Zoos in the 21st Century
Catalysts for Conservation?

edited by
ALEXANDRA ZIMMERMANN
North of England Zoological Society
MATTHEW HATCHWELL
Wildlife Conservation Society
LESLEY A. DICKIE
Zoological Society of London
CHRIS WEST
Royal Zoological Society of South Australia

CAMBRIDGE UNIVERSITY PRESS
Cambridge, New York, Melbourne, Madrid, Cape Town,
Singapore, São Paulo, Delhi, Tokyo, Mexico City

Cambridge University Press
The Edinburgh Building, Cambridge CB2 8RU, UK

Published in the United States of America by Cambridge University Press, New York

www.cambridge.org
Information on this title: www.cambridge.org/9780521618588

First published 2007
Reprinted 2009

A catalogue record for this publication is available from the British Library

ISBN 978-0-521-85333-0 Hardback
ISBN 978-0-521-61858-8 Paperback

Contents

Contributors

DAVID E. ANDERSON
San Francisco Zoo,
San Francisco, CA, USA

ANNE BAKER
The Toledo Zoo,
Toledo, OH, USA

ANDREW BALMFORD
Conservation Science Group
Department of Zoology,
University of Cambridge,
Cambridge, UK

JEFFREY P. BONNER
Saint Louis Zoo,
St. Louis, MO, USA

SARAH CHRISTIE
Zoological Society of London,
London, UK

WILLIAM CONWAY
Wildlife Conservation Society,
Bronx, NY, USA

LESLEY A. DICKIE
Zoological Society of London,
London, UK

LEE DURRELL
Durrell Wildlife Conservation Trust,
Jersey, Channel Islands

JOHN E. FA
Durrell Wildlife Conservation Trust,
Jersey, Channel Islands

DAVID A. FIELD
Zoological Society of London,
London, UK

DEAN GIBSON
Duke University Primate Center,
Durham, NC, USA

JOHN A. GWYNNE
Living Institutions,
Wildlife Conservation Society,
Bronx, NY, USA

MATTHEW HATCHWELL
European Office,
Wildlife Conservation Society,
Bronx, NY, USA

BENGT HOLST
Copenhagen Zoo,
Frederiksberg, Denmark

MICHAEL HUTCHINS
The Wildlife Society,
Bethesda, MD, USA

ANDREA S. KATZ
Duke University Primate Center,
Durham, NC, USA

NIGEL LEADER-WILLIAMS
Durrell Institute of Conservation and
Ecology,
University of Kent, Canterbury, UK

JIMIN LEE
Center for Biodiversity and Conservation,
American Museum of Natural History,
New York, NY, USA

JOHN C. M. LEWIS
International Zoo Veterinary Group,
Keighley, West Yorkshire, UK

MATTHEW LINKIE
Durrell Institute of Conservation and Ecology,
University of Kent, Canterbury, UK

GEORGINA M. MACE
Institute of Zoology,
Zoological Society of London,
London, UK

ANDREA MANICA
Department of Zoology,
University of Cambridge,
Cambridge, UK

BRIAN J. MILLER
Wind River Foundation,
Watrous, NM, USA

JACKIE OGDEN
Disney's Animal Kingdom and Disney's Animal Programs,
Lake Buena Vista, FL, USA

INGRID PORTON
Saint Louis Zoo,
St. Louis, MO, USA

RICHARD P. READING
Denver Zoological Foundation,
Denver, CO, USA

ALEX RÜBEL
Zoo Zürich, Zurich,
Switzerland

KIM R. SAMS
Walt Disney World Public Affairs,
Lake Buena Vista, FL, USA

EVA L. SARGENT
San Francisco Zoo,
San Francisco, CA, USA

ROBERT J. SMITH
Durrell Institute of Conservation and Ecology,
University of Kent, Canterbury, UK

MARK R. STANLEY PRICE
Durrell Wildlife Conservation Trust,
Jersey, Channel Islands

ACHIM STEINER
United Nations Environment Programme,
Nairobi, Kenya

ELEANOR STERLING
Center for Biodiversity and Conservation,
American Museum of Natural History,
New York, NY, USA

BETH STEVENS
Disney's Animal Kingdom and Disney's Animal Programs,
Lake Buena Vista, FL, USA

MIRANDA STEVENSON
British and Irish Association of Zoos and Aquariums,
London, UK

OLIVIA WALTER
British and Irish Association of Zoos and Aquariums,
London, UK

CHARLES R. WELCH
Duke University Primate Center,
Durham, NC, USA

CHRIS WEST
Royal Zoological Society of South Australia,
Adelaide, Australia

DAN WHARTON
Central Park Zoo,
Prospect Park Zoo, Queens Zoo
Wildlife Conservation Society,
Bronx, NY, USA

ROGER WILKINSON
North of England Zoological Society,
Chester, UK

SCOTT WILSON
North of England Zoological Society,
Chester, UK

TOM WOOD
George Mason University, and
Conservation and Research Center,
Smithsonian Institution,
Front Royal, VA, USA

ALEXANDRA ZIMMERMANN
North of England Zoological Society,
Chester, and wildlife Conservation Research Unit,
University of Oxford, UK

Foreword

For young people living in cities, zoos and aquariums are often their first contact with nature. If we consider that more than 600 million visitors pass through the gates of zoos throughout the world, we will quickly realize the immense potential these institutions have to become the incubators of the conservationists of tomorrow.

Zoos are vital in communicating conservation and educating the general public. They also conduct research, and generate and disseminate information and knowledge about natural systems, as well as the species that inhabit them. Thus, they enhance too our understanding of the components of biodiversity and their interactions, one of the key areas of IUCN's work.

Modern conservation zoos build capacity, and transfer technology to colleagues all over the world, to ensure the longer term contribution of zoos to biodiversity conservation, while also fostering a spirit of collaboration and cooperation much needed in our troubled world. They gather financial support for conservation in the field, which demonstrates the commitment of urban populations to maintaining the wild areas of the Earth.

Catalysts for Conservation was a very appropriate title for the symposium held in London in February 2004, on which this book is based. As a vehicle for conservation zoos play a key role in forging global communities for conservation of biodiversity and sustainable development for people. By using their skills and resources they can advocate and influence the attitudes of people to conservation.

Zoos are fun! Yet in reaching out to a significant portion of the population, zoos can also provide a means for education that no other conservation organizations have. When zoo-based and non-zoo-based organizations join their efforts, their outreach can push the conservation agenda far beyond the already important "understanding biodiversity" into the realms that link this understanding to the very human concerns of sustainable livelihoods, social equity, conservation incentives, and strengthening international agreements. All of these are key areas in IUCN's mission.

Whether by themselves or by forging alliances with other organizations, zoos are moving beyond their traditional roles as simply collections of wild species "*ex situ*"; that is, far from their natural ecosystems. In recent years we have seen zoos support and implement field conservation projects "*in situ*" that are complemented by the research and activities carried out back home. In doing so they provide a window to the natural world which can empower visitors to take positive action for nature and people, to the benefit of both. I am sure that as we move further into the 21st century the value of zoos supporting conservation in the field will grow significantly.

This book should act in itself as a catalyst and spur to define a "blueprint" and benchmark standards for 21st century zoos. It will be valuable to people working in zoos who share a powerful conservation vocation and to their colleagues and partners in the wider conservation movement.

Achim Steiner
Executive Director, United Nations Environment Program
(formerly Director General, IUCN – The World Conservation Union)

Acknowledgements

This book is the work of not only its contributing authors, but also those who participated in original initiatives leading to the thinking and ideas included in this volume. These include the *Zoo Measures Group* (2001–2003), an independent group which sought to evaluate the conservation impacts of zoos, and, later, the *Catalysts for Conservation* Symposium (2004), which brought together 24 speakers and over 200 participants. This volume has therefore benefited immensely from the generous input, advice, comments, and encouragement of many people, including: Andrew Balmford (Cambridge University), Deborah Body (ZSL), Jeff Bonner (Saint Louis Zoo), Joseph Bottril (Cambridge University Press), Onnie Byers (IUCN/CBSG), Alan Crowden (Cambridge University Press), Linda DaVolls (ZSL), Bert de Boer (Apenheul Zoo), Michael Dixon (Natural History Museum), Peter Dollinger (WAZA), Andrea Fidgett (Chester Zoo), Clare Georgy (Cambridge University Press), Jo Gipps (Bristol Zoo), Sally Jeanrenaud (IUCN), Richard Lattis (WCS), Nigel Leader-Williams (DICE), Georgina Mace (ZSL), Ed McAllister (WAZA), Gordon McGregor Reid (Chester Zoo), Sarah Price (Cambridge University Press), Kent Redford (WCS), Christian Rohlff (OGeS), Alex Rübel (Zoo Zürich), Achim Steiner (UNEP), Miranda Stevenson (BIAZA), Olivia Walter (BIAZA), Roger Wilkinson (Chester Zoo), and Scott Wilson (Chester Zoo).

PART I

Zoos entering the 21st century

I

Introduction: is there a conservation role for zoos in a natural world under fire?

CHRIS WEST AND LESLEY A. DICKIE

HOW MANY PLANETS DO WE NEED?

The environmental problems currently facing the world have been well documented: climate change, degradation of pristine environments, alteration of land, over-exploitation of natural resources, and increasing urbanization all play a role in the decline of the natural world (Smith 2003). The majority of environmental challenges identified are the consequence of large-scale processes, due in the most part to profligate consumption by the developed world; smaller scale processes in turn may push fragmenting ecosystems closer to oblivion. Since human suffering is ongoing people will attempt to alleviate their personal situation by using what resources are still around them. Forests will be cleared by slash and burn agriculture and endangered species will be illegally traded or hunted for food. It is against this backdrop that the current extinction crisis, with a rate of species loss an order of magnitude higher than that of background levels (Balmford *et al.* 1998), continues apace.

An ever increasing global population, with 6.4 billion individuals in 2005 (United Nations 2004), in need of shelter, food, and water, puts further pressure on already straining environments. The human population has nearly doubled in the past 30 years, with approximately 250 000 births a day (Pletscher and Schwartz 2000). It is this runaway fecundity that has led E. O. Wilson (2002) to ask how many planets do we really need to support this mass of humanity? He answers this by suggesting that on current resource use it may be as much as an additional two planets, though even that may be a modest prediction. Moreover, levels of consumption vary dramatically between nations, with the average individual energy consumption of the USA in particular far greater than that for much of the developing world. Although solutions are being sought on a number of fronts, with large-scale multi-country agreements to important protocols, such as the Convention on Biological Diversity and the Kyoto Protocol, there has also been a proliferation in the number of environmental and species conservation non-Governmental organizations (NGOs) in the latter part of the 20th century. Amidst these organizations, where should zoos position themselves, and how can they make a useful and meaningful contribution to the conservation of wildlife?

WHY ZOOS?

Zoos have traditionally been placed in the entertainment sphere, by both the viewing public and zoos themselves. However, there has been a sea change in how some zoos, but by no means all, view their role, repositioning themselves as conservation organizations. Unfortunately it is also clear that using conservation "sound bites" in marketing campaigns and overuse of these terms is commonplace. What do we really mean by "conservation" and how do we assess what zoos do? The development of critical and scientifically valid evaluation techniques of zoo-led conservation activities, and their impact on species security in the wild, have been largely absent to date, and allow spurious claims to continue unchallenged. In addition the general public are becoming more "welfare aware," with pressure groups linking all zoos to poor care and little conservation value. However, this is at odds to how the zoo-going visitor views the role of zoos (Reade and Waran 1996), having far greater awareness of welfare-enhancing techniques, such as environmental enrichment, and placing greater value on good care than their non-zoo-going counterparts.

The symposium *Catalysts for Conservation: A Direction for Zoos in the 21st Century*, upon which this book is based, set forth to challenge the more traditional role of zoos and to critique their current conservation practices. The authors in this book will set out to examine where zoos have progressed (Chapter 2), where their core strengths may lie (Chapters 17, 19), but also where zoos can improve their conservation performance (Chapters 9, 16, 20, 21), setting out questions, and, where possible, some tentative answers (Chapter 14).

TO WHICH COMMUNITY DO WE BELONG?

An estimated one in ten of the global population is believed to visit a zoo each year (IUDZG/CBSG 1993), amounting to 600 million people worldwide. However, these zoos will range from roadside menageries to highly developed conservation parks. The ability of these different institutions to convey conservation messages will also vary, but it is one area in which zoos have the potential to excel. Caro *et al.* (2003) demonstrated that university students became more conservation minded after completing conservation electives during their degree courses. But how do zoos reach their visitors in the relatively short space of time during a zoo visit?

Traditionally, this has been in the area of teaching the visiting public about the decline in species. More recently this has developed to include a more holistic "one world" view. This communication role is vital (Chapter 4), however critical evaluation of both teaching methods and the efficacy of interpretation in the zoo have been largely absent. Balmford *et al.* (Chapter 9) investigate whether differences in public awareness can be affected by a visit to the zoo without directed teaching, whilst Reading and Miller (Chapter 6) question whether zoos are currently able to communicate effectively to a wider audience.

DOES URBANIZATION MATTER?

The larger, more financially able zoos are inherently based in more developed societies, often societies that are, due to their industrialized nature, more remote from the natural world and considerable consumers of world resources (Venetoulis *et al.* 2004). Linking people to nature, to ideas that they are not separate from the world around them, rather that they are part of the global ecosystem, is one area in which zoos have the opportunity to have a profound influence upon their visiting public.

However, this requires that visitors to zoos must be engaged with the animals they see during a visit. How do zoos achieve this? The development of innovative and inspirational exhibits (Chapter 5) giving both children and adults the chance to experience, however briefly, a little of how a rainforest or a desert feels is vital. The vast majority of visitors to developed country zoos (who are more likely to have the financial resources to donate to conservation projects) will never see a gorilla or a tiger in the wild. By being the "front of house" for a zoological society, the exhibits must work, both in terms of the welfare of the animals housed and in generating an experience.

This has been called by some the "Disneyization" of zoos (Beardsworth and Bryman 2001); however, in a leisure environment that is increasingly competitive, zoos must be pragmatic in ensuring that visitors have a good day out, whilst balancing this against the primacy of the conservation mission.

Can zoos take the lead in environmental or "green" issues, communicating this to their visitors? Individual zoo visitors can be helped to relate to environmental and conservation concerns at an individual and local level as, unlike the majority of conservation NGOs, zoos can provide a deeply personal experience and enable visitors to feel positive and empowered – not hopeless and repelled. However, it is important that zoos should "walk the talk" in relation to environmental practices and sustainable-use issues. Environmental auditing, the ecological footprint of the zoo and carbon neutrality are all areas in which zoos could re-examine their activities. Can zoos really promote conservation awareness if they are not managing their own resources wisely? Zoos must also ensure that all their staff are aware of the conservation mission and have "bought in" to its importance, ensuring greater teamwork and a unified face (Chapter 8).

Larger zoos in the developed world, which are more likely to carry out *in situ* projects, should work in partnership with local in-country zoos and NGOs. Financially able zoos should see it as part of their conservation development to partner and assist struggling zoos to achieve good welfare standards, as well as enabling them to participate more fully in their own country's conservation initiatives. There has been recent criticism of some of the world's larger conservation NGOs in their relationships with indigenous peoples and small-scale local NGOs (Chapin 2004). Zoos should ensure that they work in partnership with indigenous and local peoples (rather than bringing with them an "imperialist" approach), thereby avoiding conflict, which ultimately hinders conservation activity. Zoos may also find a role in developing sustainable and environmentally friendly ecotourism projects. Issues around the sustainability of human livelihoods and their impact on wildlife cannot be ignored by organizations that view conservation of wildlife as their primary goal. Ultimately, the fate of many species is linked to that of the local inhabitants, ensuring that zoos must be aware of human actions and the often difficult conditions of local peoples in relation to their field activity, finding ways in which, at the very least, activities they carry out do not make the lives of local people economically more difficult without providing alternative sources of income. However, it must also be considered that local peoples may have little interest in some conservation issues, as with some individuals in developed nations, and that ultimately protection of critically endangered species may involve exclusion of local peoples from resource areas. How do zoos and other conservation NGOs fulfil their species conservation aims under these circumstances?

The zoos of the future must also grasp their responsibilities in relation to legislation, lobbying at every level to alter conservation-unfriendly laws (Chapter 3) or to promote an awareness of biodiversity loss issues at government level. They must also embrace and use existing legislation which supports their aims. Public advocacy on a national and international scale has not been an arena in which zoos have traditionally been active. The recent European Association of Zoos and Aquariums (EAZA) Bushmeat campaign was an example of how zoos, when working collectively, can promote enhanced awareness of a conservation issue, whilst also galvanizing public support through petitions (Chapter 19). Over 1.9 million signatures were collected and presented, the largest petition ever received by the European Parliament.

WHAT VALUES AND ETHICS DO WE SHARE?

The overriding mission of the modern zoo must be the conservation of endangered species; however, this does not mean that animal welfare considerations can be ignored. It is a reality that zoos operating in different parts of the world are subject to different local attitudes to the use of animals. Can world zoos find a common approach? The ethical responsibilities of zoos are currently under debate at regional level, for example through EAZA, and within individual zoos. Subjects such as examples of the use of flight-restraint methods and management of euthanasia are areas that require open and honest debate if progress is to be achieved. Additionally the subject of sourcing of animals for captivity is still unresolved (Chapter 15), with wide disparity in ethical standpoints even within the same regional associations. The use of animal dealers, although legal, is condoned by some zoos, whilst regarded as unethical by others. Whilst legislation must be strictly adhered to (Chapter 3) zoos must be continually re-examining their behavior.

It is also incumbent upon zoos that they facilitate or lead discussion of conservation issues and ethics in their local constituencies, rather than catching up with the thoughts of the wider community. Responsible zoos must be outspoken in defending their existence and by educating the public as to the differences between conservation, welfare, and rights (Chapter 7).

WHERE SHOULD ZOOS TARGET THEIR RESOURCES?

Whilst species declines have been well documented, for some of the most enigmatic animals conservation indicators such as accurate numbers (e.g., Asian elephant, *Elephas maximus*, Blake and Hedges 2004) or threat status (gorilla, *Gorilla* sp., Harcourt 1996) are still somewhat open to debate. Where then should zoos focus their funds? Only a small proportion of the vertebrate fauna currently threatened can be preserved as assurance populations in captivity. Regional collection plans have been initiated to try to make best use of cage space, targeting species that are conservation dependent. However, for some of the most high profile orders, the most commonly housed species are ones that have been held in captivity historically, or are easy to breed or easy to source from the wild, not the ones that are most conservation dependent.

Captive breeding as a long-term method to preserve endangered species has received criticism (Rabinowitz 1995, Snyder *et al.* 1996), yet Snyder *et al.* (1996) also recognize that for many species it has made the difference between extinction and survival, and others have countered the arguments expressed (Gippoliti and Carpaneto 1997) suggesting that zoos have made profound advances in breeding some species. Whilst zoos can use high profile breeding successes to publicize their work it should also obligate them to effectively examine the coordinated breeding programs that are currently ongoing at a regional or international level. Does it really make a difference if a species is accorded European Endangered Species Program (EEP) or Species Survival Plan (SSP) status? Are the managed breeding programs of zoos self-sustaining (Chapter 10), and, if they are not, can zoos identify why? This is one area in which the development of research standards in zoos and linkages between zoos and outside researchers may prove especially fruitful (Chapter 12).

The issue of reintroducing an endangered species is discussed by Stanley Price and Fa (Chapter 11). This much vaunted goal of zoo breeding programs has had some high profile successes, such as the Golden Lion Tamarin project (*Leontopithecus rosalia*) and Arabian oryx (*Oryx leucoryx*), but in the current arena, are we using these examples because we truly believe that this is a viable option for many species, or is it a convenient marketing tool in a hostile media environment that has little time for complicated discussion, opting rather for "feel good" stories of returns to the wild? The problems associated with reintroductions are many: financial costs, disease risk, inability of reintroduced animals to adjust to the wild, and losses of life in early release phases. Some of these difficulties (disease risk, mortality) can also be associated with translocation of individuals in the wild (Cunningham 1996) and the efficacy of translocations or reintroductions may be species dependent (Tenhumberg *et al.* 2004). How best then can zoos support and bolster wild populations?

Should zoos focus their limited conservation funds on "hotspots," areas of high biodiversity and endemism (Myers *et al.* 2000). This approach, however, has received some criticism in relation to small-scale, local levels (Reid 1998, Smith *et al.* 2001, Kati *et al.* 2004, though see Balmford 1998, Brooks *et al.* 2002) and across taxonomic groups (Harcourt 2000). Should private organizations, such as zoos, buy land in hotspot areas, as suggested by E. O. Wilson (Anon 2001, Wilson 2002)?

The recent Masoala exhibit at Zoo Zürich, opened in 2003, is an example of the links between an area of exceptional conservation importance, the Masoala National Park in north east Madagascar, and the development of an innovative zoo exhibit (Chapter 14). Linking with conservation partners *in situ* (WCS, Care International) allowed Zoo Zürich to create a world class exhibit (Rübel 2003), whilst gaining experience in conservation practice in the wild. The well-developed lines of communication between zoos, traditionally developed to exchange animals for breeding purposes, have been used to good effect in recent years in coalitions that bring together differential expertise, linking people and organizations for conservation issues. The Madagascar Fauna Group (Chapter 18) and the EAZA yearly campaigns (Chapters 17 and 19) are examples of such "coalitions for conservation."

THE 21ST CENTURY ZOO?

The original *World Zoo Conservation Strategy* (IUDZG/CBSG 1993) represented a statement of intent, with its primary objective to identify areas in which zoos could perform conservation actions. This first strategy was relatively modest in its aims and the new *World Zoo and Aquarium Conservation Strategy* (WAZA 2005) is far more explicit and challenging in its form, asking questions such as "what is the unifying philosophy and purpose of zoos?" and "how can they help conservation in the wild?" It is envisaged that the momentum that was generated by the *Catalysts* symposium will be ongoing and work in tandem with *The World Zoo and Aquarium Conservation Strategy* (WZACS), (WAZA 2005), with this book a part of this ongoing process to reposition zoos and their areas of expertise.

It is imperative that zoos should continue this debate and they should not shy away from open and thoughtful dialog, engaging in this discussion both in private and public. The final chapter of this book highlights a way of moving this process forward. However, this will only happen if zoos take up the challenge, recognizing that they must continue to change or become an anachronism. Zoos can become a force of benefit to conservation, perhaps one of the most productive and powerful forces to emerge in the 21st century, but only if they really wish it. Will global zoos take up the challenge?

References

Anon (2001). Buying land for biodiversity. News & Comment. *Trends in Ecology and Evolution*, **16(8)**, 429.

Balmford, A. (1998). On hotspots and the use of indicators for reserve selection. *Trends in Ecology and Evolution*, **13(10)**, 409.

Balmford, A., Mace, G. M., & Ginsberg, J. R. (1998). The challenges to conservation in a changing world: putting processes on the map. In *Conservation in a Changing World*, eds. G. M. Mace, A. Balmford, & J. R. Ginsberg. Cambridge: Cambridge University Press.

Beardsworth, A. & Bryman, A. (2001). The wild animal in late modernity: the case of the Disneyization of zoos. *Tourist Studies*, **1(1)**, 83–104.

Blake, S. & Hedges, S. (2004). Sinking the flagship: the case of forest elephants in Asia and Africa. *Conservation Biology*, **18(5)**, 1191–1202.

Brooks, T. M., Mittermeier, R. A., Mittermeier, C. G. *et al.* (2002). Habitat loss and extinction in the hotspots of biodiversity. *Conservation Biology*, **16(4)**, 909–923.

Caro, T., Borgerhoff Mulder, M., & Moore, M. (2003). Effects of conservation education on reasons to conserve biological diversity. *Biological Conservation*, **114**, 143–152.

Chapin, M. (2004). A challenge to conservationists. WorldWatch Institute. www.worldwatch.org.

Cunningham, A. A. (1996). Disease risks of wildlife translocations. *Conservation Biology*, **10(2)**, 349–353.

Gippoliti, S. & Carpaneto, G. M. (1997). Captive breeding, zoos and good sense. *Conservation Biology*, **11(3)**, 806–807.

Harcourt, A. H. (1996). Is the gorilla a threatened species? How should we judge? *Biological Conservation*, **75**, 165–176.

Harcourt, A. H. (2000). Coincidence and mismatch of biodiversity hotspots: a global survey for the order, primates. *Biological Conservation*, **93**, 163–175.

IUDZG/CBSG (IUCN/SSC) (1993). *The World Zoo Conservation Strategy: The Role of the Zoos and Aquaria of the World in Global Conservation*. Brookfield, IL: Chicago Zoological Society.

Kati, V., Devillers, P., Dufrene, M., Legakis, A., Vokou, D., & Lebrun, P. (2004). Hotspots, complementarity or representativeness? Designing optimal small-scale reserves for biodiversity conservation. *Biological Conservation*, **120**, 471–480.

Myers, N., Mittermeier, R. A., Mittermeier, C. G., da Fonesca, G. A. B., & Kent, J. (2000). Biodiversity hotspots for conservation. *Nature*, **403**, 853–858.

Pletscher, D. H. & Schwartz, M. K. (2000). The tyranny of population growth. *Conservation Biology*, **14**(6), 1918–1919.
Rabinowitz, A. (1995). Helping a species go extinct: the Sumatran rhino in Borneo. *Conservation Biology*, **9**(3), 482–488.
Reade, L. S. & Waran, N. K. (1996). The modern zoo: how do people perceive zoo animals? *Applied Animal Behaviour Science*, **47**, 109–188.
Reid, W. V. (1998). Biodiversity hotspots. *Trends in Ecology and Evolution*, **13**(7), 275–280.
Rübel, G. A. (2003). Conservation through cooperation: the Masoala Rainforest at Zoo Zürich. In *Proceedings of the EAZA Conference 2003*. Leipzig: European Association of Zoos and Aquaria, pp. 78–81.
Smith, D. (2003). *The State of the World Atlas*. London: Earthscan Publications Ltd.
Smith, T. B., Kark, S., Schneider, C. J., Wayne, R. K., & Moritz, C. (2001). Biodiversity hotspots and beyond: the need for preserving environmental transitions. *Trends in Ecology and Evolution*, **16**(8), 431.
Snyder, N. F. R., Derrickson, S. R., Beissinger, S. R. *et al.* (1996). Limitations of captive breeding in endangered species recovery. *Conservation Biology*, **10**(2), 338–348.
Tenhumberg, B., Tyre, A. J., Shea, K., & Possingham, H. P. (2004). Linking wild and captive populations to maximise species persistence: optimal translocation strategies. *Conservation Biology*, **18**(5), 1304–1314.
United Nations (2004). World population prospects: the 2004 revision population database. http://esa.un.org/unpp/.
Venetoulis, J., Chazan, D., & Gaudet, C. (2004). *Ecological Footprint of Nations 2004, Sustainability Indicators Program*. www.Redefining Progress.org.
WAZA (2005). *The World Zoo and Aquarium Conservation Strategy: Building a Future for Wildlife*. Liebefeld-Bern: WAZA.
Wilson, E. O. (2002). *The Future of Life*. London: Abacus.

2

Entering the 21st century

WILLIAM CONWAY

The last Carolina parakeet (*Conuropsis carolinensis*) seen in nature was sighted in 1904 when the Bronx Zoo was 5 years old and London Zoo was 78. The very last Carolina parakeet, "Incas," died on 21 February 1918 at the Cincinnati Zoo. He was one of a flock of 16 captured in Florida in the late 19th century, as many were, and bought by Cincinnati for $2.50 a bird. Subsequently, London Zoo attempted to buy the last two of these parakeets, "Incas" and his cage mate, "Lady Jane," for $400. The species was easily bred in captivity. Famed ornithologist Robert Ridgway, who thought it important to predict the species' extinction, owned a small flock which hatched chicks. He thought this of no importance. Neither Ridgway nor any zoo attempted to propagate the numerous captive birds (Cokinos 2000).

When Incas and Lady Jane died, zoos thought little about conservation. Despite the successful 20th century struggle by the Bronx Zoo's William Hornaday to stop the slaughter of fur seals, migratory and plume birds, the breeding and reintroduction of the American bison (*Bison bison*) by the Zoo was an anomaly and so was its New York Zoological Society's field research. Zoos were an archipelago of territorial entities; islands on which species after species, some already rare, would, like Christmas lights, appear, prosper, and soon disappear: thylacines (*Thylacinus cynocephalus*), passenger pigeons (*Ectopistes migratorius*), giant tortoises (*Geochelone* and *Aldabrachelys* spp.), winking on, off, and disappearing. Perhaps it is relevant that 90% of historical avian extinctions have occurred on islands (Diamond 1984). In any event, wildlife no longer seems abundant.

Human beings directly influence 83% of Earth's land surface, and 98% of the area where it is possible to grow rice, wheat, or maize (Sanderson *et al.*

2002). The World Conservation Union (IUCN) now considers that 23% of all evaluated species of vertebrates are threatened: mammals 23%, birds 12%, reptiles 51%, amphibians 31%, and fishes 40% (IUCN 2006). Almost all big terrestrial animals, zoo favorites, are in trouble: big cats, storks and cranes, parrots, pythons, antelopes, monkeys, great apes, wild cattle, tapirs, elephants, and rhinos.

Here, in London, wildlife loss seems far away. It isn't. An immense silent slaughter is taking place close to home. In the UK, domestic cats are allowed to kill about 20 million birds a year (Churcher and Lawton 1989); in the US, cats kill over a billion small mammals and hundreds of millions of birds each year (Coleman *et al.* 1997). Throughout the world, tall communications towers slay migratory birds by the millions. From 1957 to 1994, one tower in Wisconsin killed 121 563 birds of 123 species. There are over 70 000 such towers in the US and thousands more in planning (Winegrad 2000, Holden 2001, Anon 2003).

Although yearly human increase has slowed to about 70 million, if each additional person requires just one acre of land to provide lifetime resource needs, humanity must annually find an additional 283 000 km^2. This is an area 17% larger than the UK or twice the size of New York State. Four cheerless trends seem clear. Alteration of terrestrial habitats will continue to accelerate. In the humid tropics, massive hunting for bushmeat snuffing-out wildlife will continue even where habitat remains (Bennett *et al.* 2002, Milner-Gulland *et al.* 2003). So terrestrial wildlife will continue its decline. Overfished marine resources will continue to shrink, aquaculture and shoreline habitat destruction will expand, and marine wildlife will diminish. By 2050, some expect that few large marine species will remain and most coral reefs will be degraded (Jenkins 2003). Lawlessness, invasive species, global warming, and ecological succession will reduce the wildlife-carrying capacity of many protected areas. Economic and social conflict, combined with short-sighted attempts at poverty alleviation, will divert support from wildlife conservation and cause further loss (Robinson and Bennett 2002). There is little reason to expect that any terrestrial wild animal which competes with humans in some way, real or invented, or is larger than a two-gallon bucket, will survive for long without increased protection and knowledgeable care. We live in an age of extinction.

So, how are zoos and aquariums (I'll call both "zoos") to contribute to the survival of wildlife in the 21st century, to provide feasible solutions, a vision for the future? Zoo conservation lessons are not mandatory; no passing grades are required. No one attends unless they wish to, nor are zoos money-makers. They have little economic clout. But their live animals make them unique and their survival is both the stimulus and the solution. They are the key to what zoos can do that other conservation organizations cannot.

To begin with, zoos must recognize that traditional, highly diverse, international zoo animal collections maintained by frequent importations are unsustainable. In the 21st century, we must not only do better, we must do differently. We must reinvent the zoo and its vision for the future.

Zoo conservation education potentials are large, unique, and unrealized. Although zoos attract more than 650 million people each year, their conservation impacts are lessened by two peculiarities: most are perceived as local education and recreation services, not as part of an urgent national and global environmental obligation, and zoo education mostly emphasizes children, not decision-makers.

Despite the fact that almost all zoos are in population centers, where opinions and policies are made, they find it difficult to connect attendees and conservation action – beyond urging financial contributions and setting forth the effects of thoughtless consumption. It is time for focused legislator, lawyer, and conservation support exhibits at least as specialized and imaginative as those for children, and also for far more vigorous efforts to reach the people where zoo wildlife actually lives. Yes, creating familiarity with wild animals in urban children, inspiring interest, is a basic zoo education function. But it is hard to take it very seriously in the context of overseas conservation solutions to mounting extinctions. Today's children will not be decision-makers until billions more children have been born and much more wildlife has been lost. Helping New York six-year-olds to learn about African monkeys may not help African monkeys.

These dilemmas are stimulating much study, such as that of "The Ocean Project" by AZA aquariums (Vescolani and O'Callaghan 2004). However, personal observation convinces me that zoos do fill a yawning experiential gap for city dwellers. They help to produce that primordial potion of human affiliation and concern, that "biophilia" (Wilson 1984), which drives so much of global conservation and gives life to vague notions of biodiversity and ecological literacy. Yet, they act too weakly and their messages tend to ignore unpleasant problems.

For example, well-meaning groups are campaigning to save superabundant invasive animals they know from control while ignoring their destruction of other species they don't – and believing that *this* is conservation. Consider those preventing the control of house cats globally, or alien mute swans in Maryland, or elephants in South Africa's Kruger Park, animals whose destructive proliferation, exempt from science-based management, is destroying other species – and habitats. How many zoos are addressing this mix of ethics and ignorance by teaching the rudiments of wildlife carrying-capacity, the obligatory unpleasantnesses of sustaining wildlife in human-constricted spaces? Mike Hutchins (Chapter 7) addresses this critical problem.

Live animal exhibits are a zoo's heart and its organizing principle. They are fundamental in helping people to understand and care. The best exhibits are not only cognitive tools, but also powerful esthetic experiences almost magically connecting their municipal visitors with the beauty of distant nature. Surveys at the Bronx Zoo's JungleWorld, World of Birds, and Congo Gorilla Forest show that wonderful wild creatures in realistically beautiful settings can affect people very deeply. Today, advanced zoo arts and sciences are resulting in inspirational exhibit bull's eyes, as well as a few victims of "friendly fire" misaimed by the biologically oblivious or esthetically impaired. The advent of interactive electronics and video theater enables previously undreamed of interpretative presentations. John Gwynne shows examples (Chapter 5).

Yet, the most exciting, the most compelling, challenge of 21st century zoo exhibition is beyond beautiful exhibits and beyond making wild animals more understandable to people who cannot help them. It is to employ the new zoo skills in the prodigious tasks of wildlife perpetuation; to make clear to everyone the urgency and vision of the zoo's conservation effort. Zurich Zoo's "Masoala" exhibit, helping to support wildlife in Madagascar, and Bronx Zoo's "Congo Gorilla Forest," which now provides more than $1 million a year to African forest conservation (Conway 2003), are two examples. But zoo animal exhibitry only matters if zoos can sustain their animals.

Zoos can breed the vast majority of species they exhibit. Even in the early 1990s, they bred 85.7% of the IUCN-listed threatened species of mammals that they held in captivity, 87.8% of the birds, and 88.8% of the reptiles. However, the total number of threatened taxa in zoos was only 878 of the 2015 listed by IUCN at that time and far fewer were bred long-term (Conway 1995, Rahbek 1993). Today, fewer than 200 species of threatened mammals are sustainably propagated in the world's zoos.

There are only 163 species in 109 AZA Species Survival Plans (SSPs) (Krantz 2004). There are no more than six or seven reasonably "secure" species of Old World monkeys – populations probably independent of further importations (C. McCann, pers. comm.). There are not 60 secure species of exotic birds. Robert Ridgway would feel right at home. "Incas" and "Lady Jane" would not. Although there are 281 species in AZA

Population Management Plans, they are predicted to lose 10% of their founding gene diversity in 2 years. On average, current SSP populations will lose 10% in 40 years (Earnhardt *et al.* 2001, Krantz 2004). Yes, zoos can breed rare animals, but they do not do so sustainably. To do so will require inclusive agreed-upon regional collection plans that consider all zoo space, a major increase in collaborative propagation, and more space at least as good for propagation as exhibition. Today, all zoo populations are small, many are bizarre. Of 38 AZA zoos which have clouded leopards (*Neofelis nebulosa*), 19 have only one sex. Yet, space and collaboration are critical. No one collection can sustain secure populations of much of anything. Founding gene diversity in a population of 50 will decline 3.6 times as fast as one of 200 (Earnhardt *et al.* 2001).

The needed propagation effort requires zoo-wide standards of care and priority, with which WAZA and every regional zoo association are already struggling. But, most of all, it requires the recognition that this is the right thing to do; an understanding, a vision, that zoos are not simply theaters housing passing shows for community education and recreation but conservation centers empowering their communities to join with others in responding to human-caused global extinction and helping to sustain a legacy of life.

So, zoos must allot and obtain more space, more public space and more breeding centers, for serious breeding and research, and they must make a special commitment to creatures *in extremis*, those with no near-term hope of habitat return. What could make our mission clearer or more urgent, to our visitors?

In 1986, I calculated that all of the zoo animal spaces in the world could fit within New York's 212.7 km^2 Borough of Brooklyn (Conway 1986). Eighteen years later, this calculation is still valid. Little new space has been acquired. Here is another calculation: zoos sustain approximately 600 Amur tigers (*Panthera tigris altaica*), each of which would require the prey base in at least 20 km^2 to survive in nature – 12 000 km^2 of tiger space and food obtained from *human habitat*, for a change, and more than 50 times the actual space of all zoo exhibits combined. Despite all the problems, all the caveats, this says something about the potentials of a room at the zoo, about buying time for a few wonderful species that will otherwise be lost. So, what percentage of your zoo have you designated for serious on-going propagation and conservation research – and the exhibition of that?

How curious it is that zoos are more likely to seek endowments to maintain buildings than to maintain the species in them, or in nature. How surprising that we do not employ our masterful exhibit technology to sustain real species with the creativity we apply to synthetic habitats.

The most useful, the most challenging, of zoo exhibitions for our new century are of two kinds: first, that of vanishing species in breeding programs tied to the support of their populations in nature (Conway 1988, 1989, 1995) and, second, support for the almost dodos (*Raphus cucullatus*): Przewalski horses (*Equus przewalskii*), Père David deer (*Elaphurus davidianus*), Arabian oryx (*Oryx leucoryx*), California condors (*Gymnogyps californianus*), black-footed ferrets (*Mustela nigripes*), Vancouver Island marmots (*Marmota vancouverensis*) – the growing legion of the nearly gone whose most viable "habitats," for some period of time, may be in captivity (Conway 1999). Their public stewardship is a unique responsibility for the preservation of future options. It is saving seeds. If zoos still had a small population of Carolina parakeets, thylacines, pink-headed ducks (*Rhodonessa caryophyllacea*), golden toads (*Bufo periglenes*) or dodos, is there really any doubt that each would one day become the stimulus for compelling new efforts in habitat restoration and reintroduction? Yes, the history of reintroductions from captive propagation is discouraging (Beck *et al.* 1994), but hardly more so than preservation in nature. Besides, attempts are multiplying and we are getting better at it (Maunder *et al.* 1999).

Regardless, if we seek to sustain small wildlife populations, in or out of zoos, we must strengthen curatorial sciences, genomics, assisted reproduction, and veterinary programs. The ability to move sperm and ova rather than animals between parks and zoos will become increasingly important (Asa 2001, Goodrowe 2001). The management of reproductive cycles and the posthumous reproduction of genetically valuable animals is insurance against catastrophic loss while genomics is key in the maintenance of small populations in both zoos and restricted parks. At least as urgent is the care of wildlife populations in the face of disease. As William Karesh (WCS) observes, human and wildlife diseases are not separate nations, "There is only one health." Ebola, West Nile fever, and severe acute respiratory syndrome (SARS) have made this clear. Tightening veterinary regulations could make the collaborative management of wildlife reserves, as well as of zoos, impossible by prohibiting translocations and animal exchanges. The only response is convincingly authoritative zoo veterinary science and animal care standards.

Uniquely, among conservation organizations, zoos deal with animals close-up, in sickness and in health, and in very small populations – experience which applies to the needs of more and more wildlife in nature as well as zoos. Many species are destined to survive, if at all, in small, disjunctive, populations in areas of limited carrying-capacity and no room for growth – yet recent work suggests that unmanaged wild populations must support about 7000 adult vertebrates to ensure long-term persistence (Reed *et al.* 2003). So survival may sometimes depend upon an intensity of care natural to zoos but rarely available in developing nations or through conventional conservation organizations. Some species will be viable only as part of a metapopulation maintained by translocations or reintroductions. Many will survive only in undersized protected areas where curatorial management and veterinary aid are available – zoo skills. If the major task of conventional conservation organizations is to save functioning ecosystems, perhaps a major task for zoo expertise and endangered species breeding is to help save non-functioning ecosystems (Conway 1999). Thus, a clear direction for 21st century zoos is to aid the maintenance of wildlife populations in parks and reserves.

Nevertheless, in conservation, captive propagation and reintroduction are last resorts. Zoos must increase support for studies essential to sustaining wildlife in the wild (Conway *et al.* 2001). The *AZA Annual Report on Conservation and Science*, for 2000–2001, notes that member institutions participated in 2230 conservation projects in 94 countries which ranged from breeding vanishing butterflies to anti-bushmeat campaigns and park support. EAZA and other regional associations are making similar efforts and coalitions of zoos are focused on specific areas, such as the Madagascar Fauna Group, which Lee Durrell discusses (Chapter 18). What about "zoo reserves"?

Zoo acquisition or stewardship of unprotected habitat, lands or waters, containing important wildlife communities as "zoo reserves" is a concept whose time may have come (Conway 1998, 2003). The basic idea is to protect more habitat and provide local incentives for conservation where costs would be supported, in part, by animal off-takes for zoo exhibits. I visualize 100 or more zoo and aquarium coalition reserves located in important wildlife areas tied to the fabric of SSPs, European Endangered Species Programs (EEPs), and their brethren. Some might develop as ecotourist locations as well as venues for research and training for indigenous conservationists and rangers. They could be powerful assets in regional education efforts, addressing the task of achieving a shared vision of the value of wildlife; provide local employment in less developed areas; and become a constructive alternative to costly *ex situ* breeding programs.

Zoo ecology is complex, its resources modest, its habitat fragmented. Zoos vary enormously in their capabilities and are only now forging international standards and a commonality of purpose. But that this effort is underway is demonstrated by the collaborative production of the new *World Zoo and Aquarium Conservation Strategy* (WAZA 2005). Yet, zoo futures are manifold and uncertain. They include specialized collections with a clear conservation mission, such as the International Crane Foundation, the Wildlfowl Trust, or the World Center for Birds of Prey, and such focused collections as the Durrell Wildlife Conservation Trust and the Arizona-Sonora Desert Museum. But zoos might also devolve into amusement parks exhibiting semi-domesticates as ornaments among thrill rides and game shows with no conservation significance. However, we could reinvent zoos as conservation parks – diverse and reflecting the special opportunities and obligations of place – but with a clear and collaborative vision of stopping as much extinction as possible, of preserving nature and the seeds of life.

I believe that the 21st century zoo can become a resource of hope and a call to action, a barrier to the threat of extinction and, more powerfully than ever, urban humanity's introduction to living wildlife, a promoter of environmental literacy, and a recruiting center for conservationists – a catalyst for conservation. Zoo people are stepping out beyond their fences, teaching, aiding field science, and helping protected areas. They are sustaining species which have lost their habitats and conducting campaigns to restore them – but far too slowly. Their path ahead is precipitous and they must attract more help. To do so, we need to make known a new zoo vision, a new perception, not only for zoos but about them. It must be a vision that faces up to the dismal realities of an age of extinction with feasible solutions to which we are committed. It must be enormously ambitious, and it must be now. It is no longer good enough to simply predict extinction. Down with Robert Ridgway! Up with Incas and Lady Jane!

ACKNOWLEDGEMENTS

I thank George Amato, Anne Baker, John Behler, Jim Doherty, Joanne Earnhardt, Nate Flesness, Jo Gipps, Matthew Hatchwell, Mike Hutchins, Palmer Krantz, Bob Lacy, Richard Lattis, Colleen McCann, Christine Sheppard, Patrick Thomas, and Chris Woods for helpful information and discussion.

References

Anon (2003). USA Towerkill summary: 1–3. http://towerkill.com/issues/consum.html (accessed 1 January 2007).

Asa, C. (2001). The future of reproductive biology in zoos and in conservation. *Communiqué*, February, pp. 19–21.

Beck, B., Rappaport, L., Stanley Price, M., & Wilson, A. (1994). Reintroduction of captive-born animals. In eds. G. Mace, P. Olney, & A. Feistner. Creative

conservation: interactive management of wild and captive animals. In *Proceedings of the Sixth World Conference on Breeding Endangered Species*. London: Chapman and Hall.
Bennett, E., Eves, H., Robinson, J., & Wilkie, D. (2002). Why is eating bushmeat a biodiversity crisis? *Conservation in Practice*, **3**(2), 28–29.
Churcher, P. & Lawton, J. (1989). Beware of well-fed felines. *Natural History*, **7/89**, 40–47.
Cokinos, C. (2000). *Hope is the Thing with Feathers*. New York: Tarcher/Putnam, 359 pages.
Coleman, J., Temple, S., & Craven, S. (1997). *Cats and Wildlife*. Washington, D.C.: Great Lakes Indian Fish and Wildlife Commission. Wisconsin: University of Wisconsin.
Conway, W. (1986). The practical difficulties and financial implications of endangered species breeding programs. *International Zoo Yearbook*, **24/25**, 210–219.
Conway, W. (1988). Can technology aid species preservation? In *Biodiversity*, ed. E. O. Wilson. Washington, D.C.: National Academy of Sciences Press, pp. 263–268.
Conway, W. (1989). The prospects for sustaining species and their evolution. In *Conservation for the Twenty-First Century*, eds. D. Western & M. Pearl. Oxford: Oxford University Press, pp. 199–209.
Conway, W. (1995). Wild and zoo animal interactive management and habitat conservation. *Biodiversity and Conservation*, **4**, 573–594.
Conway, W. (1998). Zoo reserves; a proposal. In *Proceedings of the AZA Annual Conference*, Tulsa Zoo & Living Museum, Oklahoma. Chicago, IL: AZA, pp. 54–58.
Conway, W. (1999). Linking zoo and field and keeping promises to dodos. In *Seventh World Conference on Breeding Endangered Species: Linking Zoo and Field Research to Advance Conservation*. Cincinnati 22–26 May 1999, eds. T. Roth, W. Swanson, & L. Blattman. Cincinnati, KY: Cincinnati Zoo, pp. 5–11.
Conway, W. (2003). The role of zoos in the 21st century. *International Zoo Yearbook*, **38**, 7–13.
Conway, W., Hutchins, M., Souza, M., Kapetanakos, Y., & Paul, E. (2001). *The AZA Field Conservation Resource Guide*. Atlanta, GA: Zoo Atlanta, pp. 1–323.
Diamond, J. (1984). Historic extinctions: a Rosetta Stone for understanding prehistoric extinctions. In *Quaternary Extinctions: A Prehistoric Revolution*, eds. P. Martin & R. Klein, Jr. Tucson, AZ: University of Arizona Press.
Earnhardt, J., Thompson, S. T., & Marhevsky, E. (2001). Interactions of target population size, population parameters, and program management on viability of captive populations. *Zoo Biology*, **20**, 169–183.
Goodrowe, K. (2001). The role of genome resource banking in wildlife conservation programs. *Communiqué*, February, pp. 13–14.
Holden, C. (2001). Curbing tower kill. *Science*, **291**, 2081.
IUCN (2006). *IUCN Red List of Threatened Species, Summary Statistics*. www.redlist.org/tables/table1.
Jenkins, M. (2003). Prospects for biodiversity. *Science*, **302**, 1175–1177.
Krantz, P. (2004). The WCMC Dilemma. AZA Director's Meeting, January 2004, unpublished, Riverbanks Zoo and Garden, Columbia, SC, pp. 1–29.

Maunder, M., Stanley-Price, M., & Soorae, P. (1999). The role of in-country ex-situ facilities in supporting species and habitat recovery: some perspectives from East Africa. In *Seventh World Conference on Breeding Endangered Species: Linking Zoo and Field Research to Advance Conservation.* Cincinnati 22–26 May 1999, eds. T. Roth, W. Swanson, & L. Blattman. Cincinnati, KY: Cincinnati Zoo, pp. 31–47.

Milner-Gulland, E., Bennett, E., & The SCB 2002 Annual Meeting Wild Meat Group. (2003). Wild meat: the bigger picture. *Trends in Ecology and Evolution,* **18**(7), 351–357.

Rahbek, C. (1993). Captive breeding – a useful tool in the preservation of biodiversity? *Biodiversity and Conservation,* **2**, 426–437.

Reed, D., O'Grady, J., Brook, B., Ballou, J., & Frankham, R. (2003). Estimates of minimum viable population sizes for vertebrates and factors influencing those estimates. *Biological Conservation,* **113**, 23–34.

Robinson, J. & Bennett, E. (2002). Will alleviating poverty solve the bushmeat crisis? *Oryx,* **36**(4), 332.

Sanderson, E., Malanding, J., Levy, M., Redford, K., Wannebo, A., & Woolmer, G. (2002). The human footprint and the last of the wild. *BioScience,* **52**(10), 891–904.

Vescolani, B. & O'Callaghan, P. (2004). Turning hope into action. *Communiqué,* January, pp. 39–42.

WAZA (2005). *The World Zoo and Aquarium Conservation Strategy: Building a Future for Wildlife.* Liebefeld-Bern: WAZA.

Wilson, E. O. (1984). *Biophilia: The Human Bond with Other Species.* Cambridge, MA: Harvard University Press.

Winegrad, G. (2000). New study documents towers killing birds. *Bird Calls,* **4**(1), 1.

3

How do national and international regulations and policies influence the role of zoos and aquariums in conservation?

BENGT HOLST AND LESLEY A. DICKIE

INTRODUCTION

The ongoing, wide-scale loss of biodiversity is a consequence of human activity, and the development of laws to tackle it is the consequence of the need to regulate human behavior. It is self-evident that the more people that exist on the planet, the more legislation is required to manage their interactions, both between themselves and with the environment in which they live (Pletscher and Schwartz 2000). A great variety of legislation now exists to govern the use, welfare, and protection of animals, some of which has relevance for zoos, particularly in regard to the movement of animals. Historically, zoos have been consumers of wildlife, although there is no evidence that collecting animals for zoos has led to significant declines in any species (Conway 1995). Sourcing animals for zoos was a relatively straightforward arrangement: an animal dealer would be contacted, an order placed, and at some specified time the animal would be delivered from the wild to the zoo. The previous sentence appears flippant but these activities should be seen in the context of the time period, and even Gerald Durrell, that doyen of zoos and conservation, as a young man earned a living procuring animals for captive collections (Botting 1999). However, the environmental landscape was significantly different in the past. Animals in the wild appeared bountiful and indeed the majority of species were at no significant risk from human activity. Little care or caution was paid to the viability of populations, animals were simply "unprotected public property" (Pletscher and Schwartz 2000), zoos made wish-lists or took whatever species were on offer, and collection planning remained an unknown topic. Only later did the startling decline of species become apparent.

Now, at the beginning of the 21st century, 25% of all mammals, 12% of birds (IUCN 2006) and more than a third of amphibians (Stuart *et al.* 2004) are threatened with extinction. While sustainable harvesting of species from the wild should still be viewed as acceptable, and for much of the human population an intrinsic part of daily life, it is apparent that stringent legislation must be formulated, developed, implemented, and maintained if global conservation aims are to be met. Much of this legislation will have a direct effect upon zoos, or is specifically formulated to regulate the activities of zoos. It is therefore fundamental that zoos should be aware of the existing legislation, and fully understand where it impacts upon their goals, both positively and negatively.

This chapter will discuss the ethical and legal responsibilities of zoos in society and some of the supra-national regulations that have an impact upon the zoo community. This will not cover the entire panoply of legislation, a feat requiring an entirely separate book, but will discuss how zoos should approach legislative issues in relation to conservation aims. Non-legally binding published guidelines by organizations such as IUCN and regional zoo associations will also be discussed.

ZOOS AS PART OF SOCIETY

Keeping wild animals for exhibition has been commonplace throughout the ages. Over time, however, the motivation for keeping and exhibiting animals has changed. Historically, large collections of animals were often totemic of power and prestige, with the collections of Montezuma and the Ming dynasty of China (Mazur 2001) particularly expansive. "Zoos" as we know them only came into being in the late 18th and early 19th centuries, often as the result of initiatives by newly established scientific societies interested in studying the exotic animals they held. The flourishing of zoos in Europe at this time coincided with greatly increased urbanization and the headlong drive towards industrialization. A growing disconnection from the natural world probably played a role in the popularity of a visit to the zoo. Latter-day urbanization too has been linked to an increased public interest in wildlife (Bandara and Tisdell 2003). This fascination with zoos began to change in the 20th century with a growing awareness of animal welfare and environmental degradation. While Plutarch (AD 46–130) advocated compassion for animals (Stevens and McAlister 2003), such concern did not become widespread until relatively late in human history. When it did, again it is probably linked to the disconnection of some sections of humanity from natural processes. What also changed was the growing awareness of human impacts on the natural world. Rachel Carson's *Silent Spring* (1962) was highly influential in changing the view of the relationship between modern humans and their environment. From the late 1960s onwards, anti-zoo groups became more vociferous and visitor numbers in many zoos declined. Zoos responded to pressure, internal and external, to change, with conservation becoming a more prominent part of the expounded mission.

It is clear that zoos do not operate in a vacuum. They are, as they have been since their modern inception, part of the social landscape, providing recreational activities and a learning environment in many cities, but also responding and changing with society, sometimes following, sometimes leading, the changes. Their cultural value and ongoing popularity is often underestimated[1] and zoos themselves perhaps do not address their amenity value in sufficient detail either. It is unusual for a capital city not to have a zoo, and often the zoo is the green heart of an otherwise concrete urban sprawl. They contribute meaningfully to their local economies and many rural zoos provide significant employment in their local area.

This imposes upon zoos a pronounced level of social responsibility. In a sense, they are "pillars" of their local community and must inspire the trust of that community. No more is this trust relationship greater than in relation to animal welfare and the acquisition and disposition of animals. Animal welfare legislation is well developed across continents (although different cultural values shape the disparities between countries) and international treaties and conventions determine the movement of animals across borders. Society looks to zoos to abide rigorously by laws set out to protect biodiversity and the welfare of individual animals. However, society as a whole – and even the zoo community – has yet to realize the potential influence that zoos around the world may be able to exert on the framing of legislation. The potential contributions of zoos to conservation advocacy

[1] In Britain and Ireland in 2005 the British and Irish Association of Zoos and Aquariums reported 18.2 million visits from a subset of their total membership. This does not take into account non-BIAZA-licensed zoos. This zoo visitation figure is greater than that recorded for the attendances of professional football matches in the same time period (M. Stevenson, pers. comm.). In addition in Australia, a country with a population of approximately 20.3 million, there were 13.5 million zoo visits in 2005, far greater than for any organized sport in a country that is commonly associated with a love of sports (C. West, pers. comm.).

have not yet been fully developed either, even though collaborative campaigns (see Chapter 19) have made some progress in that respect.

LEGISLATING FOR WILDLIFE

Today there is an overwhelming array of conventions and regulations whose sole purpose is to protect wildlife and regulate wildlife trade, constituting a complex legal framework with which zoos must comply. Historically, much of the legislation in relation to wildlife trade was designed to meet economic objectives, and restrictions focused upon protecting species were limited (Roe *et al.* 2002). The Lacey Act of 1900 was a notable exception, being prompted by the burgeoning demand for feathers for use in the fashion industry. The present conventions and regulations focus primarily on animal welfare, animal health, and conservation of species (Convention on Biological Diversity, International Animal Health Code, International Air Transport Association Regulations, CITES, etc.).

In addition many countries also have specific legislation for zoos such as the European Community Zoos Directive (99/22/EC) (EC 1999). Within Europe the interpretation of this Directive is subject to local cultural values and the speed and/or flexibility of in-country legislative frameworks. As a result, despite having been in existence for some years already, it has not yet been translated into the national legislation of all EU members. In addition, some observers feel that the Zoos Directive is a "missed opportunity" (Rees 2005) in that it does not sufficiently clarify the activities that are required of zoos, and is therefore weakened as a tool for ensuring that zoos play their part in implementing the Convention on Biological Diversity (CBD 1992). In tandem with the European Directive, some countries have well-developed legislation aimed at maintaining good standards in zoos. One example is the UK Secretary of State's Standards for Modern Zoo Practice (DETR 2000), which states that zoos should demonstrate *measurable* performance in conservation, research, and education. However, as with the EC Directive, there is some leeway in interpretation.

The CBD is the highest-level legal instrument currently in place governing biodiversity and its conservation. Article 9 of the Convention requires contracting partners to undertake *ex situ* measures where appropriate and therefore explicitly suggests a role for zoos in conservation. It is important that zoos take full advantage of the mandate that the convention confers upon them, since it defines and formalizes their role in conservation in a way that no other document can or does. It also points towards a new and perhaps key function that zoos in developed countries can play in assisting their colleagues in the biodiversity-rich developing world, in that it states that *ex situ* conservation should take place preferably in the country of origin of the biological component. For this to occur, the majority of developing world captive facilities must improve dramatically. This resonates with the *World Zoo and Aquarium Conservation Strategy* (WAZA 2005) that encourages better-resourced zoos to "mentor" zoos that are financially or technically less able at present. This means that, in future, the staff of WAZA member zoos working on field conservation projects should not simply drive past local zoos on the other side of the road, but rather should engage to find out if their assistance may be required or welcomed.

In addition to these international conventions and regulations, there are IUCN Guidelines including the IUCN/SSC Guidelines for Re-introductions (1998) and the Technical Guidelines on the Management of *ex situ* Populations for Conservation (2002). Prepared by IUCN Specialist Groups with the purpose of guiding people who work with specific conservation issues, the guidelines are not regulatory, however there is a moral responsibility to use these documents. A promising development is that many wildlife authorities and conservation organizations, including zoo and aquarium associations, have made these guidelines mandatory, thus giving them a status on a par with national and international regulations.

The IUCN Technical Guidelines on the Management of *ex situ* Populations for Conservation (2002) make explicit statements concerning the powerful and increasing conservation role that captive facilities can play (Maunder and Byers 2005). The vision of these guidelines is "*To maintain present biodiversity levels through all available and effective means including, where appropriate,* ex situ *propagation, translocation and other* ex situ *methodologies.*" *Ex situ* methods are thus clearly linked to other methodologies, indicating that there are no "silver bullet" approaches to conservation, rather a variety of techniques that work in conjunction with each other. The guidelines go on to state that "*The management of* ex situ *populations must minimise any deleterious effects of* ex situ *management, such as loss of genetic diversity, artificial selection, pathogen transfer and hybridisation, in the interest of maintaining the genetic integrity and viability of such material.*" The impact of international conventions has a regulatory effect in direct relation to this last statement. The possibility of unintentional artificial selection and hybridization in captive populations calls for proactive and rigorous management practices; the related loss of genetic diversity can be prevented only through exchange of genetic material. It is thus important to permit the transfer of animals between zoos, and between zoos and the wild. Zoos should be mindful of the risk of pathogen transfer; however, the possibility of disease transmission is carefully considered in the veterinary regulations included in several conventions and directives. Finally, the *ex situ* guidelines also place a further responsibility on the managers of *ex situ* populations. The guidelines state, "*Those responsible for managing* ex situ *populations and facilities should seek both to increase public awareness, concern and support for biodiversity, and to support the implementation of conservation management, through education, fundraising and professional capacity building programmes, and by supporting direct action* in situ." In other words, it is not sufficient for zoos to practice sound husbandry to comply with these guidelines. They are urged also to implement education plans for biodiversity conservation and to link their *ex situ* programs with *in situ* conservation activities. While some zoos may find this challenging, these stipulations are an essential aspect of achieving the full potential of the 21st century zoo.

Regulations are thus an important instrument for policy- and decision-makers within zoos, who can exercise some degree of influence in implementing guidelines and regulations that will further increase the contribution of zoos to conservation. Zoos meet their conservation obligations through a suite of activities, some of which are listed here: development of collection plans geared towards meeting conservation objectives; development of husbandry guidelines to ensure sound management of *ex situ* populations, thereby ensuring their genetic and ecological fitness; development and participation in *ex situ* breeding programs; conducting conservation education programmes leading to greater conservation awareness, and thereafter action, on the part of their visitors; establishment of links to *in situ* conservation projects; fundraising for conservation activities, and so on.

How national and international regulations influence, and potentially limit, the conservation opportunities for zoos can be illustrated by examining the meta-population management of black lion tamarins (*Leontopithecus chrysopygus*). The black lion tamarin is endemic to a small area in the western region of Sǎo Paulo State, Brazil. The population is highly fragmented and translocations of animals between the different forest fragments to improve genetic diversity are planned. Analyses (Ballou and Valladares-Padua 1991) have shown, however, that these translocations are not sufficient to ensure the necessary degree of genetic diversity within and between populations. It is necessary to include also the captive population of black lion tamarins in these transfers. Based on this realization, a meta-population management plan has been formulated, allowing animals from all small populations to be freely exchanged, whether captive or wild. The conservation program for the black lion tamarin thus includes an *ex situ* breeding program with the involvement of several zoos, a reintroduction program and a translocation program. All these defined conservation actions are dependent on transporting animals back and forth across borders and/or within the same country. These movements are regulated by both CITES (1973) and the World Animal Health Organization (OiE) regulations.

The CITES Convention describes thoroughly the criteria used for the issue of export and import permits. Since both the Convention and zoos are working for conservation of biodiversity they should co-operate closely on this topic. Unfortunately, this is not always the case. There is always room for a certain level of interpretation of the regulations, and far too often things can go wrong, with the result that conservation efforts are delayed or even stopped altogether. In some instances this may be due to authorities being unwilling to acknowledge the role of zoos in conservation, or, more likely, simply failing to understand that role. Some even consider zoos as the main consumers of wildlife, even though zoos account for less than 1% of all CITES imports from the wild. With reference to animal transfers for conservation purposes, the misinterpretation by local authorities of international conventions and regulations remains a serious problem. In these cases it is not the regulations themselves that are restrictive, but the ways in which they are interpreted. Finally, local bureaucracy and red tape often delay the issue of permits significantly. Zoos can and must, if they have carried out all their obligations correctly and ethically, place responsibility for the failure of the regulations fairly on the shoulders of the authorities when these obstacles are contradictory to the ultimate goals of the conventions; that is, to conserve viable populations of animal and plant species.

However, individual zoos are not blameless in this respect. Too often zoos have no ethical framework in place and many create a great deal of publicity about their conservation activities rather than concentrating on doing good conservation work in the first place. Furthermore, some zoos still release "surplus" animals into the wild, calling these dubious exercises "re-introductions." Activities such as these are more akin to public relations stunts rather than sound conservation actions. In these cases the IUCN Re-introduction Guidelines (1998) are ignored, as are the opinions of other zoos and conservation NGOs. Zoos must be honest about their actions and not use "conservation" as a justification for such activities. The consequence of the irresponsible actions of the few may be the distrust of all zoos by national and regional authorities charged with implementing biodiversity legislation. While some zoos continue on this irresponsible course the majority will find it difficult to convince the authorities about their new role as serious conservation bodies.

Zoos should also be aware that legislation changes in response to external developments, and the spectre of emerging infectious diseases (EIDs) serve as an example of how new environmental challenges can lead to rapid legislative changes. The H5N1 variant of avian influenza, although endemic in many countries in Asia for a number of years, only reached Europe in 2005 (DEFRA 2006a). In response, an EC Directive (2005/94/EC) on the control of avian influenza was brought into force in December 2005 (EC 2005). The route of infection into Europe is still equivocal, with speculation ranging from agricultural imports of infected poultry to migration of infected wild birds (DEFRA 2006c). Although the risk of human infection remains slight, and there is as yet no conclusive evidence of human-to-human transmission or of a human variant, the poultry industry in Europe could be devastated if there were a pan-European outbreak in wild birds, with major economic consequences.

At the time of writing (2006) there is an EU ban on the import of wild-caught birds (2005/759/EC and 2005/760/EC as amended by 2006/79/EC) and all quarantine procedures are being carefully scrutinized (DEFRA 2006b, Dimmock *et al.* 2005). While some view this as problematic for conservation, driving the trade underground (Cooney and Jepson 2006, Gilardi 2006, Roe 2006), it is welcomed by others (RSPB 2005). The large-scale trade in wild-caught birds has a negative impact upon wild populations (Beissinger 2001), the welfare of transported birds is often poor (Knights 1991, Knights and Currey 1993) and, due to lax quarantine, the risk of importing infections to disease-free zones is likely (Alexander 2006). Non-Governmental organizations from around the world are lobbying the EU to ban permanently the trade in wild-caught birds on the grounds of animal welfare and disease risks. Enhanced and more rigorous welfare legislation is also being developed (e.g., Animal Welfare Bill, UK). Zoos that rely routinely upon wild bird imports to stock their aviaries may find the ban impeding. However, it should be welcomed as an independent conservation measure. Importation for defined conservation aims should not be affected as long as all quarantine procedures are followed and the legitimate import of wild-caught birds for conservation purposes is catered for in the developing legislation, with derogations[2] in the EC Avian Influenza Directive

[2] It should be noted that the derogations in this legislation came about due to the close co-operation between the drafting MEP and the British and Irish Association of Zoos and Aquariums.

relating to zoos and conservation imports. It may also bring sharper focus to zoos in relation to their management of some bird species and their ongoing need to import.

What the above demonstrates is the need for close dialogue between the zoo community and the authorities, as well as the active participation of zoos in the formulation of such legislation in order to secure conservation-friendly regulations for the benefit of both zoos and legislative bodies. It is easy for zoos to criticize current legislation, but if zoos want changes to that legislation that lead to improved conservation outcomes they need to become more proactive and engaged. Zoos should:

1. Make themselves heard and participate actively in the decision-making processes wherever possible. It takes time and effort but it will bring dividends. Zoos as yet have been poor at actively participating in politics, often hesitant to be seen as political organizations. Yet the future of biodiversity is intimately tied to the political process. Zoos ignore that process at their peril.
2. Not use conservation as an excuse for keeping animals in zoos, but use the animals they do manage to improve their conservation efforts. Zoos do not have to find excuses for their existence. There are many good reasons for having zoos all over the world, and conservation is one of them.
3. Act as responsible conservation partners and ensure that their external partners do the same. Zoos must follow all legislation and IUCN guidelines whenever relevant.
4. Communicate openly and honestly about their conservation efforts and make their data available. *Ex situ* conservation management relies to a major extent on the exchange of animals between zoos, and between zoos and the wild. In order to communicate effectively, zoos must have access to a common animal database with all relevant data. Concealing data will complicate the matter and elicit rumours of poor, or even unethical, management.
5. Support the development of new conservation tools (international animal registration systems, conservation planning tools, evaluation tools, etc.) and not simply wait until they are delivered by external bodies. Development costs money and effort, and all zoos can contribute one way or another.
6. Consider themselves as part of a national/regional/global zoo community and provide their support to its conservation efforts. It is

unacceptable for zoos to consider themselves as isolated entities with no major links to other similar institutions. Responsible zoos depend on each other, no more so than in relation to conservation. Cooperation is essential.

CONCLUSION

International conventions and regulations are necessary to protect wildlife and ensure good animal welfare. None of them are negative towards zoos. On the contrary, some of the more important conventions explicitly enshrine the necessary role of zoos in conservation. More commonly, local interpretation fails. Such misinterpretations must be challenged, but this can only happen if zoos can convince the authorities of their emerging role in conservation and act collectively to do so. Zoos have a past reputation as consumers of wildlife and as simply a nice place to visit. That has changed in recent years but it is up to zoos to prove it through open dialogue and high ethical standards in all their activities. It is not enough for only a few zoos to exemplify these new standards. The majority must follow the same path before we can expect all misperceptions to be things of the past.

Getting the dialogue right will sometimes be difficult but should be pursued. It is a huge task that calls for cooperation from all sides, but also a task vital for the successful fulfilment of future conservation efforts. Aristotle (384–322 BCE) put it far more succinctly, *"Even when laws have been written down, they ought not always to remain unaltered."* Zoos should be mindful of this and seek to improve poor legislation where it fails conservation.

References

Alexander, D. (2006). *An Assessment of the Risk of Introduction of Highly Pathogenic Avian Influenza into the European Union by the Importation of Wild Caught Captive Birds.* Horsham: RSPCA.

Ballou, J. & Valladares-Padua, C. (1991). Population extinction model for the lion tamarins in protected areas. In *Leontopithecus: Population Viability Analysis*, eds. U. Seal, J. Ballou, & C. Valladares Padua. Apple Valley, MN: CBSG/IUCN, pp. 79–94.

Bandara, R. & Tisdell, C. (2003). Comparisons of rural and urban attitudes to the conservation of Asian elephants in Sri Lanka: empirical evidence. *Biological Conservation*, **110**, 277–342.

Beissinger, S. R. (2001). Trade in live birds: potentials, principles and practices of sustainable use. In *Conservation of Exploited Species*, eds. J. D. Reynolds, G. M. Mace, K. H. Redford, & J. G. Robinson. Cambridge: Cambridge University Press.

Botting, D. (1999). *Gerald Durrell: The Authorised Biography*. London: Harper Collins.
Carson, R. (1962). *Silent Spring*. Boston, MA: Houghton Mifflin Company.
CBD (1992). Convention on Biological Diversity, 5 June 1992, United Nations, International Environmental Laws – Multilateral Treaties, 992:42 01/043.
CITES (1973). Convention on International Trade in Endangered Species of Wild Fauna and Flora. United Nations Environment Program. Washington, 3 March 1973.
Conway, W. G. (1995). Zoo conservation and ethical paradoxes. In *Ethics on the Ark: Zoos, Animal Welfare and Wildlife Conservation*, eds. B. G. Norton, M. Hutchins, E. F. Stevens, & T. L. Maple. Washington and London: Smithsonian Institution Press, pp. 1–9.
Cooney, R. & Jepson, P. (2006). The international wild bird trade: what's wrong with blanket bans? *Oryx*, **40**, 18–23.
DEFRA (2006a). In *HPAI H5N1 Situation in Europe and Potential Risk Factors for the Introduction of the Virus to the United Kingdom*, eds. M. Sabirovic, S. Hall, J. Wilesmith, N. Coulson, & F. Landeg. International Animal Health Division, 1A Page Street, London, SW1P 4PQ, United Kingdom. Version 1, Released 06 July 2006, p. 26.
DEFRA (2006b). Government response to the Independent Review of Avian Quarantine. http://www.defra.gov.uk/animalh/diseases/pdf/avianquarantine-govnresponse.pdf.
DEFRA (2006c). Wild birds' role in HPAI crisis confirmed. FAO Newsroom (press release). Food and Agriculture Association. http://www.fao.org/newsroom/en/news/2006/1000312/index.html).
DETR (2000). *Secretary of State's Standards on Modern Zoo Practice*. H. M. Government: Department of the Environment, Transport and the Regions.
Dimmock, N., Bradley, A., Lightfoot, N., Russell, T., Scott, P., & Wathes, C. (2005). Report of the Independent Review of Avian Quarantine: a report to the Chief Veterinary Officer and Secretary of State for Environment, Food and Rural Affairs on UK quarantine arrangements for captive birds. http://www.defra.gov.uk/animalh/diseases/control/avianquarantine/independentreview/report.pdf.
EC (1999). Council Directive 1999/22/EC of 29 March 1999 Relating to the Keeping of Wild Animals in Zoos. Official Journal of the European Communities. 09/04/1999.
EC (2005). Council Directive 2005/94/EC of 20 December 2005 on Community measures for the control of avian influenza and repealing Directive 92/40/EEC.
Gilardi, J. (2006). Captured for conservation: will cats save wild birds? A response to Cooney & Jepson. *Oryx*, **40**, 24–26.
IUCN (1998). Guidelines for re-introductions. Prepared by IUCN/SSC Re-introduction Specialist Group. Gland, Switzerland and Cambridge, UK: IUCN.
IUCN (2002). Technical guidelines on the management of *ex situ* populations for conservation. Prepared by IUCN/SSC Conservation Breeding Specialist Group. Gland, Switzerland and Cambridge, UK: IUCN.

IUCN (2006). *Red List of Threatened Species: A Global Species Assessment*, eds. J. E. M. Baillie, C. Hilton-Taylor, & S. N. Stuart. Gland: IUCN.

Knights, P. (1991). *A Study of the Trade in Wild Caught Birds in Argentina*. London: Environmental Investigation Agency.

Knights, P. & Currey, D. (1993). *Investigation into the Trade in African Grey Parrots from Ghana*. London: Environmental Investigation Agency.

Maunder, M. & Byers, O. (2005). The IUCN technical guidelines on the management of ex situ populations for conservation: reflecting major changes in the application of ex situ conservation. *Oryx*, **39**(1), 95–98.

Mazur, N. (2001). *After the Ark: Environmental Policy Making and the Zoo*. Melbourne: Melbourne University Press.

Pletscher, D. H. & Schwartz, M. K. (2000). The tyranny of population growth. *Conservation Biology*, **14**(6), 1918–1919.

Rees, P. A. (2005). The EC zoos directive: a lost opportunity to implement the convention on biological diversity. *Journal of International Wildlife Law and Policy*, **8**, 51–62.

Roe, D. (2006). Blanket bans – conservation or imperialism? A response to Cooney and Jepson. *Oryx*, **40**, 27–28.

Roe, D., Mulliken, T., Milledge, S., Mremi, J., Mosha, S., & Grieg-Gran, M. (2002). Regulation of the international trade in wildlife. Making a killing or making a living? Wildlife trade, trade controls and rural livelihoods. *Biodiversity and Livelihoods Issues*, **6**, 21–31.

RSPB (2005). Why the RSBP wants a permanent ban on wild bird trade. RSPB species policy department. http://www.rspb.org.uk/policy/wildbirdslaw/banbirdtrade/whyban.asp.

Stevens, P. M. C. & McAlister, E. (2003). Ethics in zoos. *International Zoo Yearbook*, **38**, 94–101.

Stuart, S. N., Chanson, J. S., Cox, N. A., Young, B. E., Rodrigues, A. S. L., Fischman, D. L., & Waller, R. W. (2004). Status and trends of amphibian declines and extinctions worldwide. *Science*, **306**, 1783–1786.

WAZA (2005). *The World Zoo and Aquarium Conservation Strategy: Building a Future for Wildlife*. Liebefeld-Bern: WAZA.

PART II

The challenge of changing behavior

4

Conservation education in zoos: an emphasis on behavioral change

ELEANOR STERLING, JIMIN LEE, AND TOM WOOD

The rationale behind the creation of zoological collections has evolved over time, influenced by shifting historical situations and cultural values. In the long history of zoos, there have been significant changes in the audiences they serve, the methods they use to display animals and impart knowledge, and the messages they seek to communicate. Early animal collections, often termed "menageries," mainly functioned as displays of prestige and power and as entertainment. In the 18th and 19th centuries, venues opened to the public, and zoos increasingly took on the purpose of advancing scientific knowledge and education. In the last few decades, zoos have continued to pursue research and provide education, but under the banner of a new primary goal: to advance conservation objectives. Truly successful conservation education should influence people's attitudes toward wildlife and conservation and inspire behavior changes that are consistent with better stewardship of the environment. Today's zoos face the challenge of effectively acting as vehicles for conservation messages while retaining the entertainment value that attracts people in the first place.

The purpose of early collections of exotic animals, dating back to the ancient menageries of Mesopotamian kings and Egyptian pharaohs (Seidensticker and Lumpkin 1991), was to impress others – the "audience" for these displays included state and personal enemies and neighbors, fellow people of high social standing, and the masses over whom they ruled or whom they outranked. The audience for European zoos in the 19th century was expanded to welcome the general public, while still aiming to impress neighbors and rivals (Baratay and Hardouin-Fugier

2002). During that century, institutions such as the Zoological Society of London placed a strong emphasis on learning and education, establishing a library and important publications. It also provided educational opportunities for various groups, including public lectures starting in 1870, courses of instruction for schoolteachers in 1910, and group admittance to university students in the 1930s (Bullough and Hamilton 1976). The message conveyed by zoos started to include the idea that "we are cultured peoples who value knowledge."

Individuals and institutions endorsed the importance of zoos' educational role throughout the 19th century (see Link 1883, Schufeldt 1889). However, the extensive and organized type of educational programming and exhibition design we see today would not become standard for many years. The setting in which animals are displayed influences our perception of their value. Designs for zoo displays in even the most advanced establishments in the 19th and early 20th centuries usually failed to take into account animals' behavioral and psychological needs, and therefore sent strong subliminal messages about the dominance of humans over nature. Animals were often placed one species per cage, with the cages arranged taxonomically, like a visual encyclopedia. When the famous animal dealer Carl Hagenbeck revolutionized animal display design in the early 1900s, his barless exhibits that presented visitors with staged illusions of wild habitats containing mixed species helped collection managers to think about individual animals and their welfare and audience members to better understand animals and their interactions with their environment (Hancocks 2001).

Conservation efforts during the 17th and 18th century were predominantly undertaken by royalty to exclude the commoners, and education came in the form of edicts. Although it did not rise to the fore for many decades, a fledgling conservation philosophy was established early in the history of American zoos, with the efforts of pioneers such as William Temple Hornaday and the formation of the New York Zoological Park (Bronx Zoo). The conservation-based philosophies that guide today's leading zoos gained traction in the 1960s and 1970s, reflecting the public's growing awareness of environmental and animal welfare issues. As ethically questionable practices of the past and the idea of the zoo as a site intended primarily for entertainment rapidly became unacceptable, zoo professionals increasingly embraced conservation ideas (Curry-Lindahl 1965, Lovejoy 1974). Major changes were made to animal display and care methods. In 1980, the Association of Zoos and Aquariums (AZA) declared that conservation was a top priority. In an effort to dissociate themselves from the traditional image of zoos and emphasize the focus on conservation some institutions in the United States and Europe have even toyed with changing their names to better reflect their new mission. In 1993 the venerable New York Zoological Park (Bronx Zoo) was renamed the International Wildlife Conservation Park. (The name change was criticized in the media and ignored by the public, resulting in a reversal years later.)

The idea of public education as a key objective of zoos did not become mainstream until the mid to late 20th century. A 1978 survey found that only 4% of United-States-based zoos and aquariums responding had educational departments prior to 1950, while 77% did in 1977 (Hensel 1978). As educational programs matured and conservation education became a top priority, there was a shift in their target audience. In the 1970s, educators were spending the majority of their time working with elementary-level school children (Hensel 1978). Programming typically emphasized animal adaptations, introduced the idea of endangered species, and was meant to guide children toward becoming environmentally aware adults. Some people felt that zoos needed to expand their educational reach both in concept and audience; for example, with programs aimed at adults, who were more likely to be in a position to take immediate and effective action (Lovejoy 1974). Such questioning led many zoos to re-focus their programming to target parents, teachers, and university students in the 1980s (Tompson 1989).

Today, zoos are expected to offer a variety of educational opportunities and to engage in conservation efforts. In a world where natural areas are shrinking and populations are increasingly urbanized, for many people zoos have become the most accessible place to get in touch with wildlife (Conway 1969, Lovejoy 1974, Tunnicliffe 1995a, b). The range of educational experiences offered includes animal displays (particularly in naturalistic exhibits), interpretive graphics and text, publications, live animal demonstrations, interactive exhibits or activities, technology-assisted programs, formal education programs, opportunities for communicating with staff and docents, and outreach beyond the walls of the zoo. A 1989 study of visitor attitudes before and after zoo visits found that zoos with an educational focus that displayed animals in authentic environments had a positive impact on visitors' attitudes toward wildlife, while more traditional zoos increased their fear of or indifference to wildlife (Kellert and Dunlap 1989). More recent efforts have focused on involving the zoo-going public as participants in exhibits, as research shows that interactive, participatory exhibits and programs increase learning (Bielick and Doering

1997, Borun and Dritsas 1997). Even the species of animals on exhibit and the materials for sale in gift shops can contribute to the education of zoo visitors.

The mission of many modern zoos is to advance conservation work, and in large part that means educating visitors, informing the public's understanding of conservation matters, and inspiring meaningful changes in people's attitudes and actions. Recent decades have seen the goal of educational programs redefined to emphasize behavior change in their target audiences, a resolve to better evaluate the success of educational initiatives, and a continuation of efforts to broaden the audience. This important transitional era for education in zoos and aquariums resembles the concurrent changes in our approach to education as a society.

PARALLEL TRANSITIONS IN EDUCATION: ZOOS AND UNIVERSITIES

The transitional thinking about educational practices that zoos and aquariums initiated in the 1970s parallels a larger reform effort directed toward more hands-on experiential learning in undergraduate education in the United States. In The Carnegie Foundation's frequently cited report, *Scholarship Reconsidered, Priorities of the Professoriate* (1990), Ernest Boyer chronicled the shifting priorities of teaching, research, and service within and outside universities in post World War II culture. Importantly, the Foundation studies revealed "growing social separations and divisions on campus, increased acts of incivility, and a deepening concern that the spirit of community has diminished." This disconnect between students' lives and their academic pursuits led to a call for revisions in the scholarship of discovery, integration, application, and teaching that are now the hallmark of progressive educational reform. This reform provided background for the debate about "ways of knowing," opening the door to valuable educational opportunities outside the traditional classroom. Active learning processes that provide development and understanding beyond exposure to disconnected facts, and learning environments that promote a wide range of learned thinking, from valuing to critical thinking, are now working their way into the educational system.

This is significant to the educational reform movement in zoos and aquariums in several ways. First, zoo educators are products of our colleges and universities, and they provide learning opportunities to the public based on their understanding of how education works. Second, primary and secondary school teachers are developing a better understanding of the educational opportunities zoos and aquariums represent. And third, zoos and aquariums are excellent places to introduce issues requiring local, national, and international civic engagement.

The transitions all progressive educators have faced during the past 30 years are represented in the emergence of "new zoos" discovering how we learn and what should be learned. These are civic issues, and zoos and similar institutions have the opportunity to influence the direction of public discourse, rather than follow it. In thinking about conservation education programs in zoos, it is significant to consider not whether students and the general public should be educated in zoos, but what they should learn and how zoos can best help them to learn. A better understanding of this question is found by investigating general educational reform in universities as it has paralleled reform in zoos and aquariums. For example, Boyer asked "How can knowledge be responsibly applied to consequential problems?" Zoos and aquariums are not at a loss for identifying problems, be it the bushmeat crisis or local environmental issues, and providing knowledge about those problems. However, we would suggest that important learning only starts with knowing facts about these issues, and improves when people make connections to their values and modify their behavior based on critical review of available information.

EVALUATING SUCCESS IN REACHING THE PUBLIC WITH CONSERVATION MESSAGES

George Rabb, in a 1994 article on the changing role of zoological parks in conserving biodiversity, wrote that "[c]onservation is a challenge to all of us, individually and collectively, to change the way we relate to and act with respect to the rest of the world . . . conservation brings about change in human behaviors to a more harmonious and sustainable relationship with nature . . . ." Clearly, behavior change across various audiences needs to be a fundamental goal of conservation education programs in zoos, and we should be working together to identify the most important methods for eliciting conservation behavior.

Education researchers in the international zoo community are now stressing the need for research that documents behavioral change and civic engagement of visitors. This literature has started to emerge in Australia (Hopkins 1992, Woodside and Kelly 1995), Great Britain (Broad 1996, Esson and Cowan 1998, Esson 2001), and in the United States (Swanagan 2000, Stoinski *et al.* 2001, 2002), but does not yet adequately describe the effectiveness of zoo-based education programs, nor does it generally define what should be learned, and how that learning can be responsibly applied by citizens.

Recognizing the need for more thorough assessment of the overall impact zoos and aquariums have on the visitors, the AZA recently initiated a Multi-Institution Research Project (MIRP) to "investigate the overall impact of visits to zoos and aquariums on visitors' conservation-related knowledge, attitudes, affect and behavior" (Dierking *et al.* 2002). This comprehensive initiative represents a real opportunity to examine the transition that accredited zoos are making from simple recreation-based collections of animals to centers for conservation, research, and learning. With over 200 accredited institutions and over 100 million visitors to AZA institutions annually, this study will allow sampling of institutions of different sizes, geographic regions, and types.

A recent literature review on visitor learning in zoos and aquariums in preparation for the MIRP initiative (Dierking *et al.* 2002) suggests that the past three decades of research in zoos has "superficially" described the interests, knowledge, and beliefs of the visiting public, and those studies tended to investigate public perception of animals, with little attention paid to the overall conservation messages conveyed. In general, research on the impact of conservation messages on zoo visitors is in its infancy (Swanagan 2000, Dierking *et al.* 2002). Similar to conservation education programs in general (Jacobson 1999), few conservation education programs in zoos receive formal evaluations (Dierking *et al.* 2002). Only 50% of AZA institutions conduct evaluations, and fewer publish their results in peer-reviewed journals (Stoinski *et al.* 1998).

The reasons for this dearth of research are many and varied. Zoo educators face the perennial questions of why, how, and what they should evaluate. While it is fairly obvious that evaluation can help to improve programs and staff performance, as well as help in planning future programs, evaluations are problematic. A major stumbling block in assessing education in general and informal learning in particular is how to measure results (Kellert and Dunlap 1989). In part because of expediency, it was previously assumed that merely presenting students with animals in a zoo environment results in improved knowledge and attitudes towards nature (Gutierrez de White and Jacobson 1994). While conservation knowledge is sometimes positively correlated with environmentally responsible consumer actions, voting, and engagement in conservation-related volunteerism, it is difficult to isolate the source of inspiration for behavior change. The zoo-going public has a higher affinity for conservation messages than the general public and might have changed their behavior even without the zoo experience.

Our visitors are already conservation-minded, so we are sometimes measuring minute differences, with visitors moving from doing pretty well to doing really well on changing conservation behaviors. This implies that with many of our programs, we are reaching relatively few of the people who could make a major change in behavior. We might have a greater overall impact if we focused on directly effecting change in people who are not currently committed to conservation behaviors. Alternatively, we could engage the audience that is committed to conservation and work with them to reach others who are not.

Also, multiple sources over time inform visitor actions and construction of their understanding of the world. Any evaluation of a single experience needs to incorporate information on the context – the previous knowledge and experiences of the visitor. Traditional evaluation methods define an educational experience as a single visit or a single issue, again in part because these experiences are easier to evaluate than multiple visits and/or programs across institutions. Often, the evaluations reflect only immediate reactions to an educational experience and do not consider later changes in knowledge or behavior. Very few zoo education evaluation studies incorporated follow-up techniques. Yet those few studies show that people may not immediately absorb what they learned, and that they understand messages better after reflection, discussion with others, and information from other sources. Behavioral change in particular improved with follow-up reinforcement of messages about action (Falk and Dierking 1992, Dierking *et al.* 2002).

We are beginning to examine many active learning processes that are both educational and experimental that can lead to behavior change. The emerging field of conservation psychology provides a framework for better understanding of motivations for conservation behavior (Saunders 2003). While previous paradigms assumed that behavior change is predominately reinforcement driven (Monroe and DeYoung 1993), researchers are now identifying alternative perspectives on motivations for change. For instance, people are more likely to change behavior if they see a clear role for themselves, and feel that this role is not optional but critical to the success of an initiative (Kaplan 1990, Folz 1991). Motivations such as commitment, altruism, and intrinsic satisfactions can help to promote more durable behavior change than reinforcement-dependent changes. Similarly, researchers are finding that behavioral competence and avoidance of feelings of helplessness are critical to effective learning situations. This translates into not introducing new concepts and new techniques simultaneously. Situations that reduce feelings of novelty help individuals to focus their attention on the main messages (Falk *et al.* 1978). Finally, consistency of message across an entire institution is important in fostering behavior change. For instance, if a zoo education program emphasizes the importance of recycling or of composting, and then the overall institution has no recycling or obvious waste separation in its facility, the message often falls short of effecting change.

Often zoo education programs focus mostly on raising awareness, hoping that behavior change will follow. However, the exact kind of behavior change is not articulated, leaving the zoo-going visitor in a state of anxiety about a problem and helplessness about the solution. A substantial number of visitors come in to the zoo or museum experience aware that there is a biodiversity problem, but thinking that it is someone else's job – government, industry, etc. – to fix it. Effective programs incorporate relevant information on skills or actions (Monroe and DeYoung 1993) and do not tell people what to do, but rather give them choices.

Preliminary research shows that zoo visitors have a clear interest in hearing more about the tangible things they can do to help as individuals. Obviously, one difficulty is in identifying appropriate messages that are relevant to their lives. In an aquarium, it is easy to talk about eating sustainably harvested fish. In botanical gardens, appropriate messages include composting, lawn eradication, avoidance of chemicals in gardens/lawns, and planting native species in gardens.

Zoo-based messages are sometimes harder to pinpoint, but have included encouraging visitors to help create and improve habitats for wildlife, choose appropriate pets, limit the negative effects of their travel, donate money to conservation organizations, talk to others about their experience, contact elected officials, vote, recycle, pick up garbage, volunteer, and minimize resource use in general. Effective programs can influence what people watch on TV, which articles they choose to read in the newspaper or magazines, how they make consumer choices, what civic groups they work with, and encourage visitors to reinforce conservation messages with visits to like institutions. Researchers have found that aquarium visits, for instance, inspired individuals to go to other similar facilities, such as zoos, museums, parks, and other aquariums (Dierking *et al.* 2002). Clearly, it is important when planning conservation education programs to identify the roles of nearby related institutions in the overall educational goals and to design programs that take on the post-visit experience (books, materials, websites, public lectures, take-home activities), extending the time and space for the messages (Adelman *et al.* 2000).

As one example, in the 1990s the Bronx Zoo and the American Museum of Natural History in New York City simultaneously built exhibits focusing on African Rainforest habitats. Over the planning of the exhibits, the curators and designers shared ideas for how to reinforce messages in the two exhibits, knowing that there is a high percentage of cross-over visitors between the two institutions. The Congo exhibit, which is described by Gwynne (Chapter 5), is a terrific example of an exhibit focusing on a conservation message. Visitors are immersed in the landscape, interacting as scientists, field workers and conservationists for a day. They learn about gorillas, their habitat, and conservation concerns. They also have the opportunity at the end of the exhibit to choose where the Zoo should invest their entrance fees. For instance, they can pick which species or which conservation strategy to invest in.

A few miles away, the Hall of Biodiversity at the American Museum of Natural History covers four themes: what is biodiversity, why is biodiversity important, what threatens biodiversity, and what we are doing to stem the biodiversity crisis? The extensive Central African Rainforest diorama shows a pristine forest grading into a forest showing more signs of human use. A unique feature of the hall is the biobulletins, high-definition videos that we update quarterly and news flashes that we update weekly. A website extends the experience, as do evening programs and books and magazines. Clearly, building messages across exhibits, formal education, and public programs within and between institutions will help us achieve our goal of changing or reinforcing conservation-friendly behaviors. Both the Congo exhibit and the Biodiversity Hall underwent extensive evaluation – formative and summative of visitor knowledge. To date, neither of these institutions has undertaken a follow-up survey to see what the cross-over learning has been and how their efforts might have contributed to behavior change subsequent to the visits.

The increased interest in how we learn and what we learn over the past 30 years has also stimulated the cooperation of universities and zoos and aquariums to examine common education goals. Given that there are long-standing relationships between research institutions such as the Smithsonian Institution's National Zoo, the Wildlife Conservation Society and the Zoological Society of London and many universities, the educational connections between the smaller zoos and aquariums and universities was set to emerge. One example of this cooperation is the establishment of an educational curriculum, requiring complex collaboration well beyond the educational objectives of single exhibits or programs.

EDUCATING EXISTING ZOO PROFESSIONALS

It has been noted that conservation and education goals of zoos are constrained by the administrative and financial context in which zoos operate. Mazur and Clark (2001) suggest "specialized skills of zoo staff are not always channeled productively into policy reform." This is a reasonable criticism of a complex administrative structure and professional training needs to keep pace with reform. We believe this is best achieved by providing interdisciplinary education of zoo professionals at an advanced curriculum level.

Recognizing the complexities and challenges faced by zoos in meeting conservation and educational goals, in the spring of 2003, George Mason University and the AZA cooperatively announced the establishment of a Master of Arts in an Interdisciplinary Studies degree in Zoo and Aquarium Leadership. Within the first year, the program had admitted over 40 students with positions in zoos and wished to broaden their knowledge of the zoo industry as they were promoted. The core of the curriculum is based on the AZA Board of Regents' response to industry needs and the strengths of the University in administration, environmental science, and education. Ultimately, this partnership represents a serious commitment to provide knowledge of education, collection management, and administration skills to employees who will provide future leadership to zoos and aquariums.

Many other programs have interesting partnerships, such as the training programs for zoo professionals run at Durrell Wildlife Conservation Trust, and more and more non-traditional partnerships arise as zoos reach out to new audiences as exemplified by the collaborative project between the Wildlife Conservation Society and the Zurich Zoo, described by Hatchwell and Rübel (Chapter 14).

In closing, it is clear that conservation education has evolved over the years, sharpening the message towards significant action, and zoo education programs have paralleled these general efforts. Some zoos are developing innovative programs in this arena, and our vision – perhaps our challenge – is for the role of conservation education to increase, especially the emphasis on fostering behavior change and civic engagement.

To summarize a few of the recommendations that emerged from this preliminary survey: first, zoo education programs could develop such that people come to expect conservation messages and not just recreation during their visit. People should depend on zoos as a resource for information on how change can happen across individuals and institutions. Zoos could further interest the public in conservation of species beyond the traditional charismatic fauna.

Second, where possible, zoos could think across institutions (zoos, aquariums, museums, botanical gardens, and universities) to design *and evaluate* programs that will reinforce messages about what individuals can do.

Zoos could be evaluating more throughout the development and implementation of education initiatives. This would include not only incorporating front end research in the early design phase, but also formative evaluation while programs are developing, summative evaluation once programs are in place, and follow-up evaluation with participants well after their engagement in programs to assess behavior changes. Zoos could incorporate more sophisticated statistics and analyses, no matter when the evaluation takes place.

More zoos could develop conservation education programs that are responsive to a diverse population and identify alternative ways of reaching new audiences who do not currently share our conservation values, and who could make major behavioral changes that would benefit themselves and conservation efforts.

Zoos could collaborate more closely with individuals and institutions working towards better understanding motivations and incentives for behavioral change, and identify ways to incorporate their findings into our programs. They could also focus more carefully on the link between raising knowledge and changing values, attitudes, and behavior. Avoiding too many assumptions about links between these areas and fostering effective research will be important steps to help us to allocate limited resources.

Finally, zoos could publish their results, sharing not just what works, but what did not work. We, the authors, were asked by the organizers to undertake a critical view of zoo conservation education from the viewpoint of an external expert. The dearth of published evaluations prevented us from doing so. Substantial interesting work continues to lie buried in unpublished evaluations of programs and exhibits. Lessons learned can be very valuable within and across institutions and should be more widely disseminated and not just available to people in the know.

With that in mind, we look forward to the results of the ongoing Multi-Institution Research Project and other initiatives such as those summarized in this volume. We urge the zoo community to continue to focus on raising knowledge, affect, and attitudes towards conservation, but also to ensure that these result in our ultimate goal, that of civic engagement in conservation behavior.

References

Adelman, L. M., Falk, J. H., & James, S. (2000). Assessing the National Aquarium in Baltimore's impact on visitors' conservation knowledge, attitudes and behaviors. *Curator*, **43**, 33–61.

Baratay, E. & Hardouin-Fugier, E. (2002). *Zoo: A History of Zoological Gardens in the West.* London: Reaktion Books.

Bielick, S. & Doering, Z. D. (1997). An assessment of the "Think Tank" exhibition at the National Zoological Park (Report No. 97–1). Washington, D.C.: Smithsonian Institution.

Borun, M. & Dritsas, J. (1997). Developing family-friendly exhibits. *Curator: The Museum Journal*, **40/3**, 178–196.

Boyer, E. L. (1990). *Scholarship Reconsidered, Priorities of the Professoriate.* San Francisco: Jossey-Bass, The Carnegie Foundation for the Advancement of Teaching.

Broad, G. (1996). Visitor profile and evaluation of informal education at Jersey Zoo. *Dodo*, **32**, 166–192.

Bullough, W. S. & Hamilton, F. (1976). The role of education. In *The Zoological Society of London 1826–1976 and Beyond (The Proceedings of a Symposium held at The Zoological Society of London on 25 and 26 March 1976)*, ed. L. Zuckerman. London: Academic Press, pp. 223–231.

Conway, W. G. (1969). Zoos: their changing roles. *Science*, **163**, 48–52.

Curry-Lindahl, K. (1965). Conservation of nature – a duty for zoological gardens. *International Zoo Yearbook*, **5**, 100–102.

Dierking, L. D., Burknyk, K., Buchner, K. S., & Falk, J. H. (2002). *Visitor Learning in Zoos and Aquariums: A Literature Review.* Annapolis, MD: Institute for Learning Innovation.

Esson, M. (2001). Does conservation education travel well? Transferring skills from Jersey Zoo to institutions in the developing world. *Dodo*, **37**, 80–87.

Esson, M. & Cowan, K. (1998). Cross-curricular activities: a richness of opportunity for zoo educators. *Dodo*, **34**, 115–124.

Falk, J. W., & Dierking, L. W. (1992). *The Museum Experience.* Washington, D.C.: Whalesback Books.

Falk, J., Martin, W., and Balling, J. (1978). The novel field trip phenomenon: adjustment to novel settings interferes with task learning. *Journal of Research in Science Teaching*, **15**, 468–472.

Folz, D. H. (1991). Recycling program design, management and participation: a national survey of municipal experience. *Public Administration Review*, **51**(3), 222–223.

Gutierrez de White & Jacobson (1994). Evaluating conservation education programs at a South American zoo. *Journal of Environmental Education*, **25(4)**, 18–22.

Hancocks, D. (2001). *A Different Nature: The Paradoxical World of Zoos and Their Uncertain Future.* Berkeley: University of California Press.

Hensel, K. A. (1978). Education in zoos and aquariums – trends and projections. In *American Association of Zoological Parks & Aquariums Annual Conference Proceedings 1978*, pp. 117–124.
Hopkins, C. (1992). Zoo education into the 1990s. *International Zoo Yearbook*, **31**, 99–103.
Jacobson, S. (1999). *Communication Skills for Conservation Professionals.* Washington, D.C.: Island Press.
Kaplan, S. (1990). Being needed, adaptive muddling and human-environment relationships. EDRA 21. In *Proceedings of the Twenty-First Annual Conference of the Environmental Design Research Association*. Oklahoma City, OK: EDRA.
Kellert, S. R. & Dunlap, J. (1989). *Information Learning at the Zoo: A Study of Attitude and Knowledge Impacts.* Philadelphia: Zoological Society of Philadelphia.
Link, T. (1883). Zoological gardens, a critical essay. *American Naturalist*, **17**, 1225–1229.
Lovejoy, T. E. (1974). The functions of zoological gardens. *Annual Report/Jersey Wildlife Preservation Trust*, **11**, 14–15.
Mazur, N. A. & Clark, T. W. (2001). Zoos and conservation: policy making and organizational challenges. In *Species and Ecosystem Conservation: An Interdisciplinary Approach*, eds. T. W. Clark, M. J. Stevenson, K. Ziegelmayer, & M. B. Rutherford. Yale University Press: Yale School of Forestry & Environmental Studies.
Monroe, M. & DeYoung, R. (1993). Designing programs for changing behavior. In *American Association of Zoological Parks & Aquariums Annual Conference Proceedings 1993*, pp. 180–187.
Saunders, C. D. (2003). The emerging field of conservation psychology. *Human Ecology Review*, **10**, 137–153.
Schufeldt, R. W. (1889). Zoological gardens: their uses and management. *The Popular Science Monthly*, **34**, 782–791.
Seidensticker, J. & Lumpkin, S. eds. (1991). *Great Cats.* Emmaus, PA: Rodale Press. 240 pp. [Introduction by J. Seidensticker and S. Lumpkin; Introduction to the living cats by J. Seidensticker; captions by S. Lumpkin and J. Seidensticker.]
Stoinski, T. S., Lukas, K. E., & Maple, T. L. (1998). A survey of research in North American zoos and aquariums. *Zoo Biology*, **17**, 167–180.
Stoinski, T. S., Ogden, J. J., Gold, K. C., & Maple, T. L. (2001). Captive apes and zoo education. In *Great Apes and Humans: The Ethics of Coexistence*, eds. B. B. Beck, T. Stoinski, M. Hutchins, T. L. Maple, B. Norton, A. Rowan, E. F. Stevens, & A. Arluke. Washington, D.C.: Smithsonian Institution Press.
Stoinski, T. S., Allen, M. T., Bloomsmith, M. A., Forthman, D. L., & Maple, T. L. (2002). Educating zoo visitors about complex environmental issues: should we do it and how? *Curator*, **45**, 129–143.
Swanagan, J. S. (2000). Factors influencing zoo visitors' conservation attitudes and behavior. *Journal of Environmental Education*, **31**, 26–31.
Tompson, C. G. (1989). Hope for the future: strategies for effective conservation education. *Zoo Biology*, **8** (Suppl. 1), 171–175.

Tunnicliffe, S. D. (1995a). What do zoos and museums have to offer young children for learning about animals? *Journal of Education in Museums*, **16**, 16–19.
Tunnicliffe, S. D. (1995b). Zoo talk: the content of conversations of family visitor groups whilst looking at live animals. In *AZA Annual Conference Proceedings*. Chicago, IL: AZA, pp. 645–647.
Woodside, D. P. & Kelly, J. D. (1995). The development of local, national and international zoo-based education programmes. *International Zoo Yearbook*, **34**, 231–246.

5

Inspiration for conservation: moving audiences to care

JOHN A. GWYNNE

INTRODUCTION

We all know that "the architect is the most dangerous animal in the zoo," but if good architecture is the mirror of its function and its time, what is the vision for the zoo of the 21st century? "We shape our buildings, thereafter they shape us," said Winston Churchill. If our zoo vision is conservation, what designs shape inspiration for conservation? What designs solve dichotomy between being fun family attractions and serious science institutions? Can a zoo visit truly change visitor attitudes and stimulate conservation action?

POTENTIAL FOR INSPIRATION

More than propagation or education, might the greater importance of the energetic zoo be inspiration, the ability to inspire millions to the wonders and needs of wildlife? No film, no museum, and no TV special has a zoo's enormous ability to bring wildlife to urban populations with such potential for moving people to care about an animal. No media and few people can inspire a little girl's caring more powerfully than a real gorilla can, with its great brown eyes only inches away. With these close personal encounters zoos' potential for seduction is vast. However, a New York or London gorilla alone doesn't automatically make a connection to gorilla conservation in nature. An adjacent educational graphic rarely seems strong enough. Our audiences are highly visual, information-jammed, time-compressed. If our intent is truly to connect them to *in situ* conservation, we need to think and build *very very* differently. The focus will be more about creating a visitor's emotional experience than about architecture.

EXHIBIT EXPERIMENTATION

It may be useful to consider recent directions in zoo design. In the period 1960–70 we all explored habitat. As an example, WC's World of Birds showcased birds in nature while expanding visitor perceptions of space. Later we all learned about the value of "affect" – the emotions associated with an idea. Here, JungleWorld cloud machines and dramatic vistas seduced visitors into valuing rainforests almost for aesthetics alone. A positive emotional connection for city kids to nature was also Richard Lattis' (WCS) mantra for new "Children's Zoos."

Then we all learned that "zoo rock isn't real rock," as we began to push aside hard-edge architecture covered with an expensive lump of sculpted concrete. We learned about "environmental immersion" as landscape architecture replaced architecture and as immersion exhibits put the visitor in the same habitat as animals. A replicated montane grassland for gelada baboons in the Bronx Zoo was made fun to explore. Similarly in Seattle subconscious messages were "no wilderness, no animals" and the frail beauty of a specific ecosystem in nature. We also learned about three-dimensional graphics that invite you to touch and explore, and mechanically interactive graphics that stimulate your hands, nose, ears as well as eyes.

We also learned about the liabilities of immersion. *Seattle Weekly* (Scigliano 2003) and *Washington Post* articles (Hyson 2003) squawk "show me the animals." We know that better design, behavioral enrichment, and subtle devices such as "baboon bun warmers" hidden in the grass can all bring animals up close. The real criticism of immersion is its one-dimensionality: how do you really learn something about this grass-eating specialist, about Ethiopian endemism, about geladas' increasing need for protection, or the specifics of the possibility of our helping?

We're also learning a lot about conservation connections. A direct connection occurred in 1999 when Congo Gorilla Forest visitors used touch screens to "vote" how their own $3 exhibit surcharge would be spent in Central Africa. They selected real conservation activities (Discover, Involve, or Protect) and about a million of their dollars supports field work *in-situ* yearly. Alex Rübel's visionary new Masoala Rainforest goes a step further to connect Swiss zoo visitors and their funds directly to just one great Malagasy national park (Chapter 14). The Bronx's Tiger Mountain explores on-line enrollment touch screens to connect visitors at their homes to zoo tigers

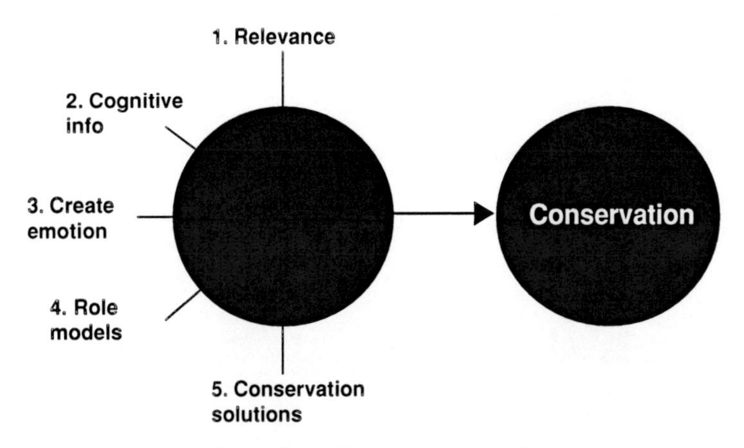

Figure 5.1 The five forces that influence inspiration for conservation

and field conservation. About 5% sign up, a much higher number than the below 1% via direct mail. Some of us are pushing further, past architecture, past landscape immersion and just interactivity into creating more powerful, even theatrical, visitor experiences that drive people further toward conservation. *Congo* was a dauntingly big project, but its choreographed one-third of a mile trail, two dozen gorillas up-close, 75 animal species, theatre with dramatically opening curtain and tunnels managed to create an experience that moved visitors. How do we do this more simply again to inspire people to care?

FINDING PATTERNS

We believe that powerful, relevant experiences are what motivate people and such experiences are the heart of inspiring people for conservation. Five forces, working in tandem, seem to influence this experience. The first, *relevance*, comes from the visitor, not from us. Next, our *cognitive* messages appear important. We need to *create emotion* and to find *role models* other than ourselves. The fifth seems to be providing *positive conservation solutions* (Figure 5.1).

Create relevance for visitors

How many of us have heard a distinguished director's exasperation that "they don't read signs?" How many have seen the director read that sign? Interpretation is produced in the tradition of what we think "they" need to know. No one's thinking that both children and the zoo director may either

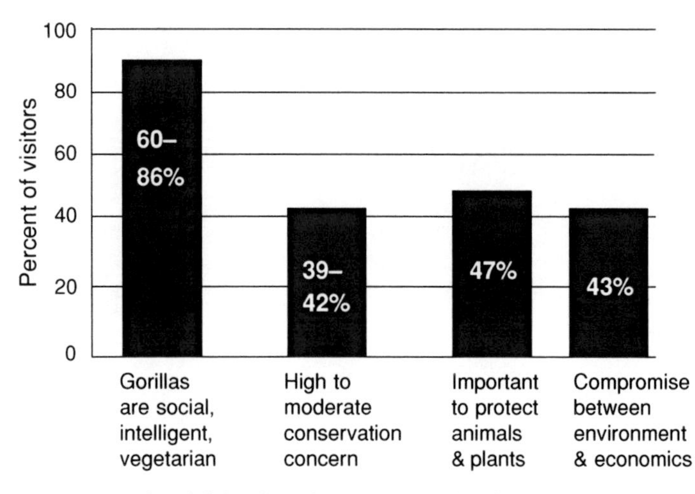

Figure 5.2 What did they know? (Source: People, Places and Design Research 1996)

already know it's a giraffe or that our curatorial information may hold zero relevance to our audiences. They are not blank slates, ready to be filled with our information.

Evaluations indicate that many US zoo visitors arrive with much higher baseline information than we anticipate. Our 1996 *Congo Gorilla Forest* audience prestudy (People, Places and Design Research, 1996) indicated 60%–86% of our New York audiences were beyond the novice stage: they already knew gorillas are social, intelligent vegetarians (obviating the need for basic social, intelligence or diet graphics); 42% cited high and 39% moderate concern for conservation (Figure 5.2).

They also bring "naive knowledge," sometimes misinformation, that pre-shapes how they approach our messages. To reach them we must find a common ground and meet them there. A small survey in 2003 found that 45% of Bronx Zoo audiences can't identify Asia on a map by its shape alone and 47% can't recognize South America without a larger world view (Fraser *et al.* 2003). These certainly are problematic data if one is trying to present substantive information to zoo audiences in the USA, such as reduction in habitat. Similarly the front-end study for *Tiger Mountain* indicated that only 26% of New York visitors believe that tigers come from Asia, while 47% incorrectly responded "Africa" (Media Transformations 2002a). The same poll indicated an interesting knowledge link. Of those who knew Asia, 52% said they would donate money; while of those who incorrectly stated Africa, only 19% were interested in donating. We probe for access points

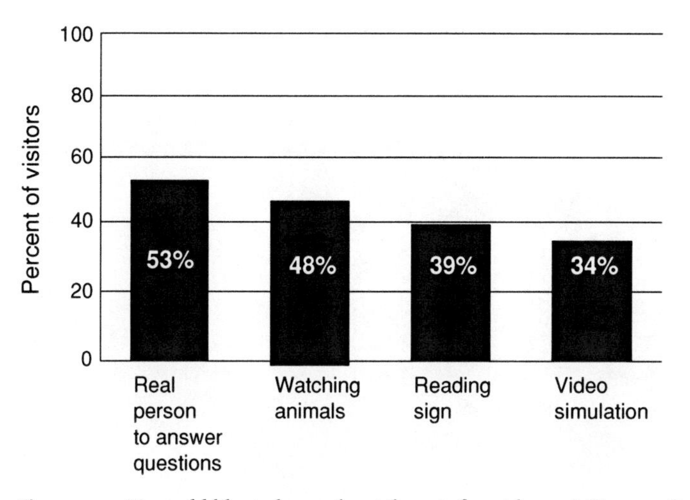

Figure 5.3 "I would like to learn about the rainforest by . . ." (Source: People, Places and Design Research 1996)

to what's relevant to our guests. Geography clearly wasn't naturally relevant to them, so the programming/design team opted to focus on tigers and conservation.

A small up-front study for a Madagascar exhibit indicates that only 42% of our English-speaking and 27% of our Spanish-speaking audiences reported having heard of Madagascar. Fewer know where it is. However, consistent with our *Tiger Mountain* findings, they may know little about Madagascar but indicate moderate-to-high compassion for its endangered species (Media Transformations 2003).

Education level influences attitudes. At the New York Aquarium the final graphic design solution for the *Alien Stingers* exhibition purposefully presents two levels of information. Design 1 has white type and carries more substantive facts. Design 2 has yellow handwritten type and provides a more casual approach, what we call the 'voice of Dr Science." This double-tier approach resulted from a formative study comparing alternative design mockups. People with graduate school (66%) or college education (57%) liked design 1 that was more fact-oriented. Those with less formal education preferred casual design 2 with handwritten short answers (People, Places and Design Research 2001). We combined both solutions to serve both audiences.

To determine visitor preferences for getting information, evaluations for *Congo Gorilla Forest* indicated slightly more visitors (53%) preferring talking with a live person, versus just watching animals (48%), sign graphics (39%),

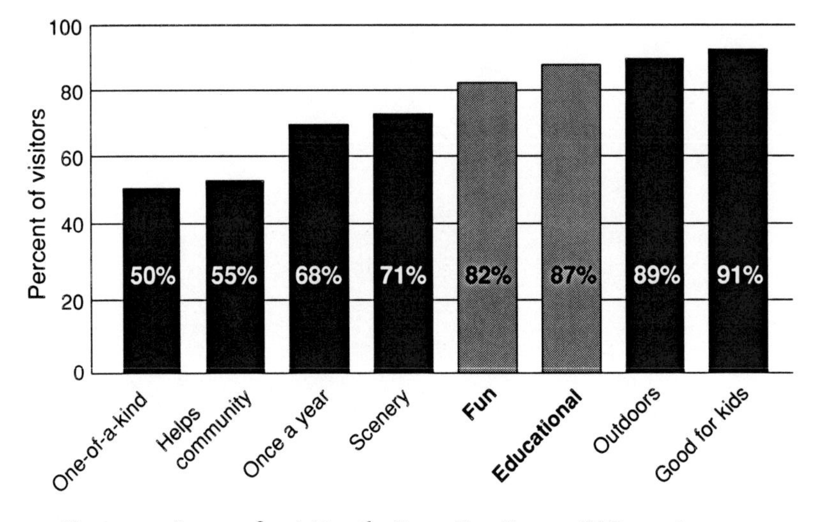

Figure 5.4 Reasons for visiting the Bronx Zoo (Source: BSG 2002)

video, and media (34%) (Figure 5.3; People, Places and Design Research 1996).

Much of our audience is highly visual. Adults tend to want the answer quickly. Children's minds are apparently more influenced by game-based inquiry, hence the proliferation of interactive learning tools for kids. Three-dimensional and interactive stations do seem to grab more of children's interest than two-dimensional graphics.

Exhibits should encourage dialogs, not monologs. While the traditional curator felt responsibility for conveying instructional information, few visitors want to reveal ignorance, especially parents to their kids. Indirect ways of getting information across are probably more effective. Keeper talks in *Tiger Mountain* have worked at creating dialog and at generating excitement about behavioral enrichment with a live tiger so close.

Some want information in new ways; the "Game Generation" (Prensky 2001) of multi-tasking for elementary-to-college students apparently attunes to different methods of information gathering – including random non-linear access, graphics first, highly digitally oriented, multiple tracks at once. How are we anticipating relevance to this new audience?

Concentrate on cognitive information

New York audiences seem to recognize the educational value of a zoo visit for their families. When asked in 2002 why they like to come to the Bronx Zoo, we saw them scoring education value at 87%, even higher than fun at 82% (BSG 2002) (but see also Chapter 6). This is consistent with US national trends. We have traditionally taken the approach that audiences may not be attracted to the zoo for education or conservation *per se*, but appear to value it once they are inside the park. But when one looks at this study, one wonders if we are underrating how much our audiences value coming to teach their kids (Figure 5.4).

Interpreters strongly advocate a rigorous process to winnow down a multitude of ideas to the exhibit's single "big idea." It is important that the bulk of visual and cognitive messages reinforce it throughout. At the Wildlife Conservation Society (WCS), we believe each "graphic" or interactive element can also really only illustrate one idea at once. (Imagine a car advertisement trying also to advertise car paint at the same time.) However, we also greatly value the visual layering of rich sub-messages. This is communication design with layers prioritized to reinforce the message. To the audience it has content and creates an easy way to learn.

"What they see is what they get." Our audiences are so highly visual that if the cognitive message isn't aligned with what they're seeing, they disconnect. Visual and cognitive messages must be in synch in order to work. At *Conservation Hall* in the New York Aquarium we expected that visitors would make conservation connections with the overhead graphics. But what are they really seeing? They see fish, not conservation. If we expect people to think conservation, we have got to devise ways to put conservation within the center of the tank without diminishing nature. As an example, we're experimenting with a video "pepper's ghost." A six-foot tall talking version of Curator Loisel will appear in the tank seeming to interact with living cichlids which he saved just prior to extinction in nature. With bubbly words he will tell its story. Fun, yes. Will visitors look? Yes. Will they understand our message? Probably – with the right balance of clear message and this interesting delivery. Will this "cheapen" science or diminish the fish? Not if we do it properly. How long will it be retained? That's the toughest to measure.

Create an emotion-based experience

The third ingredient to inspiring people is by trying to create a visitor's emotion reaction. Caring itself is affect-based and if we're trying to inspire people to conservation, one is automatically in the land of emotion.

For most of the last century, our zoo audiences saw zoo animals in places clearly created and dominated by man. At many a renowned elephant house, visitors saw a piece of architecture so monumental that it could dwarf a herd of elephants. Animals are not ennobled by such, and now we need more than ever to present them at their best, magnificently adapted to live in complex wild ecosystems. Our future exhibitions must create very different visitor experiences.

The most powerful weapon for creating positive emotion is a close encounter with a real animal doing natural behavior in a natural-looking place, especially if it is engaging visitors. How can we give animals choices and huge exhibits, while providing ways for people to "feel" a snow leopard or tiger up-close? We've successfully experimented with heating rocks on a snowy day for snow leopards, but how can we do a better job designing with the animals in mind, designing with shade, breezes, food, enrichment sites, observation stations, food rewards, and social dynamics to bring animals in "wild habitats" close-up? New "behavior enrichment" devices provide important stimulation for animals, but can we better involve visitors with such enrichments? Living animals aren't constantly active animatronics. How do we sustain visitor interest when our animal dozes at midday or audiences are attuned to seeing it hunt on nature television? Dramatic, beautiful places, theater, and backlighting help create interest when our star is resting, but can we do more?

Can video subtly meld the real with video action, or will technology in naturalistic exhibits compete with nature? Years ago we exhibited a live koala in a diurnal exhibit along with two educational videos showing koalas eating and interacting. We did not evaluate formally but got the impression that visitors left the exhibit integrating the sleepy koala with moving videos and felt satisfied by seeing a moving animal. Buoyed by this perceived tech-usefulness, at *Tiger Mountain* we used video projection and interactive touch screens outdoors in the woods to convey most of our graphics messages. Would tech be perceived by our visitors as out-of-place? Interestingly our evaluation indicated adults thinking the tech a little more appropriate for themselves than for kids, but in actuality 86% of the 1-min use is by the children (Media Transformations 2003). A chick-hatching video at Woodland Park Zoo also produced positive guest reactions . . . one parent said "it's such a hard concept for kids to understand . . . unless they see it actually come out, they just don't get it" (Woodland Park Zoo 2003).

Find role models

To amplify our mission, from what role models can we learn about creating emotion-based experiences that attract and stimulate audiences in order to amplify a message or product? One thinks of Ralph Lauren branding a lifestyle not just a shirt. Disney theme parks have had enormous success in creating consistent experiences. On Oxford Street in London, the new energy of Selfridges comes partly from selling the experience of shopping as an event, titillating shoppers with fortune tellers and light porn on the main aisle. In an era of celebrity marketing, should we be thinking of a lemur or gorilla species as a "celebrity"? This approach is a foreign world to the museum and zoo community, but what can we learn here to enhance our message? As at Selfridges, here we see a *Congo* visitor also shopping for an experience. Hopefully she's moved by her rainforest experience as we use fog machines as dramatic build-up for finding herself in the midst of gorillas.

Emphasize positive conservation solutions

Audiences already know that the natural world is in trouble. They don't know how much and don't seem to want to know. A recent book *Beyond Ecophobia* (Sobel 1999) contends that American second- and third-graders may be feeling so impotent that they distance themselves from, rather than connect to, the natural world which is so painful. They hear of ten thousand acres of rainforest lost between their recess and lunch, of accelerating loss of endangered species. With insufficient real contact with nature they don't want to keep hearing that their favorite children's book animal is in dire trouble for reasons that they didn't cause, about which they are powerless to help, and don't know anyone who can. The Bronx Zoo's up-front evaluation for *Tiger Mountain* (mostly adults) fortunately picked up more optimistic audience data: 90% said people could do something to save tigers, and 70% identified a way they could personally do something (Media Transformations 2002b).

Some of us think we need to emphasize conservation that works. This is consistent with story lines for nature programs – a *Discovery* director told us in 1999 that they could no longer sell films without "up" endings to media. But, how do we not dumb-down issues as important as the loss of Sumatra's forest or marine fisheries? If the media aren't telling it and we don't, who will? Often our living collections are the only strong visceral connection our audiences have to Sumatra or mangrove nurseries. Isn't this story our ethical obligation?

We experimented with a powerful bushmeat photo at *Congo* just prior to introducing conservation solutions. Here another film-maker, Archipelago Films, may have found one way to get real, if unpleasant, data across. WCS had very much wanted our *Congo* film to end on a hopeful note and argued with Archipelago that we didn't want to show a lot of deforestation, bushmeat, and logging. They argued (correctly, as we think now) that in order for them as film-makers to create the most emotive upbeat finale they, in fact, needed the previous downers as dramatic contrast. It worked.

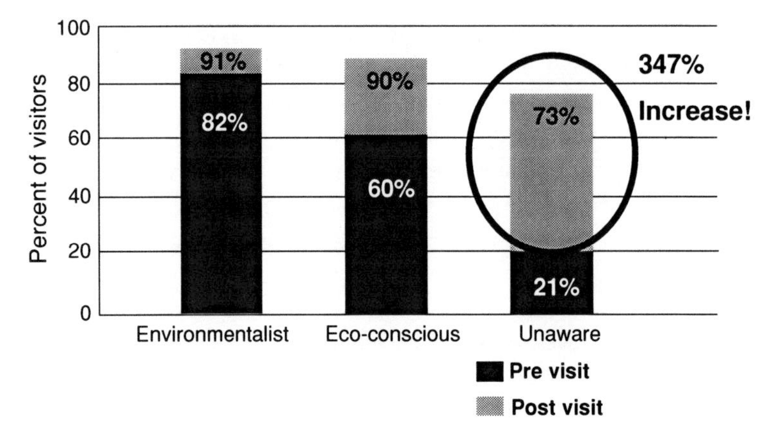

Figure 5.5 Concern for the future of African rainforests (Source: People, Places and Design Research 1996)

Regarding positive solutions, for our upcoming Madagascar exhibit we experiment with a walk-through film. A projected flash flood from upstream erosion will symbolically wipe out the animals and visitors. After a few seconds, life will begin again, music becoming optimistic, scientists and joyous local people symbolically providing support. It's the kind of affective experience that we're exploring even though it pushes us into the unexpected role of theater and film-maker. If such an upbeat solution-based finale is the key, have we stumbled upon solving our conundrum between serious science or "fun"? By emphasizing the positive, can't we be both? Zoos can be beautiful fun celebrations of life and showcase positive conservation stories in spite of occasional tough circumstances.

CHANGING ATTITUDES

We have learned we can change public opinion toward conservation, even reach new audiences. A 1996 pre-*Congo* poll indicated that our audiences were divided in concern for the future of rainforests (People, Places and Design Research 2000). The exhibit increased caring by 110% among prior environmentalists but also stimulated a gratifyingly 347% jump in concern among those previously unaware (Figure 5.5).

Interestingly *Congo* indicated there are three places in a one-third of a mile visitor tour where an *in-situ* conservation message gets across to our guests. It does not occur when they are looking at two dozen live gorillas. It happens with a graphic message of forest destruction. It happens at a theater with powerful film that "takes you" to Central Africa and it occurs at the exhibit finale where 70% of the visitors use touch screens to "vote" their exhibit entry fee directly to field conservation in Africa. Here they directed their thoughts along with their funds to Africa and apparently liked doing so: 54% of visitors said "it felt good to know I did something to help" plus 24% said "it was an enjoyable hands-on activity," plus 18% said "it helped me understand about conservation activities." Only 4% said "it wasn't very interesting." As a data cross-reference, a striking 93% said they would like the computer voting stations to remain if future changes were to be made to the exhibit (People, Places and Design Research 2000).

A NEW KIND OF INSTITUTION

How can we reinvent ourselves, so strategically sited in urban centers such as New York and London, emotional epicenters of the English-speaking world? If one's target is to create a powerful emotional experience incorporating messages and making a strong conservation connection to a distant place, it does mean we need to completely change how we build architecture. Long before we talk architecture or landscape, perhaps even collections, we need to begin with a fearless inventory of what feelings, experiences, and messages we want our visitors to get. It is not just thinking about exhibits and landscapes before architecture, but about creating meaningful emotive experiences before exhibits.

It's not easy for us, more perspiration than inspiration at first. But Nature is too important for us to give up. We need vital new ways to present her realities, so much more inspirational than fiction.

References

BSG (2002). *Growing the Bronx Zoo Visitor Base.* Bronx Zoo Group.

Fraser, J., Taylor, T., & Cave, S. (2003). *Understanding Maps, Preliminary Findings.* Bronx: Wildlife Conservation Society.

Hyson, J. (2003). No business like zoo business. Research is fine but show us the animals. *Washington Post*, 11 May 2003.

Media Transformations (2002a). *Tiger Mountain Survey Report.* Teaneck, NJ: Media Transformations.

Media Transformations (2002b). *Tiger Evaluation: Front End.* Teaneck, NJ: Media Transformations.

Media Transformations (2003). *Tiger Mountain Exhibit Summative Evaluation.* Teaneck, NJ: Media Transformations.

People, Places and Design Research (1996). *Visitors' Interests, Knowledge and Perceptions about Congo Gorilla Forest.* Amherst, MA: People, Places and Design Research.

People, Places and Design Research (2000). *Use and Perceptions of the Conservation Choices Computers*. Amherst, MA: People, Places and Design Research.

People, Places and Design Research (2001). *Formative Study #1: Mock-ups of the First Gallery of Alien Stingers, NYA*. Amherst, MA: People, Places and Design Research.

Prensky, M. (2001). *Digital Game-Based Learning*. New York: McGraw Hill.

Scigliano, E. (2003). Whose zoo. *Seattle Weekly*, 11–17 June 2003.

Sobel, D. (1999). *Beyond Ecophobia. Reclaiming the Heart in Nature Education*. Great Barrington, MA: The Orion Society.

6

Attitudes and attitude change among zoo visitors

RICHARD P. READING AND BRIAN J. MILLER

INTRODUCTION

The primary mission of zoos increasingly focuses on conservation and conservation education (Rhoads and Goldsworthy 1979, Altman 1998, Morgan and Hodgkinson 1999, Conway 2000, Swanagan 2000, Hutchins *et al.* 2003, Miller *et al.* 2004). To promote conservation effectively, we argue that zoos must do much more than simply provide information to visitors; they must help engender positive values and attitudes toward nature. Many zoos attempt just that. Unfortunately, little research has evaluated whether zoos are successful in those efforts (Rhoads and Goldsworthy 1979). Yet, many zoos alter exhibits, develop education programs, and create new graphics with the goal of increasing support for nature conservation. We encourage further research to evaluate the efficacy of these new directions and programs.

Toward that end, we initiated a study into the attitudes of visitors to the Denver Zoo. We view this study as a first step in evaluating how well our zoo addresses its conservation education mandate. We asked visitors to the Denver Zoo questions about their views on zoos in general, their reasons for visiting the zoo, and their attitudes toward the relationship between people and nature. We also asked questions to visitors about a specific conservation topic of interest to people in Colorado, namely attitudes towards gray wolves (*Canis lupus*) and wolf restoration in the state. We chose wolf restoration because it is a timely and controversial conservation topic in Colorado, especially given the recent restoration efforts to the north (Wyoming, Idaho, and Montana) and south (Arizona and New Mexico) of the state, and because we were also studying attitudes toward this topic among the general public (Meadow *et al.* 2005). We discuss our results with respect to future directions for zoos in social science research.

Values, attitudes, and knowledge

Terms such as values, attitudes, and knowledge are often used inconsistently by people outside of the social sciences. Even social scientists may define these terms differently. It is therefore important that we define our terms.

A *value* is a preferred mode of behaving (e.g., honesty) or existing (e.g., equality). Numerous values exist and social scientists classify them in various ways (e.g., Lasswell 1971, Steinhoff 1980, Kellert 1996). People possess multiple values, even on a single topic, that vary in strength and are ordered hierarchically (Rokeach 1972, Brown 1984). When values clash people usually rely upon more strongly held values (core values) over less strongly held ones (peripheral values) (Williams 1979). Thus, two people can share a common value, but still come into conflict on a relevant issue if the strength of that value varies (especially relative to other values). Values influence and are affected by attitudes, contextual factors, and knowledge (Rokeach 1972, Williams 1979, Brown 1984, Brown and Manfredo 1987).

Attitudes are affinities or aversions toward an issue or entity based on beliefs, or what a person senses and understands about how that issue or entity affects a given situation (Bem 1970, Rokeach 1972, Brown 1984). Extreme attitudes tend to be based on more simple belief systems than moderate attitudes (Bright and Barro 2000).

Context describes a person's situation; for example, frequency of exposure to an issue, customs, peer pressure, socialization by institutions, mood, and physical state (e.g., health) (Brown and Manfredo 1987, Chaiken and Stangor 1987).

Knowledge is the acquisition, comprehension, and retention of information and it depends on exposure, receptivity, perception, interpretation, and memory (Petty *et al.* 1997). Knowledge is only one of several factors influencing values and attitudes, and its influence is often relatively weak. Yet the importance of knowledge is often overestimated, especially among people who value knowledge greatly, such as educators and conservationists (Reading 1993, Kellert *et al.* 1996). When values and attitudes are strongly held, new knowledge is often selectively received and interpreted (Tessler and Shaffer 1990, Olson and Zanna 1993). In part, this results because people focus on and more easily memorize and recall information that supports existing values and attitudes. These interactions are strengthened if information is poor, ambiguous, or too complex to be easily understood (Tessler and Shaffer 1990, Olson and Zanna 1993).

Changing values, attitudes, and associated behaviors is difficult and education programs directed at doing so are rarely successful, especially when values and attitudes are strongly held (Chaiken and Stangor 1987). Providing information is important, but since knowledge is only one of several factors influencing attitudes, changing the attitudes of people with strongly held values and attitudes is difficult or impossible. Indeed, for both wolves (Bath 1989, Kellert 1990) and black-footed ferrets (*Mustela nigripes*) (Reading and Kellert 1993), the two groups of people that scored highest on knowledge of the species were ranchers and conservationists. Yet, those two groups had diametrically opposed attitudes about those species.

Values, attitudes, and behaviors can and do change over time as situations, knowledge, and experiences change (Sinden and Worrel 1979, Williams 1979). Understanding why and how values, attitudes, and behaviors change is important if zoos hope to develop more supportive publics for conservation. Inducing such change usually requires more than simply providing information. Most social scientists agree that the best chances for attitude change occur when individuals become aware of internal contradictions between different values, or between values, attitudes, and behaviors (Williams 1979, Olson and Zanna 1993, Petty *et al.* 1997). For example antilittering campaigns often succeed by playing on people's patriotism and love of nature. People will seek to reduce the discomfort they experience from these inconsistencies by changing more dissonant, peripheral values and attitudes and behaviors to better reflect core values (Williams 1979, Tessler and Shaffer 1990).

Persuading people to make such changes requires that they both receive and acquiesce to a persuasive message (Olson and Zanna 1993). Receptivity depends on several factors, including motivation, the identity of the messenger, cultural congruence of the message and receiver, the strength and frequency of the message, the clarity of the message, and the state of the recipient (Chaiken and Stangor 1987, Petty *et al.* 1997). Peer pressure can play a large role in maintaining or changing values, attitudes, and behaviors (Chaiken and Stangor 1987, Tessler and Shaffer 1990). In addition, changes are more likely to occur when alternative choices are provided that facilitate attitude or behavior change or permit people to reach the same or similar goals differently (Tessler and Shaffer 1990, Petty *et al.* 1997). For example, people are more likely to throw trash into a garbage can than on to the street when such containers are made easily available. Another effective way to change attitudes is through the social institutions that form and reinforce values (Reading 1993). All major social institutions are important, but education and religious institutions may be the most influential (Reading 1993). When a value is strongly intertwined with other values, or is the product of personal experience, it is more difficult to change (Williams 1979, Olson and Zanna 1993).

Some people argue that we should not try to influence other people's values and attitudes. Usually these people disagree with the value or attitude being promoted and strive to make it seem as though they are "taking the moral high ground." In reality, such a stance promotes the status quo. While we should respect the right of people to hold different values and attitudes, we should also recognize that since everyone believes their value system is superior (otherwise they would change it), people constantly strive to influence the values and attitudes of others.

Zoos, education, and attitude change

All zoos accredited by the Association of Zoos and Aquariums (AZA) support active education programs. AZA-affiliated zoos host millions of visitors every year (over 135 million in 2002) and are considered one of the most trusted sources of information about nature by the general public (Hutchins *et al.* 2003). As such, zoos have vast potential to educate people and promote positive attitudes toward animals and nature. Yet critics argue that zoos are not educating people (Swanagan 2000).

One of the greatest challenges to zoo education programs is that visitors come to zoos largely to be entertained or for family outings, not for educational experiences (Andereck and Caldwell 1994, Tunnicliffe 1996, Acampora 1998, Morgan and Hodgkinson 1999). However, visitors, especially parents with children, often cite education as a secondary goal of zoo visits (Kidd *et al.* 1995, Morgan and Hodgkinson 1999). Thus, one of the challenges for zoos is to develop education programs that will appeal to and reach visitors who may have little incentive to learn.

We suggest that if zoos are serious about being conservation education organizations, a second challenge of their education programs is to not only provide information, but also to develop positive values and attitudes toward animals and wildlife conservation. Furthermore, the acquired information and values should empower people to act. Unfortunately, many education programs fall far short of these goals, and instead offer chatty newsletters and arcane information. These problems can occur when a zoo (or any type of organization) lacks specific goals for their programs (Masci 2000), embraces self-promotion as the primary goal, or strives for political neutrality. Because people arrange values and associated goals hierarchically, strong self-promotion, legitimacy, power, or other goals can subvert education goals (Clark 1997). Claiming institutional neutrality supports existing policies and actions or inactions that often damage nature (see Miller *et al.* 2004).

Zoos spend millions of dollars on education programs, informative signs, and improved exhibits in efforts to improve their educational impact. Unfortunately, the vast majority of these efforts go unevaluated (Kidd *et al.* 1995). Much of the evaluation that has occurred is not encouraging. For example, Altman (1998) reviews research demonstrating that visitors rarely even read zoo signs, let alone retain the information on them. Dunlap and Kellert (1989) found that people demonstrated little increase in knowledge about animals after zoo visits. Instead, most people appear to visit zoos for recreation, primarily as family or group outings, and appear little interested in education (Tunnicliffe 1996, Acampora 1998, Morgan and Hodgkinson 1999).

Alternatively, more active education programs apparently have had more success (Kidd *et al.* 1995, Altman 1998, Hutchins *et al.* 2003) and a minority, but still large percentage, of zoo visitors have cited education as a secondary goal of their trip (Morgan and Hodgkinson 1999). Morgan and Hodgkinson (1999) call for blending education programs with enjoyable, recreational activities. Heinrich and Birney (1992) found that it was possible to impart information to visitors while entertaining them. They found that spectators of animal demonstrations received and retained much of the information provided for 6 days. Similarly, research by Kidd *et al.* (1995) indicated that petting zoos created more possibilities for learning and more favorable attitudes toward animals among small children. Altman (1998) studied the impact of animal activity on learning and found that non-stereotypic activity increases an exhibit's holding power and facilitates learning. She suggests using enrichment to both improve animal welfare and increase the chances for effective education. Hutchins *et al.* (2003) provide additional examples of effective education, at least in the short-term. Despite these apparent successes, zoo education programs could be greatly improved.

Beyond the apparent limited ability of most zoo education programs to impart information, critics argue that zoos demonstrate human mastery over wildlife, symbolize humanity's dominion over nature, and anthropomorphize animals (Marvin 1994, Acampora 1998, Swanagan 2000). Of course, even with successful education programs, people may acquire information, but still not develop more favorable values and attitudes because of the differences between knowledge, values, and attitudes described above.

Exhibit designs also likely influence values and attitudes toward animals and conservation through contextual effects. In a study by Rhoads and Goldsworthy (1979), subjects rated zoo animals in significantly more negative terms than animals in natural or semi-natural settings. Studies summarized by Swanagan (2000) found that exhibit design was important. Interest in conservation increased among visitors to more naturalistic exhibits that included an educational focus.

Research more specifically directed at values and attitudes found that zoo visits reinforce or increase people's sense of superiority over nature (Tunnicliffe 1996, Kellert 1996, 1997). And, visitors often anthropomorphize animals, stating that they are funny, ugly, disgusting, cute, ferocious, etc. (Marvin 1994, Tunnicliffe 1996). These results are particularly disturbing and point to the serious need for zoos to address how and if they influence visitors' values and attitudes. This is an area ripe for additional research and, we argue, crucially pertinent to the missions of most zoos.

METHODS

We surveyed visitors to the Denver Zoo on 45 days in 1998, 2000, and 2001. We conducted the survey in conjunction with a marketing firm (Data Marketing Associates, Inc., Phoenix, AZ) that was collecting data on potential clients. Respondents self-selected themselves to complete the survey, which we located at a well-marked kiosk on zoo grounds. We offered respondents a chance to win an automobile as an incentive for completing the survey.

The survey asked respondents seven demographic questions, three questions about leisure activities, if respondents were members of the Denver Zoo, how often the respondents visit the Denver Zoo, a description of the group that was visiting (if applicable), why they visited the zoo, what they thought about zoos in general, and how they believe that people are related to animals and nature. Surveys administered in 2001 also asked three questions about wolves: (1) the level of threat they thought wolves posed to agriculture, (2) the role they felt wolves play in maintaining ecosystem health, and (3) if they would like to see wolves restored to wilderness areas in Colorado. Specific questions and a copy of the survey instrument are available upon request.

We examined all variables for normality and checked for homogeneity of group variance using Barlett's test. We compared pair means using simple *t*-tests. Count response data were compared using Pearson's likelihood ratio chi-square tests and Yate's corrected X^2 for 2×2 comparisons. Unless otherwise indicated, all means presented ± 1 S.D. We set significance at $P < 0.05$.

To assess non-response bias, we surveyed 890 people who did not complete the survey as they exited the zoo on 4 and 5 May 1998. Of these 785 surveys were usable. Comparing groups, we found no difference in age ($t = -0.15$, df = 20 606, $P = 0.88$), race ($X^2 = 1.40$, df = 4, $P = 0.85$), membership status (Yate's $X^2 = 0.00$, df = 1, $P = 0.98$), or visitation rate ($X^2 = 4.55$, df = 3, $P = 0.21$) among respondents and non-respondents. However, non-respondents were significantly more likely to be married (Yate's $X^2 = 7.34$, df = 1, $P < 0.01$), female (Yate's $X^2 = 16.96$, df = 1, $P < 0.001$), and better educated ($X^2 = 11.82$, df = 3, $P < 0.01$) than respondents. Nevertheless, overall, we found no significant differences in responses among respondents and non-respondents to the questions on why they visited the zoo ($X^2 = 4.75$, df = 6, $P = 0.58$); what they thought about zoos in general ($X^2 = 1.01$, df = 3, $P = 0.80$); and how they believe that people are related to animals and nature ($X^2 = 6.29$, df = 5, $P = 0.28$), suggesting that non-response bias was weak.

RESULTS

We received 22 028 usable surveys (hundreds of surveys were excluded from analyses for a variety of reasons, such as missing information or failure to follow instructions). Given the large sample size, most of our analyses yielded statistically significant results. Still, just because a result is statistically significant does not mean that the difference is meaningful from a social science or management perspective. Therefore, in most cases, it is more important to look at the magnitude of the difference between respondents than whether or not that difference is statistically significant.

When asked how respondents felt about zoos in general, most (55.1%) responded that they are important for education (Figure 6.1). Almost half (49.1%) also believed that zoos are important for conserving wildlife (note that respondents often chose more than one response, so percentages add to over 100%). A smaller percentage (39.3%) felt that zoos were nice places to visit. These results contrast with the reasons people gave for why they visited the zoo. The majority of respondents (56.5%) indicated that the main reason they visited the zoo was for a family outing, followed by their desire to see animals (26.9%). Only 18.4% of respondents stated that they visited the zoo to learn about animals or wildlife conservation. Other main reasons for visiting the zoo were to do something outdoors (17.7%) or for mental relaxation (12.6%).

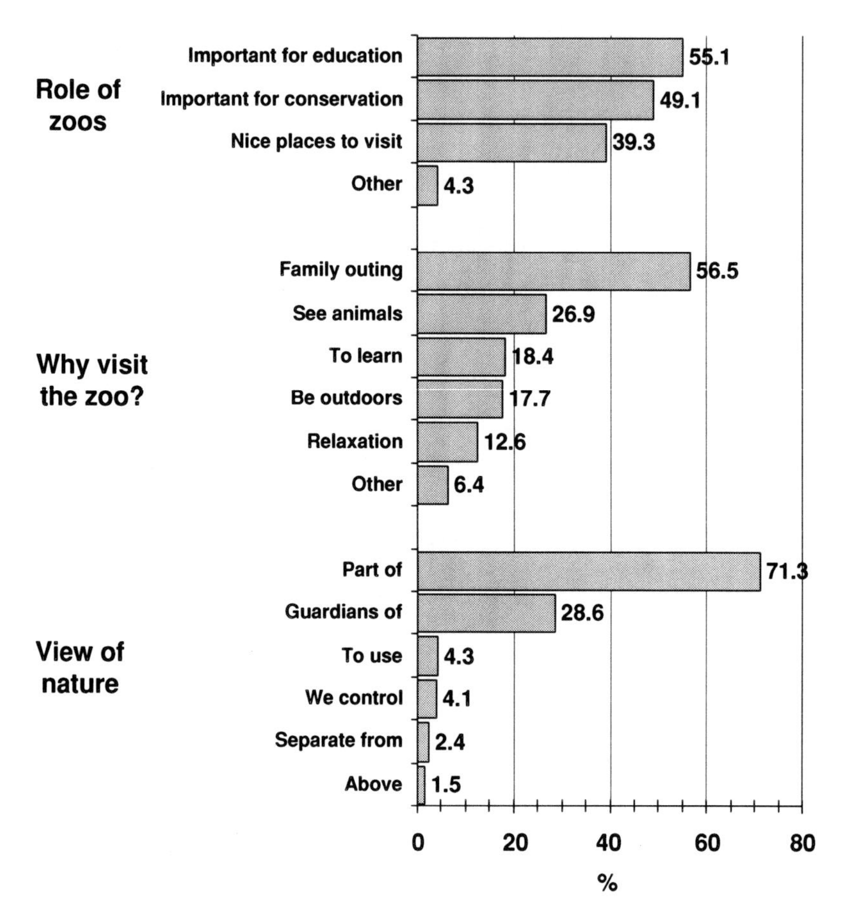

Figure 6.1 Attitudes of Denver Zoo visitors on the role of zoos, why they visit the zoo, and the relationship between people and nature

A third question asked respondents how they felt people are related to animals and nature (Figure 6.1). The vast majority of zoo visitors responded that people are a part of nature (71.3%), while 28.6% stated that people are guardians of nature. Few visitors indicated other, more dominionistic or negativistic relationships, such as nature exists for people to use (4.3%), people control nature (4.1%), people are fundamentally separate from nature (2.4%), or people are above nature (1.5%).

Attitudes toward wolves

We also asked a sub-sample of respondents three questions that addressed a more specific conservation issue that has relevance in Colorado. These

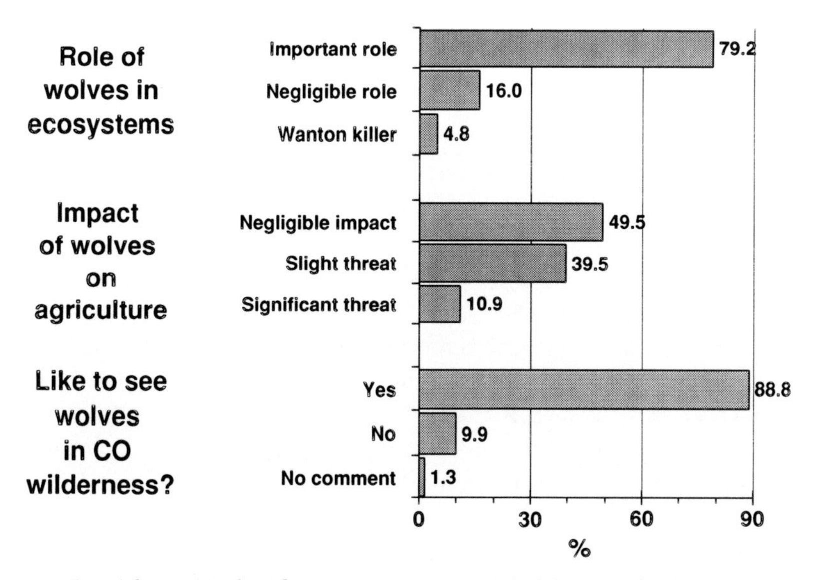

Figure 6.2 Attitudes of Denver Zoo visitors toward the role of wolves (*Canis lupus*) in ecosystems, the likely impact of wolves on agriculture, and possible wolf restoration into Colorado's wilderness areas

questions focused on attitudes toward wolves and possible wolf restoration in Colorado. We received 4237 responses to questions on wolves.

We first asked visitors about the role that wolves play in maintaining healthy ecosystems. A large majority of respondents believed that they play an important role (Figure 6.2). Only 16% of visitors thought wolves play a negligible role in maintaining ecosystem health and 4.8% felt that wolves are wanton killers of deer and elk. When asked about wolves' impacts on the agricultural economy of Colorado, just under half (49.5%) of respondents indicated that they thought wolves pose a negligible threat. Over a third (39.5%) thought that wolves would represent a slight threat to agriculture that would be somehow mitigated or compensated. Few respondents (10.9%) believed that wolves would significantly threaten Colorado's agricultural economy. Lastly, we asked if respondents would like to see wolves living in the wilderness areas of Colorado. The vast majority (88.8%) of visitors indicated that they would like to see wolves in Colorado's wilderness areas, while only 9.9% opposed the idea.

Demographic variation

We found significant variation among different demographic groups for almost all categories we analyzed for our questions that explored general attitudes toward zoos and nature. Despite the significance, the magnitude of the differences was not large in many cases (Figures 6.3a–6.5a). Respondents from different demographic groups displayed fewer significant differences in attitudes toward wolves, yet the sample size was smaller. In several instances, differences among respondents from different demographic groups were quite large for questions focused on wolves (Figures 6.6a–6.8a).

One demographic category we evaluated was ethnic descent. Visitors saying they were of non-white ethnicity generally responded quite differently from visitors who described themselves as white. Respondents of the former ethnic groups were less likely to view zoos as important for conservation ($X^2 = 70.9$, df $= 12$, $P < 0.001$; Figure 6.3a), but more likely to visit zoos to learn ($X^2 = 153.9$, df $= 24$, $P < 0.001$; Figures 6.4a). They were less likely to view people's relationship to nature as stewards (i.e., guardians) and more likely to express more dominionistic and negativistic attitudes (e.g., use, control, separate from, or above nature) ($X^2 = 232.2$, df $= 20$, $P < 0.001$; Figure 6.5a). With respect to wolves, black, Hispanic, Asian, and other non-white respondents were significantly more likely than white respondents to view wolves as wanton killers that play a negligible role in maintaining ecosystem health ($X^2 = 101.5$, df $= 8$, $P < 0.001$; Figure 6.6a) and as a species that poses a significant threat to Colorado's agricultural economy ($X^2 = 112.8$, df $= 8$, $P < 0.001$; Figure 6.7a). As such, non-white visitors were significantly more likely to oppose establishing a population of wolves in Colorado ($X^2 = 76.2$, df $= 8$, $P < 0.001$; Figure 6.8a).

Respondents less than 25 years old were less likely to view zoos as important for education, while respondents over 49 years old were more likely to indicate that zoos are important for conservation ($X^2 = 96.5$, df $= 9$, $P < 0.001$; Figure 6.3a). Perhaps not surprisingly, respondents under 25 and unmarried were less likely to visit zoos for family outings; instead visiting to see animals or be outdoors ($X^2 = 599.9$, df $= 18$, $P < 0.001$ and $X^2 = 1\,109.4$, df $= 6$, $P < 0.001$, respectively; Figure 6.4a). Younger, single visitors were more likely to view people as part of nature, while older, married visitors were more likely to view people as guardians of nature ($X^2 = 255.9$, df $= 15$, $P < 0.001$ and $X^2 = 123.9$, df $= 5$, $P < 0.001$, respectively; Figure 6.5a).

Respondents with fewer years of formal education generally thought zoos were more important for conservation and were nice places to visit than did respondents with more education ($X^2 = 63.0$, df $= 9$, $P < 0.001$; Figure 6.3a). Visitors with less education visited zoos more to see animals and to learn ($X^2 = 153.3$, df $= 18$, $P < 0.001$; Figure 6.4a) and expressed

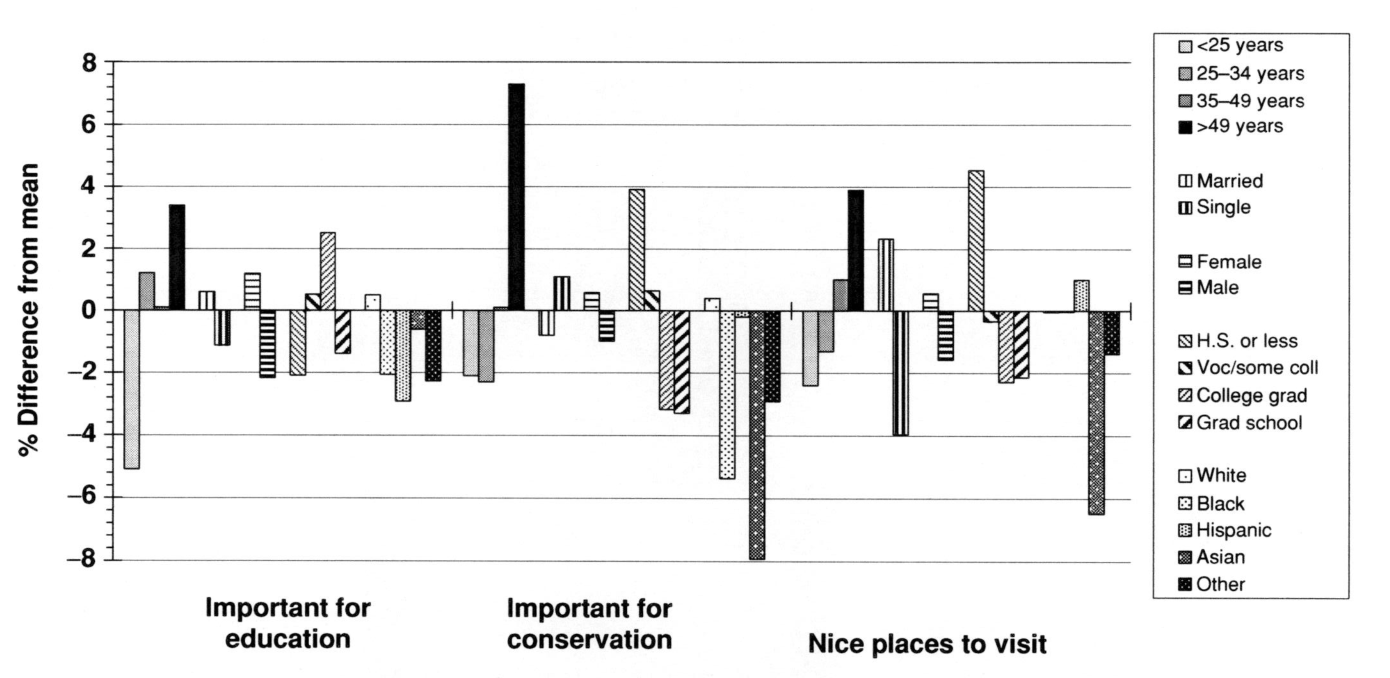

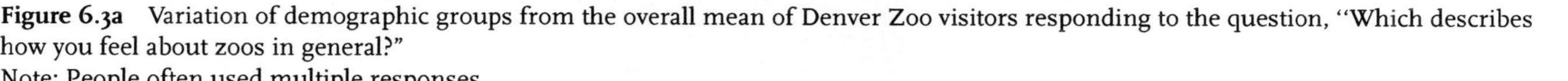

Figure 6.3a Variation of demographic groups from the overall mean of Denver Zoo visitors responding to the question, "Which describes how you feel about zoos in general?"
Note: People often used multiple responses

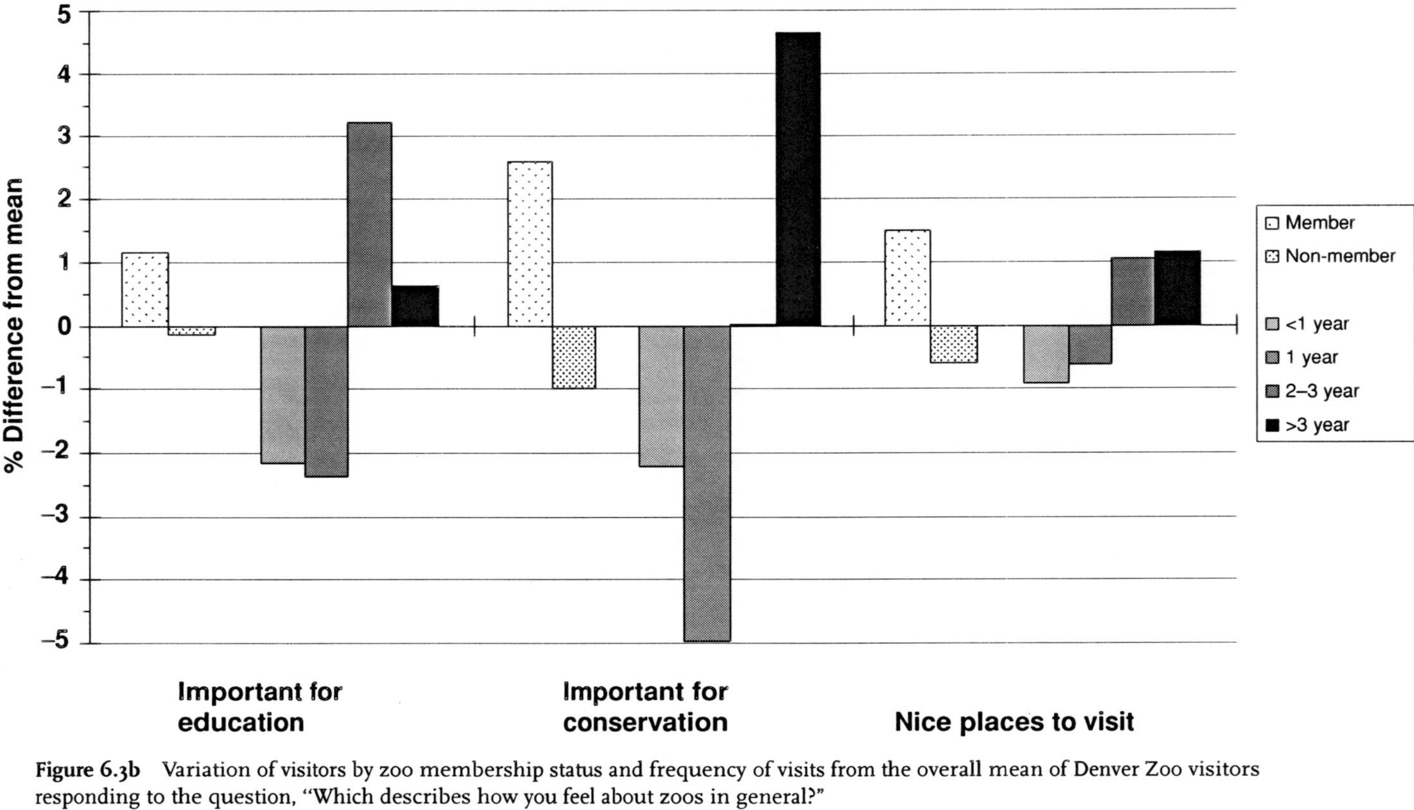

Figure 6.3b Variation of visitors by zoo membership status and frequency of visits from the overall mean of Denver Zoo visitors responding to the question, "Which describes how you feel about zoos in general?"
Note: People often used multiple responses

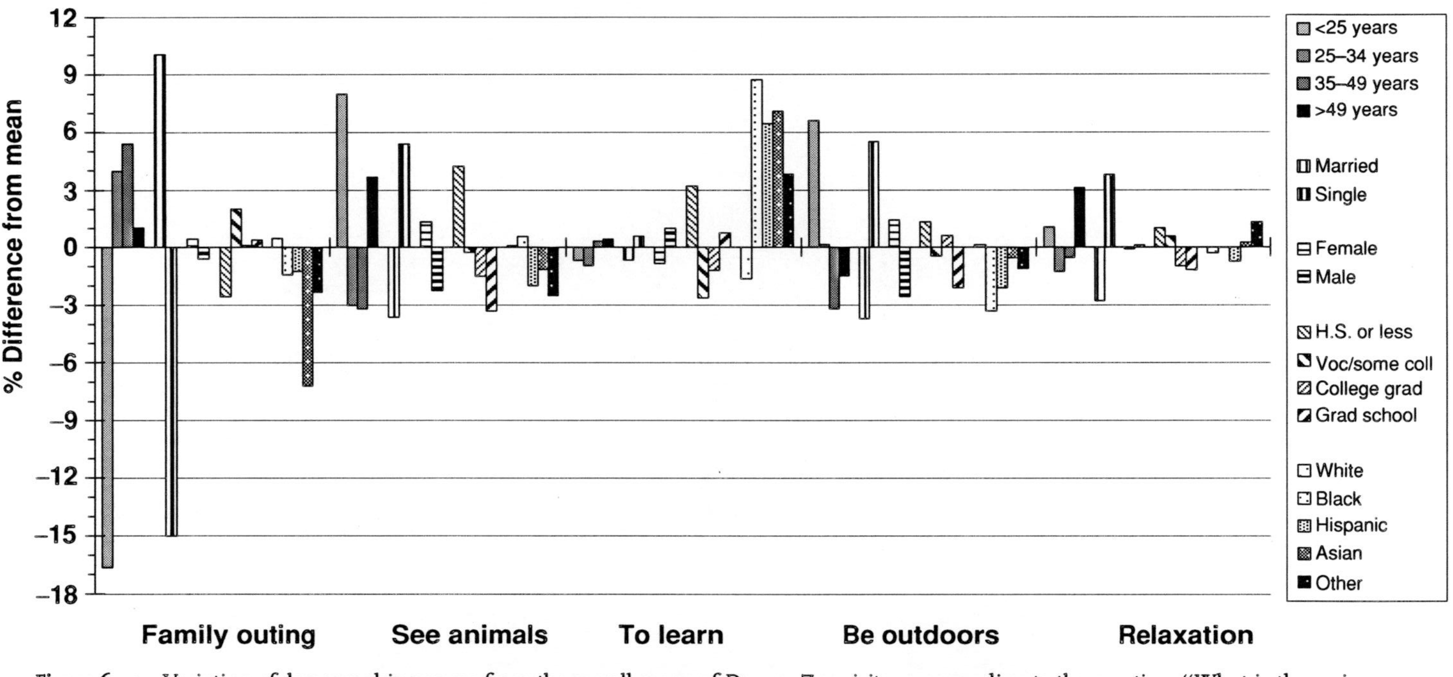

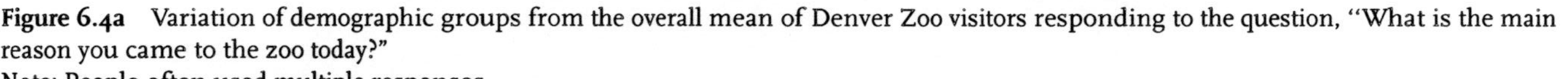

Figure 6.4a Variation of demographic groups from the overall mean of Denver Zoo visitors responding to the question, "What is the main reason you came to the zoo today?"
Note: People often used multiple responses

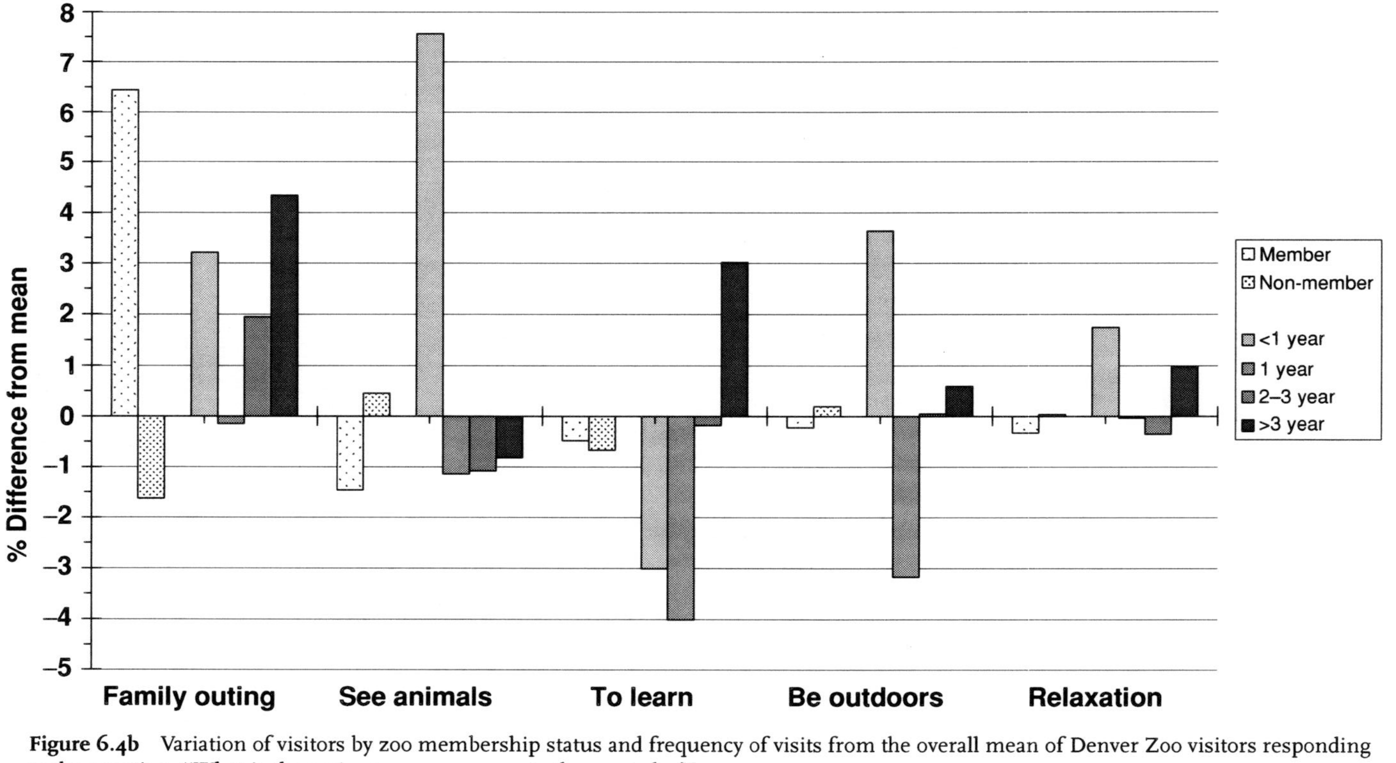

Figure 6.4b Variation of visitors by zoo membership status and frequency of visits from the overall mean of Denver Zoo visitors responding to the question, "What is the main reason you came to the zoo today?"
Note: People often used multiple responses

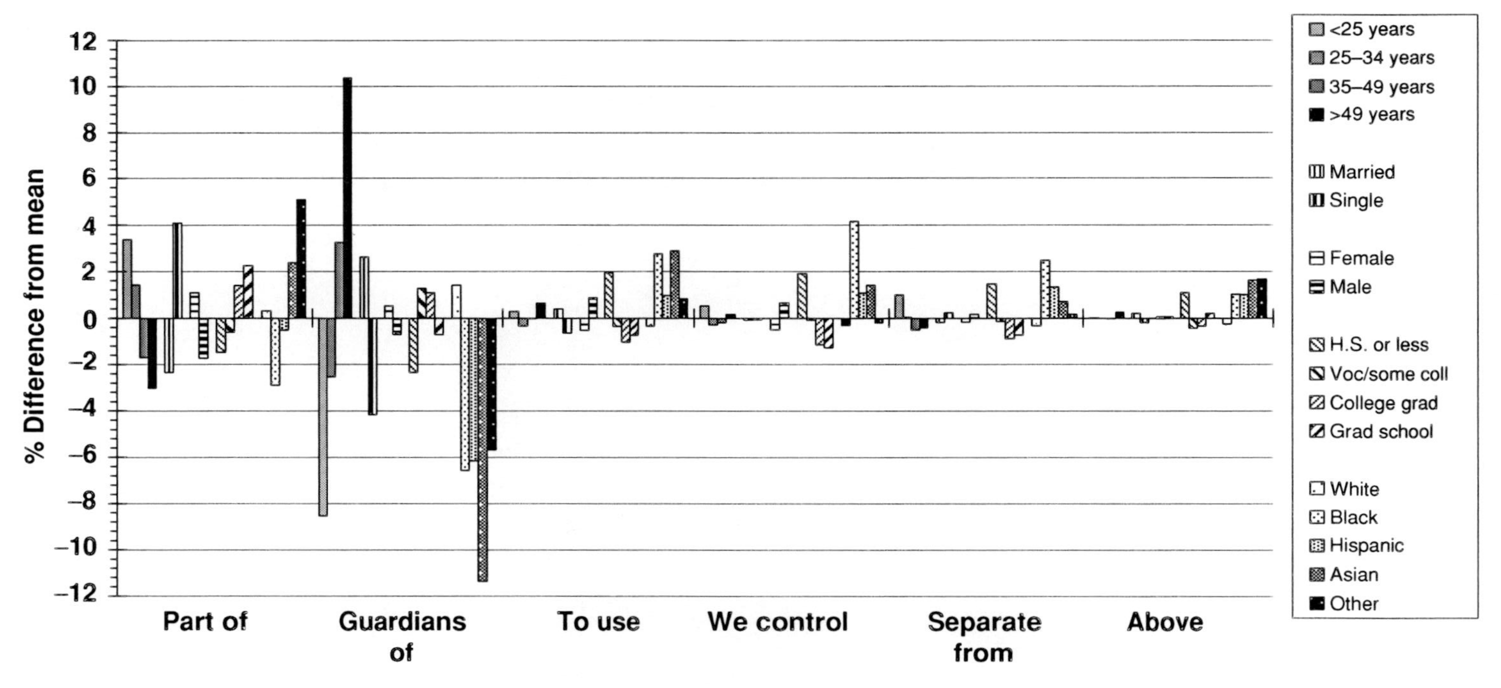

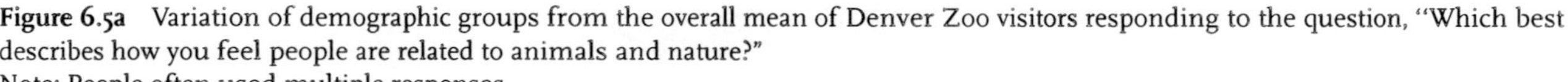

Figure 6.5a Variation of demographic groups from the overall mean of Denver Zoo visitors responding to the question, "Which best describes how you feel people are related to animals and nature?"
Note: People often used multiple responses

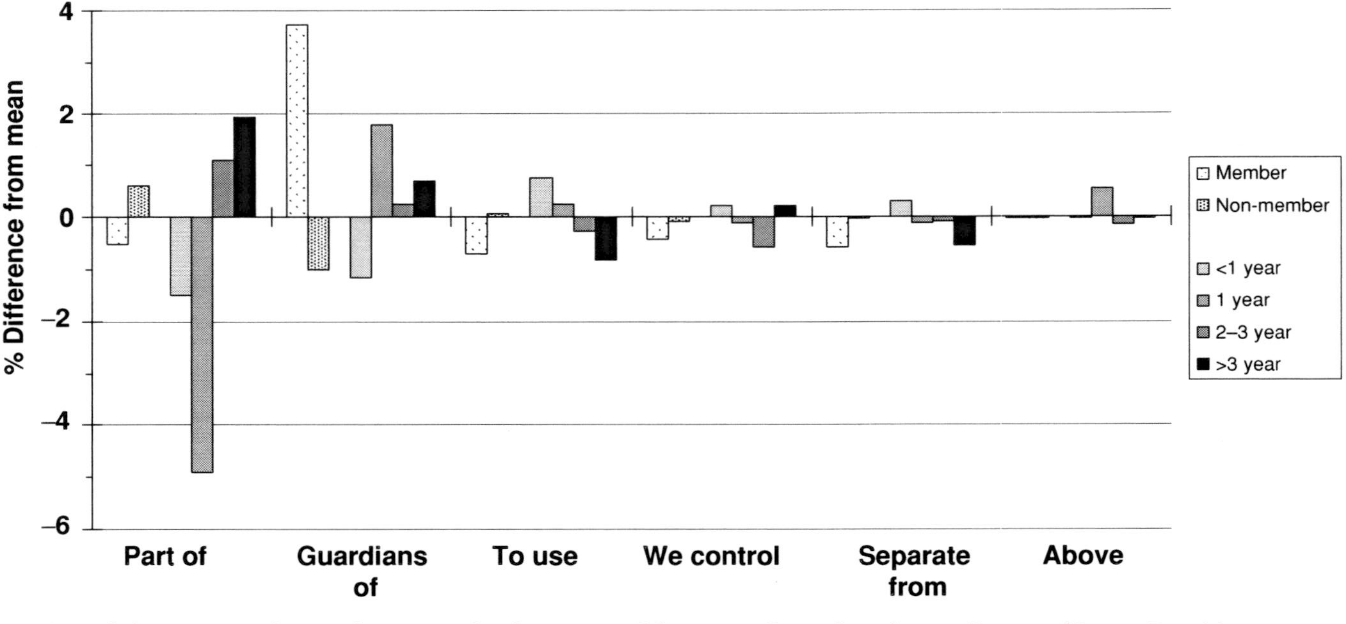

Figure 6.5b Variation of visitors by zoo membership status and frequency of visits from the overall mean of Denver Zoo visitors responding to the question, "Which best describes how you feel people are related to animals and nature?"
Note: People often used multiple responses

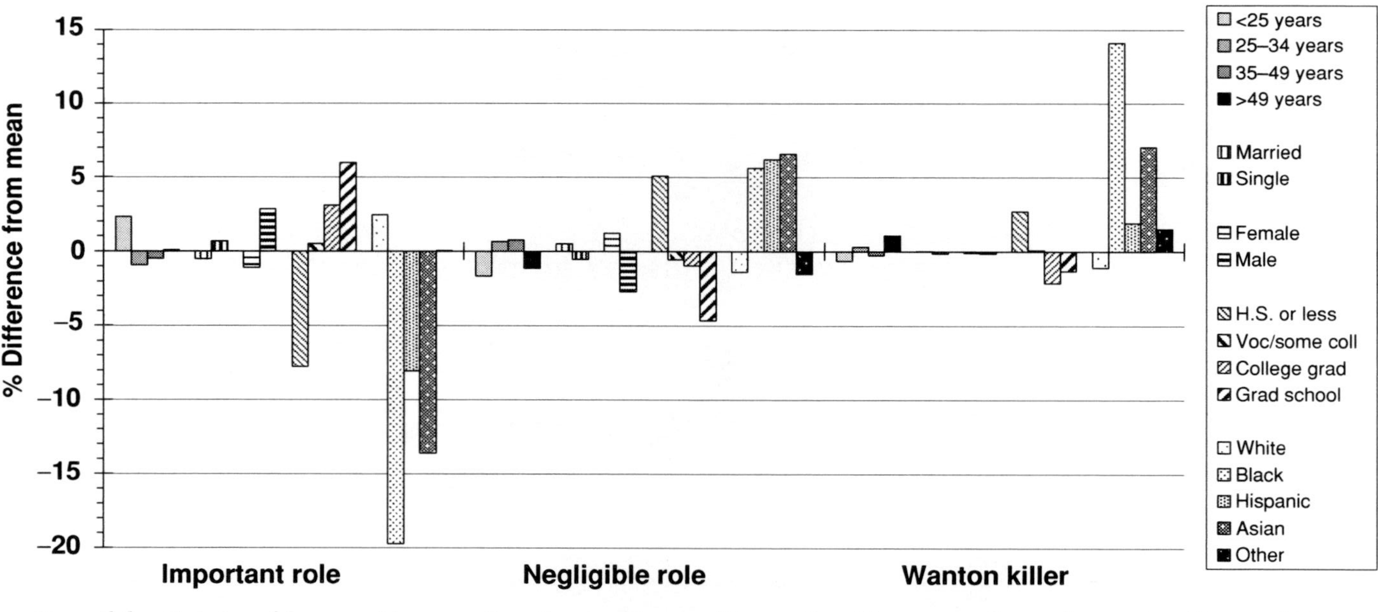

Figure 6.6a Variation of demographic groups from the overall mean of Denver Zoo visitors responding to the question, "Do you think wolves: (1) play an important role in maintaining healthy ecosystems; (2) play a negligible role in maintaining ecosystem health; or (3) are wanton killer of deer and elk?"

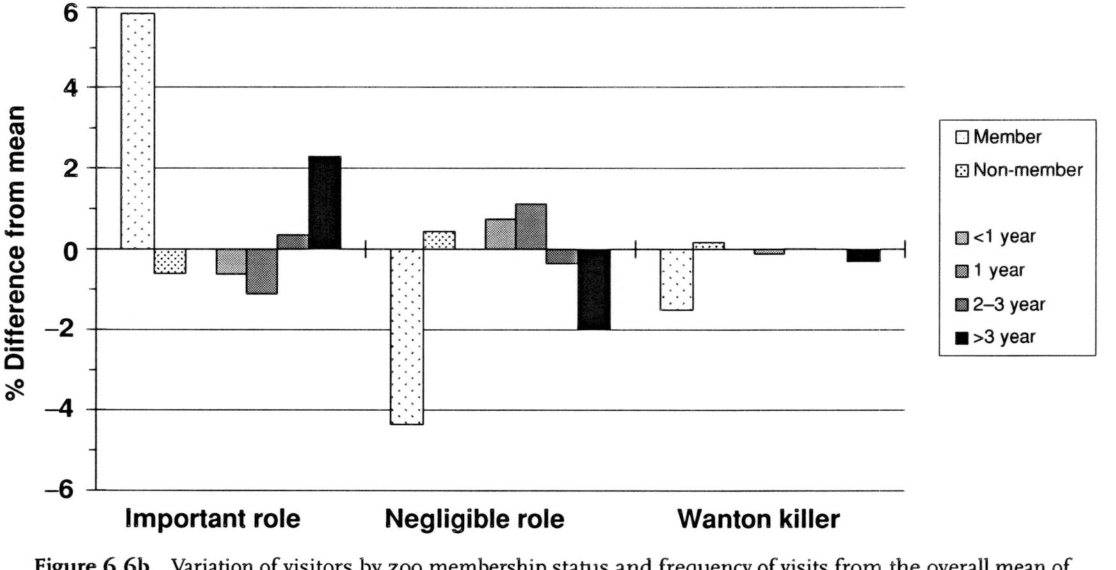

Figure 6.6b Variation of visitors by zoo membership status and frequency of visits from the overall mean of Denver Zoo visitors responding to the question, "Do you think wolves: (1) play an important role in maintaining healthy ecosystems; (2) play a negligible role in maintaining ecosystem health; or (3) are wanton killer of deer and elk?"

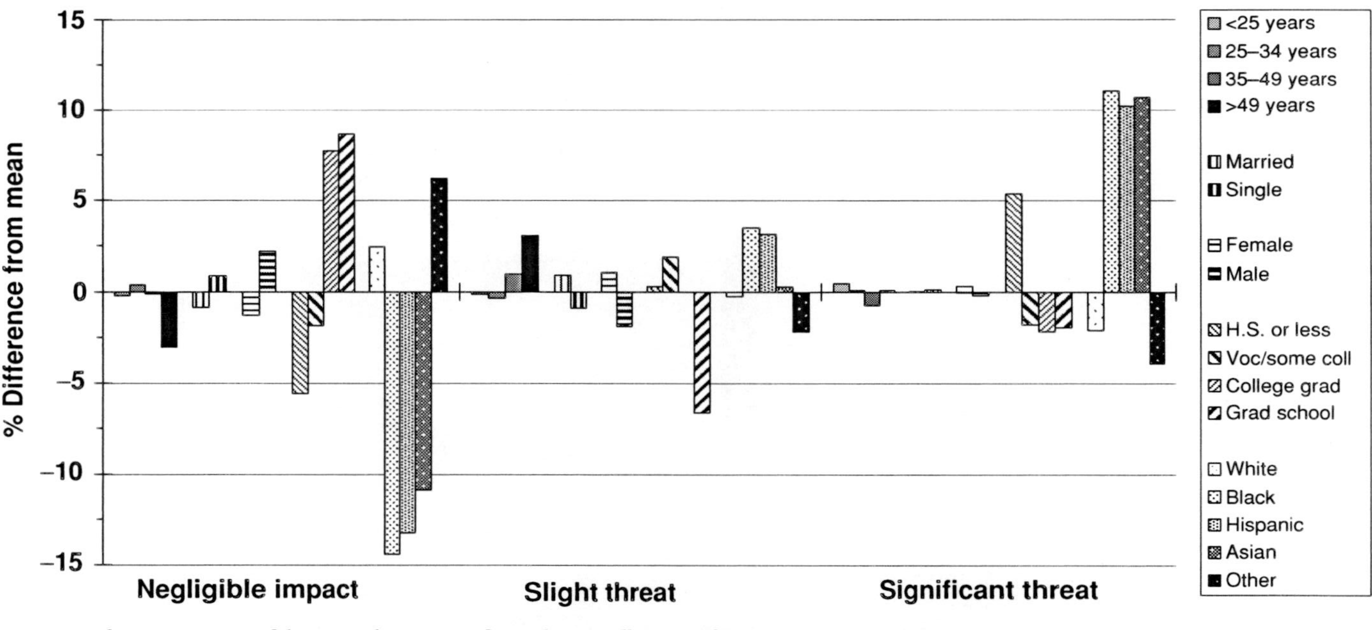

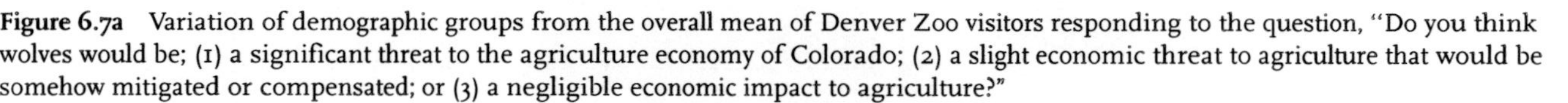

Figure 6.7a Variation of demographic groups from the overall mean of Denver Zoo visitors responding to the question, "Do you think wolves would be; (1) a significant threat to the agriculture economy of Colorado; (2) a slight economic threat to agriculture that would be somehow mitigated or compensated; or (3) a negligible economic impact to agriculture?"

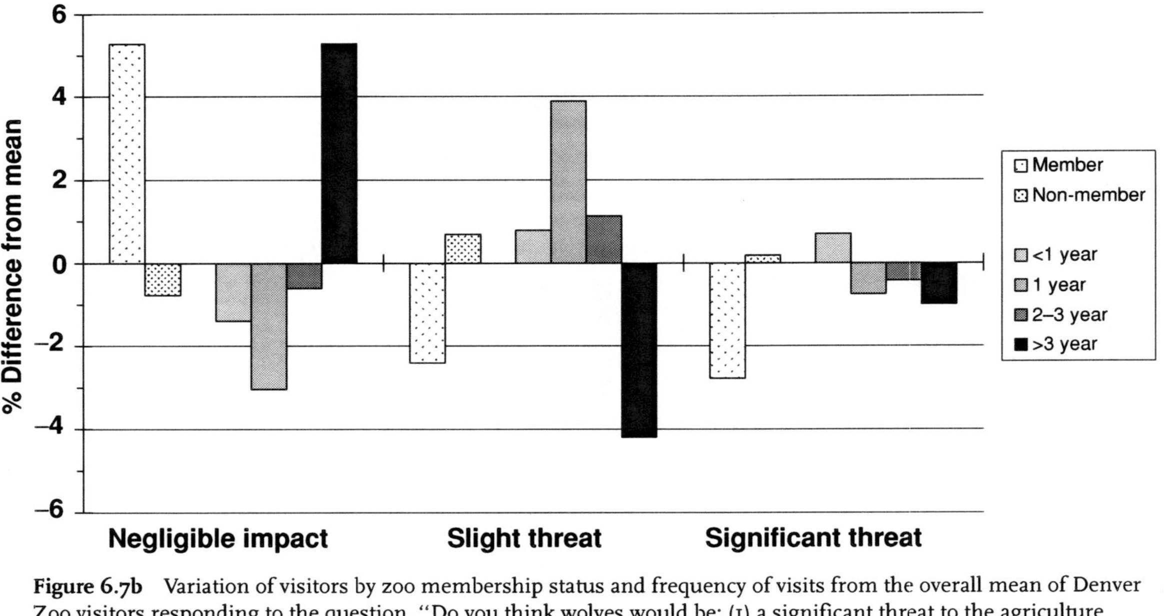

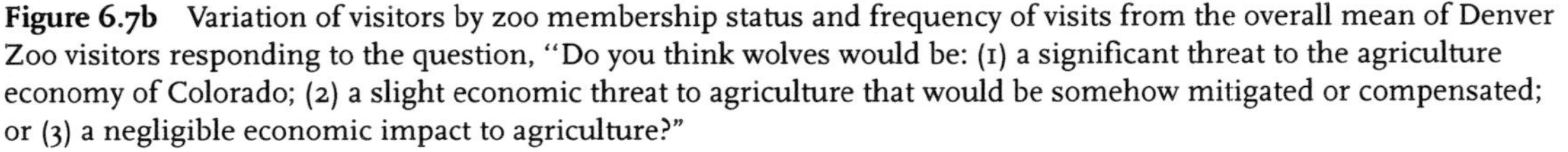

Figure 6.7b Variation of visitors by zoo membership status and frequency of visits from the overall mean of Denver Zoo visitors responding to the question, "Do you think wolves would be: (1) a significant threat to the agriculture economy of Colorado; (2) a slight economic threat to agriculture that would be somehow mitigated or compensated; or (3) a negligible economic impact to agriculture?"

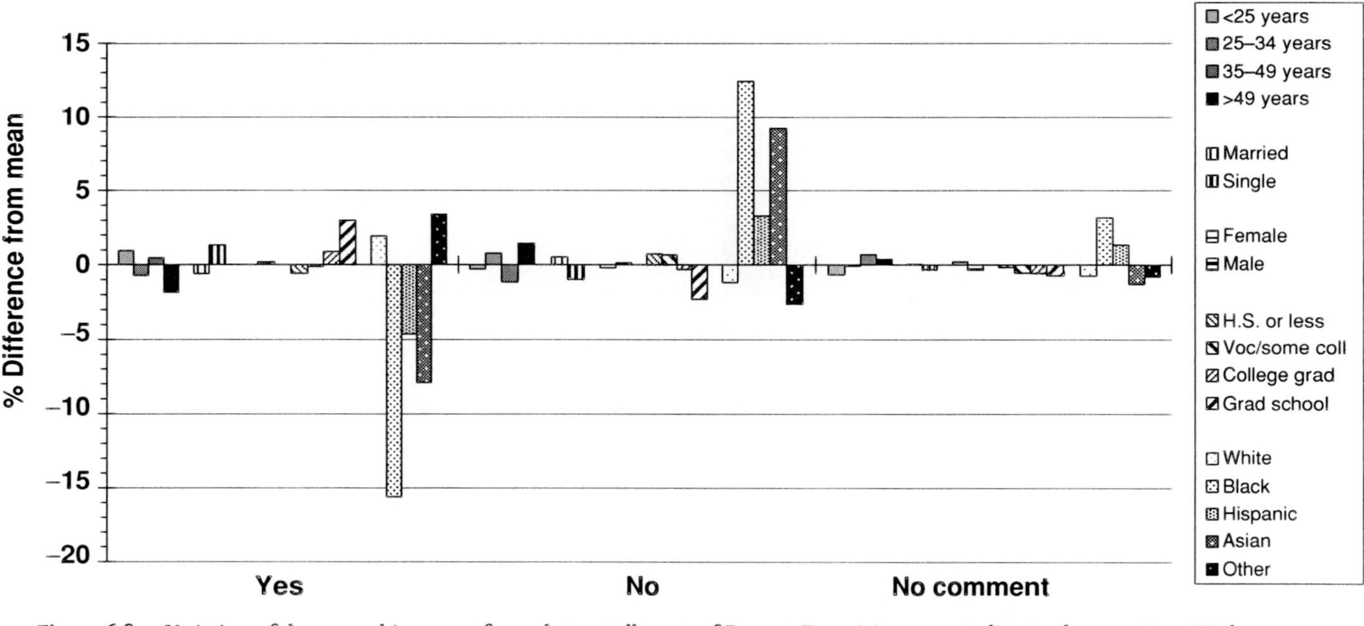

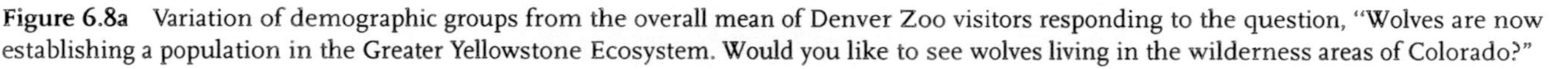

Figure 6.8a Variation of demographic groups from the overall mean of Denver Zoo visitors responding to the question, "Wolves are now establishing a population in the Greater Yellowstone Ecosystem. Would you like to see wolves living in the wilderness areas of Colorado?"

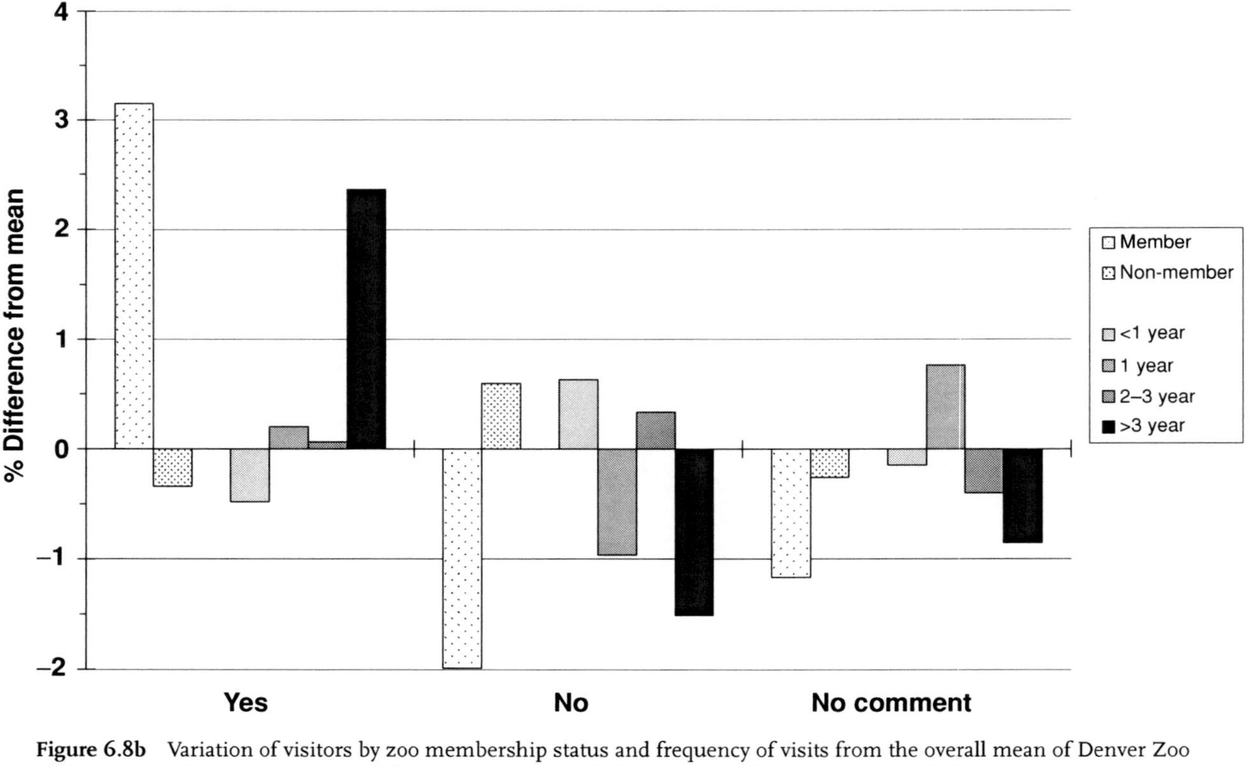

Figure 6.8b Variation of visitors by zoo membership status and frequency of visits from the overall mean of Denver Zoo visitors responding to the question, "Wolves are now establishing a population in the Greater Yellowstone Ecosystem. Would you like to see wolves living in the wilderness areas of Colorado?"

more dominionistic and negativistic attitudes toward nature ($X^2 = 277.5$, df = 15, $P < 0.001$; Figure 6.5a) than did more educated visitors. Greater formal education was also significantly correlated with the beliefs that: (1) wolves play an important role in ecosystems ($X^2 = 61.0$, df = 6, $P < 0.001$; Figure 6.6a) and (2) wolves have little impact on Colorado agriculture ($X^2 = 59.2$, df = 6, $P < 0.001$; Figure 6.7a).

Attitudes toward zoos and nature varied significantly depending on whether a respondent was a Denver Zoo member and on how many times the respondent visited the zoo. Denver Zoo members and respondents who visited the zoo frequently viewed zoos as more important for conservation than did non-members and infrequent visitors ($X^2 = 24.2$, df = 3, $P < 0.001$ and $X^2 = 121.3$, df = 12, $P < 0.001$, respectively; Figure 6.3b). Members and frequent visitors were also much more likely to visit the zoo for a family outing ($X^2 = 51.5$, df = 6, $P < 0.001$ and $X^2 = 157.3$, df = 24, $P < 0.001$, respectively; Figure 6.4b). Respondents who visit the zoo less than once per year were more likely to go to the zoo to see animals or be outdoors, while respondents who visit more than three times per year were more likely to visit to learn (Figure 6.4b). Finally, zoo members were more likely to view people as guardians of nature, while infrequent zoo visitors were less likely to view people as a part of nature ($X^2 = 38.5$, df = 5, $P < 0.001$ and $X^2 = 68.9$, df = 20, $P < 0.001$, respectively; Figure 6.5b).

Respondents who were and were not zoo members and who visited the zoo different numbers of times varied significantly in their attitudes toward wolves. A greater percentage of zoo members and frequent zoo visitors believed that wolves play an important role in ecosystems ($X^2 = 16.5$, df = 2, $P < 0.001$ and $X^2 = 7.48$, df = 8, $P = 0.49$; Figure 6.6b) and pose a negligible impact on Colorado's agricultural economy ($X^2 = 11.0$, df = 2, $P < 0.01$ and $X^2 = 21.9$, df = 8, $P < 0.01$, respectively; Figure 6.7b) than did non-members and infrequent visitors. Similarly, Denver Zoo members and frequent visitors were more likely to support wolf restoration in Colorado's wilderness areas than were non-members and less frequent visitors ($X^2 = 10.3$, df = 2, $P < 0.01$ and $X^2 = 19.4$, df = 8, $P < 0.05$, respectively; Figure 6.8b).

Discussion

Our small study only began to examine many of the important questions and issues surrounding the effectiveness of zoo education programs at influencing the values and attitudes of visitors. While we found that most visitors held positive attitudes toward zoos, toward people's relationship to nature, and toward wolves and their conservation, we could not determine the extent to which the zoo helped foster these positive attitudes toward nature or whether visitors were predisposed to such attitudes. Additional work is required. A more robust study, for example, would compare values and attitudes of zoo visitors before and after visiting the zoo (see below).

Still, this study and many previous research projects provide baseline data on which to base future studies. And some important insights can be gleaned from this work. For example, people visited the Denver Zoo primarily for family outings and other recreational reasons. Education was less important. Similarly, Kellert (1979) found that people went to zoos to educate children (36%), to recreate with friends and family (26%), to see animals (25%), and for aesthetic reasons (11%). Kellert and Berry (1980) found that people attended zoos more because they felt affection for animals than out of intellectual curiosity. Andereck and Caldwell (1994) found that visitors were oriented primarily toward education/recreation (56.3%), followed by recreation/photography (22.7%), and education (21.0%). Finally, in interviews with children, Kidd *et al.* (1995) reported that most rated zoos as fun (99%), that they learned something (47%), that they preferred unrelated activities (e.g., zoo rides) (13%), or that they were afraid of (2%), disliked (2%) or were indifferent (2%) toward animals. These results highlight the challenge that zoos face in educating visitors that primarily want to be entertained (Morgan and Hodgkinson 1999).

Another result of our study was the difference in attitudes when sorted by ethnic background. Black, Hispanic, Asian, other non-white visitors were generally more utilitarian, dominionistic, and negative toward animals and conservation than were white visitors (although all groups were still strongly supportive of nature). These findings could result from: (1) non-white ethnic groups espousing higher utilitarian, dominionistic, and negative attitudes toward animals and nature, (2) zoos attracting black, Hispanic, Asian, and other non-white visitors with a wider cross-section of viewpoints than white visitors toward animals and nature, (3) zoo education efforts working less effectively on visitors with certain ethnic backgrounds than others, or (4) a combination of these. To reach all visitors better, Kidd *et al.* (1995) suggested developing more ethnically oriented education programs and materials that are bi- or trilingual.

Previous studies of Colorado residents in general found broad and widespread support for wolves and wolf restoration to the state (Manfredo *et al.* 1993, Pate *et al.* 1996, Meadow *et al.* 2005). The higher degree of support for wolves and wolf restoration in Colorado among zoo members and frequent visitors to the Denver Zoo suggests that either the zoo is helping to foster such attitudes or people with such attitudes are more likely to become zoo members and visit the zoo frequently. Alternatively, since more zoo members than non-members view people as guardians of nature, as opposed to a part of nature, do zoos encourage more dominionistic attitudes or, again, is this a reflection of the types of people that do and do not become members? Only additional research can inform us which of these scenarios, if not all of them, are true.

Future research

Zoos serious about conservation education missions should support research that evaluates the education value of specific zoo programs and the overall zoo experience (Kleiman *et al.* 2000). We recommend combining research with planning using an adaptive management framework (Holling 1978) to continually improve performance. In the true spirit of adaptive management, changes should be based on data. Future studies should attempt to better understand the effectiveness of zoo education programs, graphics, exhibits, and the overall zoo experience in imparting knowledge to visitors and influencing their values, attitudes, and associated behaviors. The limited research conducted to date only begins to address a few of these topics. Other researchers also call for additional studies into zoo education programs. For example, Kidd *et al.* (1995) suggest conducting more long-term studies of the effectiveness of education programs. Hutchins *et al.* (2003) call for extending research into examining the impact of education programs on the welfare of the animals in a collection. Yet, we argue that research should extend much further, especially since increasingly zoos claim to embrace a primary mission of conservation education.

Studies should also examine several other aspects of zoos and educational impact. For example, zoos would likely benefit from studies comparing the efficacy of different types of graphics and of graphics versus other means of imparting information (e.g., docents). In addition, although most people would agree that naturalistic exhibits are better for the animals and for educational efforts, we still have a poor understanding of how exhibits influence human–animal relationships (Marvin 1994). Birney (1995) suggests that changes in exhibit designs are likely to influence the attitude and knowledge of visitors with less education or familiarity with an animal. Studies of different exhibit designs might help us begin to address this gap in our knowledge. We should examine the utility of more targeted education programs to reach all ethnic groups (Kidd *et al.* 1995), urban/inner city (vs. rural or suburban) residents, and people from different wealth and education categories. We also need more research into the desirability and effectiveness of combining the education and recreation functions of zoos (Heinrich and Birney 1992, Morgan and Hodgkinson 1999).

We further argue that future studies should extend beyond assessing the effectiveness of education programs to impart knowledge to visitors. While providing accurate information is important, we believe this should comprise just one part of a zoo's education goals. Arguably more important is the extent to which zoos can influence visitors' values, attitudes, and associated behaviors. For example, do immersion exhibits decrease the chances that an exhibit experience will lead to dominionistic attitudes toward animals? Related to this, what types of exhibits and what aspects of exhibits engender more positive values and attitudes toward animals and conservation? One way to begin addressing this issue is by comparing attitude change among visitors to different types of exhibits. Despite the lack of data, we believe that most, if not all, zoo education programs remain too diffuse and undirected to induce value, attitude, or associated behavioral changes, which are difficult to accomplish in any case (Chaiken and Stangor 1987, Petty *et al.* 1997). Indeed, the same could be said for educational programs of many NGOs and government agencies. Studies should therefore address the types of education programs and information in those programs that are most persuasive. And since zoos are cultural institutions that likely help shape values and attitudes, and since a dominant value of Western culture society is dominion over nature, zoos should carefully monitor educational messages and their context to maximize the development of less dominionistic and more positive values and attitudes toward animals and wildlife conservation.

Pedagogical literature exists to help guide the development of new zoo education programs, graphics, and exhibit designs, but directed research promises to help us develop more effective programs more rapidly. We recommend conducting both latitudinal (i.e., before and after studies of different people) and longitudinal (i.e., before and after studies of the same individuals) research to assess the effectiveness of (1) education programs, (2) exhibits, and (3) the overall zoo experience. Studies should extend well after zoo visits to assess the degree to which the impacts, if any, created lasting changes to values, attitudes, or behaviors. Ideally, such research would include multiple methods (Clark *et al.* 1999), containing both quantitative (i.e., statistically analyzed data on large numbers of people) and qualitative (i.e., in-depth analyses of few people) studies to increase our confidence in the results.

Finally, it is important to re-emphasize the fact that changing values and attitudes and associated behaviors is often difficult, especially when well developed. As such, zoos should focus on areas that offer the most hope of imparting knowledge and persuasive arguments that will affect change. Most likely this will be among children and visitors with poorly developed values and attitudes toward animals and nature conservation. Reaching out to these segments of society might prove the most fruitful and efficient, not to mention fulfilling. Indeed, the development of strong education programs could help zoos lead the way toward improving programs at other institutions that focus on conservation education.

Acknowledgements

The Denver Zoological Foundation funded this research. Data Marketing Associates, Inc. provided assistance in implementing the survey and inputting the data. Tammy Hill assisted with pre-testing and survey design. Bill Loessberg, Dennis Smith, Brian Klepinger, and Clayton Freiheit facilitated several aspects of this work.

References

Acampora, R. R. (1998). Extinction by exhibition: looking at and in the zoo. *Human Ecology Review*, **5**, 1–4.

Altman, J. D. (1998). Animal activity and visitor learning at the zoo. *Anthrozoös*, **11**, 12–21.

Andereck, K. & Caldwell, L. (1994). Motive-based segmentation of a public zoological park market. *Journal of Park and Recreation Administration*, **12**, 19–31.

Bath, A. J. (1989). The public and wolf restoration in Yellowstone National Park. *Society and Natural Resources*, **2**, 297–306.

Bem, D. J. (1970). *Beliefs, Attitudes, and Human Affairs*. Belmont, CA: Brooks/Cole.

Birney, B. A. (1995). Children, animals, and leisure settings. *Society and Animals*, **3**, 171–187.

Bright, A. D. & Barro, S. C. (2000). Integrative complexity and attitudes: a case study of plant and wildlife species protection. *Human Dimensions of Wildlife*, **5**, 30–47.

Brown, P. J. & Manfredo, M. J. (1987). Social values defined. In *Valuing Wildlife: Economic and Social Perspectives*, eds. D. J. Decker & G. R. Goff. Boulder, CO: Westview Press, pp. 12–23.

Brown, T. C. (1984). The concept of value in resource allocation. *Land Economics*, **60**, 231–246.

Chaiken, S. & Stangor, C. (1987). Attitudes and attitude change. *Annual Review of Psychology*, **38**, 575–630.

Clark, T. W. (1997). *Averting Extinction*. New Haven, CT: Yale Press.

Clark, T. W., Reading, R. P., & Wallace, R. L. (1999). Research in endangered species conservation: an introduction to multiple methods. *Endangered Species Update*, **16**, 96–102.

Conway, W. (2000). The changing role of zoos and aquariums in the 21st century. *Communiqué*, January, 11.

Dunlap, J. & Kellert, S. R. (1989). *Informal Learning at the Zoo: A Study of Attitude and Knowledge Impacts.* Report to the Zoological Society of Philadelphia funded by the G. R. Dodge Foundation.

Heinrich, C. J. & Birney, B. A. (1992). Effects of life animal demonstrations on zoo visitors' retention of information. *Anthrozoös*, **5**, 113–121.

Holling, C. S. (1978). *Adaptive Environmental Assessment and Management.* New York: John Wiley & Sons.

Hutchins, M., Smith, B., & Allard, R. (2003). In defense of zoos and aquariums: the ethical basis for keeping wild animals in captivity. *Journal of the American Veterinary Medical Association*, **223**, 958–966.

Kellert, S. R. (1979). Zoological parks in American society. In *Proceedings of the Annual AZA Conference*, pp. 88–126.

Kellert, S. R. (1990). Public attitudes and beliefs about the wolf and its restoration in Michigan. Internal Report, Yale University School of Forestry and Environmental Studies, New Haven, CT.

Kellert, S. R. (1996). *The Value of Life: Biological Diversity and Human Society.* Washington, D.C.: Island Press.

Kellert, S. R. (1997). *Kinship to Mastery: Biophilia in Human Evolution and Development.* Washington, D.C.: Island Press.

Kellert, S. R. & Berry, J. K. (1980). *Knowledge, Affection, and Basic Attitudes Toward Animals in American Society: Phase III.* Washington, D.C.: US Government Printing Office.

Kellert, S. R., Black, M., Rush, C. R., & Bath, A. J. (1996). Human culture and large carnivore conservation in North America. *Conservation Biology*, **10**, 977–990.

Kidd, A. H., Kidd, R. M., & Zasloff, R. L. (1995). Developmental factors in positive attitudes toward zoo animals. *Psychological Report*, **76**, 71–81.

Kleiman, D. G., Reading, R. P., Miller, B. J., Clark, T. W., Scott, J. M., Robinson, J., Wallace, R., Cabin, R., & Fellman, F. (2000). The importance of improving evaluation in conservation. *Conservation Biology*, **14(2)**, 1–11.

Lasswell, H. D. (1971). *A Pre-View of the Policy Sciences.* New York: Elsevier.

Manfredo, M. J., Vaske, J. J., Haas, G. E., & Fulton, D. (1993). *The Colorado Environmental Poll.* CEP #2. Fort Collins, CO: Human Dimensions in Natural Resources Unit, Colorado State University.

Marvin, G. (1994). Review essay. *Society and Animals*, **2**, 191–199.

Masci, D. (2000). Zoos in the 21st century. *CQ Researcher*, **10**, 355–368, 370–372.

Meadow, R., Reading, R. P., Phillips, M., Mehringer, M., & Miller, B. J. (2005). The influence of persuasive arguments on public attitudes toward a proposed wolf restoration in the southern Rockies. *Wildlife Society Bulletin*, **33(1)**, 154–163.

Miller, B., Conway, W., Reading, R. P., Wemmer, C., Wildt, D., Kleiman, D., Monfort, S., Rabinowitz, A., Armstrong, B., & Hutchins, M. (2004). Evaluating the conservation mission of zoo, aquariums, botanical gardens, or natural history museums. *Conservation Biology*, **18**, 1–8.

Morgan, J. M. & Hodgkinson, M. (1999). The motivation and social orientation of visitors attending a contemporary zoological park. *Environment and Behavior*, **31**, 227–239.

Olson, J. A. & Zanna, M. P. (1993). Attitudes and attitude change. *Annual Review of Psychology*, **44**, 117–154.

Pate, J., Manfredo, M. J., Bight, A. D., & Tischbein, G. (1996). Coloradan's attitudes toward reintroducing the gray wolf into Colorado. *Wildlife Society Bulletin*, **24**, 421–428.

Petty, R. E., Wegener, D. T., & Fabrigar, L. R. (1997). Attitudes and attitude change. *Annual Review of Psychology*, **48**, 609–674.

Reading, R. P. (1993). Toward an endangered species reintroduction paradigm: a case study of the black-footed ferret. Unpublished Ph.D. Dissertation, Yale University, New Haven, CT, USA.

Reading, R. P. & Kellert, S. R. (1993). Attitudes toward a proposed black-footed ferret (*Mustela nigripes*) reintroduction. *Conservation Biology*, **7**, 569–580.

Rhoads, D. L. & Goldsworthy, R. J. (1979). The effects of zoo environments on public attitudes toward endangered wildlife. *International Journal of Environmental Studies*, **13**, 283–287.

Rokeach, M. (1972). *Beliefs, Attitudes, and Values: A Theory of Organization and Change*. San Francisco, CA: Jossey-Bass.

Sinden, J. & Worrel, A. (1979). *Unpriced Values: Decisions Without Market Values*. New York: John Wiley & Sons.

Steinhoff, H. W. (1980). Analysis of major conceptual systems for understanding and measuring wildlife values. In *Wildlife Values*, eds. W. W. Shaw & E. H. Zube. Center for Assessment of Noncommodity Natural Resource Values, Institute Series Report Number 1, pp. 11–21.

Swanagan, J. (2000). Factors influencing zoo visitors' conservation attitudes and behavior. *Journal of Environmental Education*, **31**, 26–31.

Tessler, A. & Shaffer, D. R. (1990). Attitudes and attitude change. *Annual Review of Psychology*, **41**, 479–523.

Tunnicliffe, S. D. (1996). Conversations within primary school parties visiting animal specimens in a museum and zoo. *Journal of Biological Education*, **30**, 130–142.

Williams, R. M., Jr. (1979). Change and stability in values and value systems: a sociological perspective. In *Understanding Human Values: Individual and Societal*, ed. M. Rokeach. New York: The Free Press, pp. 15–46.

7

The animal rights-conservation debate: can zoos and aquariums play a role?

MICHAEL HUTCHINS

INTRODUCTION

There is growing public concern about the treatment of non-human animals in our society, and this has resulted in the rapid growth of membership in and support of animal rights and welfare organizations (Beckoff 1998). Similarly, there is strong public support for the conservation of endangered species and their habitats, and this has resulted in an equally rapid increase in membership in and support of organizations that identify wildlife and habitat conservation as their primary missions (Street 2004).

However, a survey conducted by the Association of Zoos and Aquariums suggests that the public is very confused about the concept of animal rights and its relationship to wildlife conservation. Preliminary results indicate that many laypersons perceive the two ideologies to be roughly equivalent (Mellen 2003); but, what are the differences between the animal rights and environmental/conservation ethics? Are these two worldviews always compatible?

ANIMAL RIGHTS AND CONSERVATION: A PRIMER

Animal rights and conservation are ethical perspectives; that is, viewpoints intended to guide us in our behavior toward animals and nature. Ethical questions are questions about right and wrong; that is, rather than focusing on "what is," which is the realm of science, ethicists focus on "what ought to be" done (White 1981). It should be noted, however, that ethical viewpoints are not absolute and our conceptions of right and wrong can vary with context (situational ethics) and can also change over time. It should also be noted that the knowledge gained from scientific study does influence our ethical discussions. An ethical decision based on ignorance or incorrect information is unlikely to stand up to scrutiny. The purpose of ethics is therefore not to seek absolute truth or perfect resolution; rather, it is a continual process of debate and reconsideration. The process of ethical debate can eventually result in consensus building, as more and more people are compelled to accept one position over another, or conversely, it can result in an extreme polarization of views. In either case, it is important that various views be critically examined, especially in cases where our decisions have important implications for society or nature. Thus, before discussing some potential sources of conflict between conservation and animal rights, it will first be necessary to examine these viewpoints in more detail.

Before comparing the philosophical foundations of animal rights and conservation, a brief caveat is necessary. In framing these discussions, I fully realize that each individual's views on any topic fall somewhere along a wide spectrum of beliefs. It is often hard to classify individuals into one category or another and, in reality, people may use terms to describe themselves, which although technically inaccurate, seem to work for them. For example, I have had conversations with people who classified themselves as animal rights advocates who, upon closer examination, were more appropriately classified as animal welfare advocates. This may be due to the complex ways in which people gather information and form opinions, and partly to people's confusion over the definition of these terms. One of the goals of this paper is to clear up some of this confusion, but it is important to note that this discussion of the animal rights-conservation debate focuses on the philosophical foundations of the animal rights and conservation movements as articulated in the academic literature.

Animal rights and welfare ethics

Animal rights proponents invoke the philosophy and language of the civil rights movement, arguing that sentient, non-human animals, like disadvantaged human groups (e.g., women, ethnic minorities), have an intrinsic and inviolate right to life, liberty, and bodily integrity. Thus any human use of non-humans that harms individuals in any way is considered unacceptable (Regan 1983).

Animal rights advocates argue that individual animals should be the focus of our ethical concerns. Furthermore, they argue that sentience – the capacity to experience pain – is the only relevant characteristic needed by animals to merit full moral consideration (Singer 1975, Regan 1983). They contend that if non-human animals have the capacity to suffer pain, then their suffering should be as important a matter of ethical concern as that of our fellow humans; that non-humans may be incapable of reason, speech, forethought or self-awareness is considered irrelevant. After all, some classes of humans (e.g., newborn infants and the severely mentally retarded) do not possess these abilities, yet are accorded rights.

The argument that human needs should take precedence over those of sentient non-humans is viewed as "speciesism" – a form of prejudice analogous to chauvinism, racism, or sexism (Singer 1975, Midgely 1983, Regan 1983). Thus, individual animals are seen as having a "right to life" and, except in very special circumstances, any attempt to kill them or to cause them to suffer pain is considered morally unjustifiable.

Animal rights differ from the concept of animal welfare. Welfare can be defined as a good or satisfactory condition of existence; thus animal welfare refers to the quality of an animal's life, whether in nature or in human care (Duncan and Fraser 1997). Like animal rights proponents and many conservationists, animal welfare advocates abhor human cruelty towards animals; they also detest unnecessary suffering or loss of life. However, many animal welfare advocates tend to be more flexible in their beliefs, arguing that non-humans can be utilized by humans, even for food, as long as pain, suffering, and loss of life are minimized. Not surprisingly, the welfare ethic does not sit well with many animal rights proponents, who see it as perpetuating aspects of the current belief system that they cannot support (Regan and Francione 1992).

Conservation ethic

Conservationists seek to ensure a future for naturally occurring biological diversity (Primack 2002). Biodiversity is defined here as the variety of organisms (species, subspecies, etc.) that inhabit particular ecosystems. The term "natural" is used here to distinguish between diversity that has occurred as the result of natural ecological/evolutionary processes (i.e., speciation, colonization, and "natural" extinction), and that which has occurred because of relatively recent human interventions (i.e., introduction of non-native invasive species, human-caused extinctions) (Aitken 1998). Aldo Leopold said that: "A thing is right if it tends to preserve the integrity, beauty and stability of the biotic community. It is wrong when it tends to do otherwise" (Leopold 1949). While change not stability is now recognized as an inherent aspect of natural systems (Callicott 1989), the dynamic systems themselves and the species that inhabit them are seen as worthy of our moral consideration (Gunn 1980, Norton 1987, Callicott 2000, Regan 2000). Thus, the biological richness of an ecosystem, as characterized by the number and variety of native species it supports, is seen as intrinsically good. Conversely, the loss of naturally occurring biological diversity, especially as the result of human activity, is seen as intrinsically bad (Aitken 1998, Primack 2002).

Proponents of the conservation ethic recognize and value the complex interdependencies of organisms within functioning ecosystems (Primack 2002). In simple terms, ecosystems consist of a source of energy (usually sunlight), a source of raw chemical materials (rocks, soil, air, water), "producers" capable of transforming and storing solar energy (usually green plants), "primary consumers," which feed on the producers (i.e., herbivores), "secondary consumers," which feed on the primary consumers (i.e., carnivores), and, finally, "decomposers," which break down the dead bodies of the producers and consumers and cycle their energy back into the system (Ricklefs 1973). According to Leopold (1949), such "food chains are the living channels which conduct energy upward; death and decay return it to the soil . . . like a slowly augmented revolving fund of life."

Conservationists have identified habitat alteration, fragmentation, and destruction as being the primary threat to wildlife populations (Ehrlich and Ehrlich 1981, Primack 2002, Wilson 2002). Therefore, a concern for wild animals should be expressed in a willingness to protect natural habitats and ecosystems. All living organisms are tied closely to the habitats in which they have evolved. It is therefore difficult to separate individual animals or species from their ecological contexts. It is equally difficult to draw a strong distinction between living and non-living components of the environment. All living organisms, whether they are bacteria or humans, are composed of non-living matter. Carnivores, for example, are as dependent on soils for their existence as they are on their prey.

CONFLICTS BETWEEN ANIMAL RIGHTS AND WILDLIFE CONSERVATION

Decisions regarding the future of wildlife and their habitats are complex. All such decisions involve the weighing of various, sometimes competing, values, and it is often excruciatingly difficult to determine which path will lead to the "greatest good." Clearly, the animal rights ethic and the conservation ethic will lead to the same decisions in many situations. For example, both ethics would consider it wrong for humans to destroy critical wildlife habitat, or to pollute it with chemicals and wastes. But, when the two viewpoints are compared, it is evident that disagreements will arise when the "rights" of individual, sentient animals come into conflict with efforts to conserve populations, species, habitats or ecosystems (Hutchins and Wemmer 1987, Norton 1987, 1995, Temple 1990, Soulé 1990, Hutchins *et al.* 1995, Vrijenhoek 1995).

Species or ecosystems do not warrant moral consideration according to the rights view, though they are considered to have "inherent value" (Feinberg 1978, Regan 1983). Ideological differences between animal rights and conservation are evident in their contrasting views of the endangered species problem. While both ethics favor saving threatened or endangered species or populations, they differ in their reasons for doing so. Regan (1983, p. 360) argues that we should conserve endangered species "... not because the species is endangered, but because the individual animals have valid claims and thus rights against those who would destroy their natural habitat, for example, or would make a living off their dead carcasses through poaching and traffic in exotic animals, practices which unjustifiably override the rights of those animals." Thus, all sentient animals, regardless of species, rarity or other considerations, are given equal moral consideration, according to the rights view.

In contrast, conservationists argue that endangered populations or species should be given special status because of their scarcity and heightened risk of final and irreversible extinction (Gunn 1980, Norton 1987, Aitken 1998, Callicott 2000, Regan 2000). That is, extraordinary efforts should be made to preserve rare populations or species, especially when an organism has become scarce due to some action on the part of humans.

The animal rights ethic is very clear on its position regarding human–animal relationships, but it is unclear with regard to pain and suffering inflicted on animals by animals. For some animal rights advocates, what predatory animals do to their prey is beyond the realm of their concern, presumably because it is done by "innocent killers" lacking in malicious intent or the knowledge of the ethical consequences of their actions (Feinberg 1978, Regan 1983). Such views open the animal rights ethic to logical criticism (Hutchins and Wemmer 1987, Callicott 1989). From an individual's standpoint, pain is pain, regardless of the intent of the predator. In addition, predation is not the only way that one animal can have a detrimental effect on another. Indeed, one weakness of a view of nature that stresses individual rights is that it fails to recognize the complex interdependencies that exist within natural ecosystems. Thus if a population of one species becomes locally over-abundant to the point that it degrades its own habitat, many other organisms may suffer as well.

So what happens in cases where conservationists recognize the need to intervene to save populations, species or ecosystems that are in danger of extinction? What if this involves harming or killing individual animals? In fact, there are many circumstances in which the "rights" of individual animals come into conflict with efforts to preserve populations, species or ecosystems (Hutchins and Wemmer 1987, Soulé 1990, Temple 1990, Hutchins 1995, Hutchins *et al.* 1995, Vrijenhoek 1995). This chapter considers those areas particularly relevant to the conservation activities of zoos.

CAPTIVE BREEDING FOR REINTRODUCTION

One form of management intervention employed by conservationists is moving endangered animals into "protective custody" for breeding and reintroduction purposes. For example, the North American black-footed ferret (*Mustela nigripes*) was nearly driven to extinction by the loss of its primary prey – the prairie dog (*Cynomys* spp.). These social rodents were the focus of a systematic control program, and as their populations declined, so did the ferrets'. Introduced diseases, such as distemper and plague, also decimated ferret populations. The species was thought to be extinct until a small population was discovered in Wyoming in the early 1980s (Miller *et al.* 1996). These animals have become the nucleus of a scientifically managed, cooperative breeding program that was intended to support the species' recovery. The cooperative effort between accredited zoos and federal and state wildlife agencies has been highly successful, and there are now more ferrets living in nature than in captivity (Lockart *et al.* 1998).

The endangered California condor (*Gymnogyps californicus*) provides another example. Only 22 wild condors remained in 1982, their numbers having been reduced by habitat loss and pollutants, such as dichlorodiphenyl-trichloroethane (DDT) and lead. The last remaining wild condors were brought into captivity in 1987. But, in this case, it was not animal rights groups that protested; it was environmental groups, such as The National Audubon Society. The breeding program has been successful and reintroduction has brought the condor back to areas in California and the American southwest (Wilcove 1999, Graham 2000). The program still faces many difficult challenges, such as power lines, lead shot used by hunters and antifreeze, all of which have caused condor deaths. However, the current population exceeds 200 and there are condors living in nature again. The Audubon Society recently apologized for its early opposition to the captive breeding program.

Conservationists do not recommend captive breeding for reintroduction as a recovery strategy in every case, and the removal of wild animals for breeding programs must be carefully considered (Koontz 1995). However, when necessary and appropriate, captive breeding for reintroduction is widely recognized as a potentially effective tool to aid in endangered species recovery (Mallinson 1995, Norton 1995, Primack 2002, Stanley Price and Soorae 2003).

However, animal rights advocates have a much different view. In fact, the animal rights ethic generally opposes the notion that wild animals may be held in captivity, even under the best of conditions. In this view, individual animals have a right to liberty and these rights are being violated when wild animals are captured and held in dedicated breeding facilities or professionally managed zoos. According to Regan (1995, p. 45), "... confining wild animals ... can be justified, according to the rights view, but only if it can be shown that it is in their best interests to do so." He further states that, "it is morally irrelevant ... to insist that captive animals serve as useful models in important scientific research or that they offer an opportunity for protecting rare and endangered species, or that they advance the interests of other individuals, whether human or non-human" (p. 46).

The reintroduction process itself appears to be a potential point of contention between conservationists and animal rights advocates. The goal of reintroduction programs is to re-establish extirpated or augment existing, though severely depleted, populations of endangered or threatened species. However, the risk to individual animals during reintroduction is considerable, and the incidence of mortality can be high, especially in a program's early stages (Beck 1995). Early in the black-footed ferret reintroduction, the mortality rate of released animals was nearly 80%. The most common cause of death was predation by coyotes (*Canis latrans*) and badgers (*Taxidea taxus*). Behavioral studies indicated that reintroduced ferrets were spending far too much time traveling above ground in daytime, thus increasing their chances of detection by predators. This problem was later solved by a combination of approaches, including pre-release conditioning of the ferrets, which were allowed to gain experience moving in and out of prairie dog burrows when confronted with danger (Miller *et al.* 1998). In order to ensure that release candidates could recognize and kill their primary food, captive-reared ferrets were also given the opportunity to hunt and kill live prairie dogs. While this experience was critical for the success of the reintroduction program, there is no doubt that it violated the "rights" of the individual prairie dogs.

The animal rights view would clearly not have allowed the black-footed ferret or California condor breeding and reintroduction programs, but at what cost? These species and many others would surely be extinct today if it were not for human intervention. The need for captive breeding for reintroduction is expected to grow over the next few decades as more and more key species are pushed to the brink of irreversible extinction.

CONSERVATION RESEARCH

Scientific research is one means by which humans gain an understanding of the natural world. Such an understanding is essential to wildlife conservation efforts. Animal rights advocates are opposed to the use of animals as research models in human biomedical research, particularly when individuals are sacrificed or caused to suffer psychological or physical pain (Regan 1983, March 1984, Rudacille 2000).

Conservation studies generally fall under the umbrella of a relatively new scientific discipline known as conservation biology (Soulé 1985). The goal of this applied science is to conserve naturally occurring biological diversity. Conservation biology is one of the most interdisciplinary of sciences, encompassing not only the biological sciences, but also economics and the social sciences. Conservation biology is the scientific foundation of the conservation ethic. Thus, its primary goal is to ensure a future for individuals, populations, species, and ecosystems. Like any other research conducted on animals, however, conservation research has the potential to violate the rights of individual animals as defined by Regan (1983). In fact the rights view would preclude all practices that "cause intentional harm." According to Regan (1983, p. 387), "This objective will not be accomplished merely by ensuring that test animals are anesthetized, or given post-operative drugs to ease their suffering, or kept in clean cages with ample food and water, and so forth. For it is not only the suffering that matters – though it certainly matters – but it is the harm done to animals, including the diminished welfare opportunities that they endure." But what are the consequences of this view?

There are many cases in which conservation-related research could prove harmful to individual animals. For example, to collect essential data on population dynamics, behavior, individual growth rates, diseases, etc., it is often necessary to capture and handle animals or to mark them for individual identification. Despite numerous precautions by scientists, animals are sometimes harmed during capture procedures. For example, some animals may suffer limb fractures or trauma due to falls and others may succumb to adverse reactions to immobilizing drugs or to shock. Still others may contract capture myopathy, an often-fatal muscular condition induced by the stress of capture and transport (Fowler 1995).

Harm that comes to individual animals during capture and handling could be considered accidental in that scientists are not harming animals deliberately. However, there are some cases where conservation research may involve deliberate harm. For example, Eaton (1972a, 1972b) studied the development of predatory behavior in captive-reared lions (*Panthera leo*) and cheetahs (*Acinonyx jubatus*). To observe predation, he released live domestic goats, which were subsequently killed and eaten by the cats. The rights view would certainly not condone such experiments because they violate the rights of the individual goats. However, despite the unfortunate consequences for individual goats, this work appears to be compatible with the more holistic conservation ethic. Humans have pushed many carnivorous species, including large cats, to the brink of extinction. One method by which conservationists hope to save some of these species is through captive breeding for reintroduction (Moore and Smith 1990). However, reintroducing captive-bred carnivores into their natural habitats poses many difficult problems, including the ability of the animals to recognize, catch, and kill their own prey (Miller *et al.* 1998). Thus, studies of this kind could prove to be essential for planning any serious reintroduction effort.

Conservation biologists have identified numerous research questions to be answered in order to fill gaps in our current knowledge of biodiversity and conservation methodology (Soulé and Orians 2001). However, the rights view would place heavy restrictions on conservation research. One implication is that information essential to conservation could not be collected except under very restricted conditions, and this might increase the probability of species extinctions (Hutchins and Wemmer 1987, Hutchins *et al.* 1995). By contrast, the more holistic conservation ethic does not oppose the use of animals in scientific research, especially if the knowledge gained were to help ensure the survival of a population, species or ecosystem. Animal rights proponents would consider this view to be "utilitarian," in that "whether the harm done to individual animals in the pursuit of scientific ends is justified depends on the balance of the aggregated consequences for all those affected by the outcome" (Regan 1983, p. 392). Since animal rights proponents find this utilitarian view unacceptable, most forms of conservation research could not be condoned.

Animal welfare advocates might argue that wildlife scientists should develop more benign methods of study. In fact, scientists themselves have taken some initiative in this regard, as few people want to inflict needless harm. For example, there has been interest in using less invasive methods to assess physical condition and diets. Physical condition can sometimes be assessed by measurements of weight, girth, blood chemistry, and other physical parameters (Franzmann *et al.* 1995). Similarly, rather than studying stomach contents, dietary preferences can sometimes be determined by watching what animals eat, by analyzing feces, or by measuring the nutritional quality and abundance of the food resources themselves (Blankenship and Satakopan 1995, Stromberg 1995). The humane treatment of animals is therefore a continuing goal. However, it may not be possible to totally avoid suffering, pain or loss of life in all instances. Conservationists seek to balance harm to individual animals against the potential gain in knowledge that may positively impact the lives of many more animals and species survival.

ANIMAL RIGHTS AND CONSERVATION REALITIES

There have been many attempts to find common ground between animal rights and conservation (e.g., Warren 1983, Ehrenfeld 1991, Callicott 1989, Varner 1998, Beckoff 2002). While both ethical viewpoints clearly share a reverence for life, they can lead to vastly different policies and practices and these differences are affecting the way that we manage and conserve wildlife, sometimes to its detriment. It is therefore particularly important that laypersons, experts, and key decision-makers fully understand the differences between these views and their implications for wildlife conservation and management policy and practice. At least with regard to some decisions affecting wildlife, it *is* important whether we classify ourselves as "animal rights advocates" or "conservationists." The two viewpoints are not always compatible, nor should they be perceived as such.

Given the fact that the animal rights ethic is often incompatible with conservation goals, is there an alternative for those of us who care about both the future of life on this planet *and* the welfare of individual animals? I believe there is, but it is not going to be easy. To accomplish this goal, we must first make conservation our highest moral imperative. And, secondarily, conservationists must also embrace the concept of animal welfare.

Animal rights proponents apparently believe that we can achieve conservation goals by focusing our attention solely on individual animals (see Regan 1983). Thus, animal rights is a highly reductionist view of nature, which seems to imply that species and ecosystems can be saved simply by preserving their component parts. In contrast, ecological science has shown us that complex interdependencies exist between various species in a biotic community, and that the whole is actually greater than the sum if its parts (Rodman 1977, Callicott 1989, 2000, Regan 2000). If so, then there is a strong justification for giving preference to endangered over common species and native species over introduced species when it comes to conservation policy.

Animal rights proponents also believe that if we simply leave nature alone, it will heal itself (see Regan 1983). However, ecologists have noted that even the largest of national parks will lose much of their biological diversity in the absence of careful management (Soulé *et al.* 1979). In fact, given the rate at which humans are altering natural ecosystems, unprecedented levels of human intervention are going to be necessary to prevent the widespread loss of biological diversity (Younghusband and Myers 1986, Coblentz 1990, Kenny 1990, Soulé 1990, Diamond 1992, Bowles and Whelan 1994, Hutchins 1995). Whether we like it or not, there are many cases in which wildlife is going to have to be managed if it is to survive.

In weighing the costs and benefits of our decisions, we must consider not only the fate of individuals, but also the fate of populations, species, and ecosystems. These "collectives" have inherent value and must be considered in our decision-making processes (Rodman 1977, Gunn 1980, Norton 1987, Callicott 1989, 2000, Rolston 1994). We must consider the fact that the extinction of a species means the "end of birth" – that is, that the potential for future individual lives will be lost forever. These future lives have value too and their potential loss is deserving of our moral consideration (Attfield 1983). And, perhaps even more importantly, because various life forms can be highly interdependent, it means that the loss of one species could mean the loss of others.

Context is also important in ethical decision making. Clearly we are in a desperate situation as far as biological diversity is concerned. The current global trends that are in play will likely mean that many species are lost in the coming decades (Wilson 2002, Rosenweig 2003). In the long-term, saving species, habitats, and ecosystems will mean finding remedies to the many human-caused problems that lead to species endangerment (Redford and Richter 2001). However, as a conservationist, I believe that we should take necessary short-term actions to conserve and/or restore as many species and habitats as possible, while at the same time trying to move toward a more sustainable society. A conservationist's work is analogous to that of an emergency room doctor's, and emergency room ethics apply. Sometimes desperate acts are going to be necessary to preserve life, even if it means causing the patient short-term harm. Of course, in the case of conservation, the "patients" are endangered populations, species, and ecosystems, and the immediate harm is being done to a subset of individual animals.

In order to preserve some semblance of nature in a human-dominated world, we are going to have to make some hard choices. However, this does not mean that conservationists should act callously and without empathy and compassion (Schmidt 1981, Hutchins and Wemmer 1987, Rolston 1992, Beckoff 2002). Conservationists and wildlife managers should recognize that individual sentient animals *are* morally considerable and we should not make these decisions lightly. When the need to control animal populations or otherwise manage wild animals becomes necessary, it should be accomplished in the most humane manner possible (Hutchins and Wemmer 1987). One problem is the current lack of effective humane alternatives. Conservation scientists must confront this issue directly and seek to refine existing methods or develop new ones.

Some conservationists argue that one reason for preserving natural processes is to protect human interests (e.g., to maintain ecosystem services, preserve aesthetic values and utility for future generations, etc.). However, others value wildlife and nature for its own sake (Callicott 2000, Regan 2000). This makes it even more difficult to gauge the rightness or wrongness of actions that simultaneously sacrifice the life of, or cause suffering in some, sentient beings, while they preserve or enhance the lives of many others. In fact, it is increasingly clear that ethical philosophy faces a severe test when it comes to the conservation problem. Wilson (1984, p. 123) said, "in ecological and evolutionary time, good does not automatically flow from good and evil from evil. To choose what is best for the near future is easy. To choose what is best for the distant future is also easy. But to choose what is best for both the near future and distant future is a hard task, often internally contradictory, and requiring ethical codes yet to be formulated."

IS THERE A ROLE FOR ZOOLOGICAL PARKS AND AQUARIUMS IN THE ANIMAL RIGHTS-CONSERVATION DEBATE?

Modern zoos and aquariums are playing an increasingly important role in educating the public about current conservation issues. In North America alone, some 140 million people visit accredited zoological institutions annually, providing a tremendous opportunity for public outreach and education (Delapa 1994). Many zoos and aquariums have structured as well as informal educational programs that cater to both children and adults. Zoo and aquarium professionals also produce many publications and frequently interact with the media and key decision-makers. In addition, these institutions are science based, share a conservation mission and have practical experience in managing wildlife on a day-to-day basis (Hutchins 2003).

Clearly, zoos and aquariums can play a critical educational role in the animal rights-conservation debate. We must do a better job of not only building an appreciation for wildlife, but in promoting a deeper understanding of the many complex challenges faced by conservationists today. It is not enough to teach people to care about animals or to be concerned about their future. In fact, building a purely emotional connection to wildlife can often backfire on conservationists. We have taught people to love animals and nature unconditionally; in fact, they love it so much that the thought of having to take a life to save others is often not acceptable. That is why zoos and aquariums need to find a way to promote what I refer to as "informed concern." If we are to save wildlife for the future, then people must begin to understand the many challenges and complexities of modern conservation, including the need to actively manage wildlife and wildlife habitats (Bowles and Whelan 1994).

Over a decade ago, conservation biologist Michael Soulé posed an important challenge to the conservation community. He said, "Conflicts between animal rights groups and management agencies are increasing in frequency and cost – the cost is being borne by endangered species and ecosystems as well as by the public that pays for expensive rescue operations and time consuming court battles. The minimization of such conflicts will require both public education and courageous leadership." (Soulé 1990).

Zoos and aquariums have a tremendous opportunity to seize this leadership role, but courage will be necessary. Zoos and aquariums, like many conservation-oriented organizations, have probably been reluctant to address these difficult issues for several reasons, including fear of the "negative public reaction that almost inevitably accompanies eradication efforts, especially for animals" (Temple 1990). However, it should be recognized that much of the research, captive breeding, reintroduction, and field conservation work done by the zoological community is also not compatible with animal rights philosophy. Indeed, if zoos and aquariums want to continue to pursue their conservation mission, they need to take the animal rights challenge very seriously (see Hutchins *et al.* 2003).

Animal rights groups have been very active in schools and in indoctrinating an entire generation of young people. They raise millions of dollars annually from concerned individuals, which they spend primarily on influencing public opinion and key decision-makers in government. Zoos, aquariums, mainstream conservation, and animal welfare organizations must join forces to help counter their inflexible, anti-conservationist views. At the same time, modern zoos and aquariums need to evolve into conservation and animal welfare organizations, doing their best to both provide quality animal care *and* contribute directly to wildlife and habitat conservation in nature (Maple *et al.* 1995, Hutchins 2003, Hutchins and Smith 2003, Hutchins *et al.* 2003).

Modern zoos and aquariums need to help the public, media, and key decision makers in government understand the hard choices that lie ahead. This can only be accomplished if people appreciate what is going to be lost if no action is taken. Zoos and aquariums provide a unique opportunity to drive home the conservation message – a message based on science and commonsense, rather than on misguided logic and pure emotion. What better time is there to impress upon people the importance of conservation ethics than when they are standing face to face with living examples of critically endangered species? While individual animals are important, a failure to act on behalf of populations, species, and ecosystems will render the debate moot – there will be no future individuals on which to focus our concerns. Like many other species before them, they will be relegated to the dustbin of extinction – victims of well-meaning, but misguided efforts to preserve individual life, liberty, and welfare at all costs.

SUMMARY AND CONCLUSIONS

1. The animal rights ethic is a poor foundation for conserving wildlife and their habitats in an increasingly complex world. Its primary weaknesses when applied to wildlife conservation are as follows. It focuses exclusively on what is good for individuals rather than on what is good for populations, species or ecosystems. It fails to recognize the complex interrelationships between species in functioning ecosystems. Species and ecosystems are morally considerable because they are necessary to sustain life. Specifically with regard to wildlife, the rights view emphasizes a hands-off approach, which in the current global context, means that virtually all forms of wildlife management would be unacceptable.
2. If we care about the future of life on this planet, then we *must* make conservation our highest moral imperative. Because conservationists are compassionate about individual animals, they should also embrace the animal welfare ethic and investigate more humane methods of population control and captive animal management.

3. Professionally managed zoos and aquariums have a role to play in the animal rights-conservation debate by educating the public about both the differences between these two worldviews and the challenges, complexities, and realities of conservation in human-dominated landscapes. When the public understands the differences between the two perspectives, it is hoped that more people will embrace the conservation ethic and understand the difficult, but crucial choices that will be necessary to secure a biologically diverse future.

ACKNOWLEDGEMENTS

I would like to thank C. West and R. Lattis for inviting me to participate in the Zoos as Catalysts for Conservation Conference. Thanks are also due to R. Allard for reading and commenting on an earlier version of the manuscript. J. Lankard provided assistance in conducting library research.

References

Aitken, G. M. (1998). Extinction. *Biology and Philosophy*, **13**, 393–411.
Attfield, R. (1983). *The Ethics of Environmental Concern*. New York: Columbia University Press.
Beck, B. B. (1995). Reintroduction, zoos, conservation and animal welfare. In *Ethics on the Ark: Zoos, Animal Welfare and Wildlife Conservation*, eds. B. Norton, M. Hutchins, E. F. Stevens, & T. L. Maple. Washington, D.C.: Smithsonian Institution Press, pp. 155–163.
Beckoff, M., ed. (1998). *Encyclopedia of Animal Rights and Welfare*. Westport, CT: Greenwood Press.
Beckoff, M. (2002). The importance of ethics in conservation biology: let's be ethicists not ostriches. *Endangered Species Update*, **19(2)**, 23–26.
Blankenship, L. & Satakopan, S. (1995). Food habits and nutrition. In *Wildlife Research and Management: Asian and American Approaches*, eds. S. H. Berwick & V. B. Saharia. Delhi, India: Oxford University Press, pp. 329–362.
Bowles, M. L. & Whelan, C. J. (1994). *Restoration of Endangered Species*. Cambridge: Cambridge University Press.
Callicott, J. B. (1989). *In Defense of the Land Ethic: Essays in Environmental Philosophy*. Albany, NY: State University of New York Press.
Callicott, J. B. (2000). On the intrinsic value of nonhuman species. In *The Preservation of Species*, eds. B. G. Norton *et al.* Princeton, NJ: The Princeton University Press, pp. 138–172.
Coblentz, B. E. (1990). Exotic organisms: a dilemma for conservation biology. *Conservation Biology*, **4(5)**, 261–265.
Delapa, M. (1994). Interpreting hope, selling conservation: 2005, aquariums and environmental education. *Museum News*, May/June, 48–49.
Diamond, J. (1992). Must we shoot deer to save nature. *Natural History*, **8**, 2–8.

Duncan, I. J. H. & Fraser, D. (1997). Understanding animal welfare. In *Animal Welfare*, eds. M. C. Appleby & B. O. Hughes. Wallingford, UK: CAB International, pp. 19–31.
Eaton, R. L. (1972a). An experimental study of predatory and feeding behavior in the cheetah (*Acinonyx jubatus*). *Zeitschrift für Tierpsychologie*, **31**, 270–280.
Eaton, R. L. (1972b). Predatory and feeding behavior in adult lions: the deprivation experiment. *Zeitschrift für Tierpsychologie*, **31**, 461–473.
Ehrenfeld, D. (1991). Conservation and the rights of animals. *Conservation Biology*, **5(1)**, 1–3.
Ehrlich, P. & Ehrlich, A. (1981). *Extinction*. New York: Random House.
Feinberg, J. (1978). Human duties and animal rights. In *On the Fifth Day: Animal Rights and Human Ethics*, eds. R. K. Morris & M. W. Fox. Washington, D.C.: The Humane Society of the United States.
Fowler, M. (1995). *Restraint and Handling of Wild and Domestic Animals*, Second Edition. Ames, IA: Iowa State University Press.
Franzmann, A. W., Cook, R., Singh, C. M., & Cheeran, J. V. (1995). Health and condition evaluation of wild animal populations. In *Wildlife Research and Management: Asian and American Approaches*, eds. S. H. Berwick & V. B. Saharia. Delhi, India: Oxford University Press, pp. 367–399.
Graham, F. (2000). The day of the condor. *Audubon*, **102(1)**, 46–53.
Gunn, A. S. (1980). Why should we care about rare species? *Environmental Ethics*, **2**, 17–37.
Hutchins, M. (1995). What do "wild" and "captive" mean for large ungulates and carnivores now and into the twenty-first century? In *Wildlife Conservation, Zoos and Animal Protection: A Strategic Analysis*, ed. A. Rowan. Grafton, MA: Tufts Center for Animals and Public Policy, pp. 1–18.
Hutchins, M. (2003). Zoo and aquarium animal management and conservation: current trends and future challenges. *International Zoo Yearbook*, **38**, 14–28.
Hutchins, M. & Smith, B. (2003). Characteristics of a world class zoo or aquarium in the 21st century. *International Zoo Yearbook*, **38**, 130–141.
Hutchins, M. & Wemmer, C. (1987). Wildlife conservation and animal rights; are they compatible? In *Advances in Animal Welfare Science 1986/87*, eds. M. Fox & L. D. Mickley. Boston, MA: Martinus Nijhoff Publishers, pp. 111–137.
Hutchins, M., Dresser, B., & Wemmer, C. (1995). Ethical considerations in zoo and aquarium research. In *Ethics on the Ark: Zoos, Animal Welfare and Wildlife Conservation*, eds. B. Norton, M. Hutchins, E. F. Stevens, & T. L. Maple. Washington, D.C.: Smithsonian Institution Press, 253–276.
Hutchins, M., Smith, B., & Allard, R. (2003). In defense of zoos and aquariums: the ethical basis for keeping wild animals in captivity. *Journal of the American Veterinary Medical Association*, **223(7)**, 958–966.
Kenny, J. (1990). Control of the wild. *National Parks*, **65**, 9–10.
Koontz, F. (1995). Wild animal acquisition ethics for zoo biologists. In *Ethics on the Ark: Zoos, Animal Welfare and Wildlife Conservation*, eds. B. Norton, M. Hutchins, E. F. Stevens, & T. L. Maple. Washington, D.C.: Smithsonian Institution Press, pp. 127–145.
Leopold, A. (1949). *A Sand County Almanac*. New York: Oxford University Press.
Lockart, M., Vargas, A., Marinari, P., & Grober, P. (1998). Black-footed ferret (*Mustela nigripes*) recovery update. *Endangered Species Update*, **15(6)**, 92–93.

Mallinson, J. (1995). Conservation breeding programmes: an important ingredient for species survival. *Biodiversity and Conservation*, **4**, 617–635.

Maple, T. L., McManamon, R., & Stevens, E. (1995). Defining the good zoo: animal care, maintenance and welfare. In *Ethics on the Ark: Zoos, Animal Welfare and Wildlife Conservation*, eds. B. G. Norton, M. Hutchins, E. F. Stevens, & T. L. Maple. Washington, D.C.: Smithsonian Institution Press, pp. 219–234.

March, B. E. (1984). Bioethical problems: animal welfare, animal rights. *BioScience*, **34(1)**, 615–620.

Mellen, J. (2003). U.S. perceptions about animal welfare, conservation, zoos, aquariums and AZA: a survey. Preliminary results. Paper presented at the 2003 AZA Annual Conference, Columbus, OH.

Midgley, M. (1983). *Animals and Why they Matter*. Athens, GA: University of Georgia Press.

Miller, B., Reading, R. P., & Forrest, S. (1996). *Prairie Night: Black-Footed Ferrets and the Recovery of Endangered Species*. Washington, D.C.: Smithsonian University Press.

Miller, B., Biggins, D., Vargasa, A., Hutchins, M., Hanebury, L., Godbey, J., Anderson, S., Wemmer, C., & Oldemeier, J. (1998). The captive environment and reintroduction: the black-footed ferret as a case study with comments on other taxa. In *Second Nature: Environmental Enrichment for Captive Animals*, eds. D. Shepherdson, J. D. Mellen, & M. Hutchins. Washington, D.C.: Smithsonian Institution Press, pp. 97–112.

Moore, D. & Smith, R. (1990). The red wolf as a model for carnivore re-introductions. In *Beyond Captive Breeding: Re-introducing Endangered Mammals to the Wild*, ed. J. Gipps. Oxford, UK: Clarendon Press, pp. 263–278.

Norton, B. G. (1987). *Why Preserve Natural Variety?* Princeton, NJ: Princeton University Press.

Norton, B. G. (1995). Caring for nature: a broader look at animal stewardship. In *Ethics on the Ark: Zoos, Animal Welfare and Wildlife Conservation*, eds. B. G. Norton, M. Hutchins, E. F. Stevens, & T. L. Maple. Washington, D.C.: Smithsonian Institution Press, pp. 102–121.

Primack, R. B. (2002). *Essentials of Conservation Biology*, Third Edition. Sunderland, MA: Sinauer.

Redford, K. & Richter, B. (2001). Conservation of biodiversity in a world of use. *Endangered Species Update*, **18(1)**, 2, 4.

Regan, D. H. (2000). Duties of preservation. In *The Preservation of Species*, eds. B. G. Norton *et al*. Princeton, NJ: The Princeton University Press, pp. 195–220.

Regan, T. (1983). *The Case for Animal Rights*. Berkeley, CA: The University of California Press.

Regan, T. (1995). Are zoos morally defensible? In *Ethics on the Ark: Zoos, Animal Welfare and Wildlife Conservation*, eds. B. Norton, M. Hutchins, E. F. Stevens, & T. L. Maple. Washington, D.C.: Smithsonian Institution Press, pp. 38–51.

Regan, T. & Francione, G. (1992). The animal "welfare" vs. "rights" debate. *The Animal's Agenda*, **12(1)**, 40–45.

Ricklefs, R. E. (1973). *Ecology*. Portland, OR: Chiron Press.

Rodman, J. (1977). The liberation of nature? *Inquiry*, **20**, 83–131.

Rolston, H. (1992). Ethical responsibilities toward wildlife. *Journal of the American Veterinary Medical Association*, **200**, 618–622.

Rolston, H. (1994). *Conserving Natural Values.* New York: Columbia University Press.
Rosenweig, M. L. (2003). *Win-Win Ecology. How the Earth's Species Can Survive in the Midst of Human Enterprise.* Oxford, UK: Oxford University Press.
Rudacille, D. (2000). *The Scalpel and the Butterfly. The Conflict Between Animal Research and Animal Protection.* Berkeley, CA: University of California Press.
Schmidt, R. H. (1981). A professional attitude toward humaneness. *Wildlife Society Bulletin*, **9(4)**, 289–291.
Singer, P. (1975). *Animal Liberation.* New York: Avon Books.
Soulé, M. E. (1985). What is conservation biology? *BioScience*, **35**, 717–734.
Soulé, M. E. (1990). The onslaught of alien species, and other challenges in the coming decades. *Conservation Biology*, **4(5)**, 233–239.
Soulé, M. E. & Orians, G. H., eds. (2001). *Conservation Biology: Research Priorities for the Next Decade.* Washington, D.C.: Island Press.
Soulé, M. E., Wilcox, B. A., & Holtby, C. (1979). Benign neglect: a model of faunal collapse in the game reserves of East Africa. *Biological Conservation*, **15**, 259–272.
Stanley Price, M. R. & Soorae, P. S. (2003). Reintroductions: whence and whither? *International Zoo Yearbook*, **38**, 61–75.
Street, B., ed. (2004). *Conservation Directory: The Guide to Worldwide Conservation Organizations.* Washington, D.C.: Island Press.
Stromberg, L. P. (1995). Vegetation sampling methods for use in wildlife habitat evaluation. In *Wildlife Research and Management: Asian and American Approaches*, eds. S. H. Berwick & V. B. Saharia. Delhi, India: Oxford University Press, pp. 135–174.
Temple, S. A. (1990). The nasty necessity: eradicating exotics. *Conservation Biology*, **4(2)**, 113–115.
Varner, G. E. (1998). *In Nature's Interests? Interests, Animal Rights and Environmental Ethics.* New York: Oxford University Press.
Vrijenhoek, R. (1995). Natural processes, individuals and units of conservation. In *Ethics on the Ark: Zoos, Animal Welfare and Wildlife Conservation*, eds. B. Norton, M. Hutchins, E. F. Stevens, & T. L. Maple. Washington, D.C.: Smithsonian Institution Press, pp. 74–92.
Warren, M. A. (1983). The rights of the non-human world. In *Environmental Philosophy: A Collection of Readings*, eds. R. Elliott & A. Gare. University Park, PA: The Pennsylvania State University Press, pp. 109–131.
White, M. (1981). *What Is and What Ought to Be Done: An Essay on Ethics and Epistemology.* New York: Oxford University Press.
Wilcove, D. (1999). *The Condor's Shadow: The Loss and Recovery of Wildlife in America.* New York: W. H. Freeman and Company.
Wilson, E. O. (1984). *Biophilia: The Human Bond With Other Species.* Cambridge, MA: Harvard University Press.
Wilson, E. O. (2002). *The Future of Life.* New York: Alfred A. Knopf.
Younghusband, P. & Myers, N. (1986). Playing God with nature. *International Wildlife*, **16(4)**, 4–13.

8

Creating a culture of conservation: a case study of a backyard approach

BETH STEVENS, JACKIE OGDEN, AND KIM R. SAMS

INTRODUCTION

At Disney's Animal Kingdom and the Living Seas at Epcot, which are both accredited by the Association of Zoos and Aquariums (AZA) and part of Walt Disney World Resort, it is part of the mission to contribute to global conservation efforts. However, success in global conservation is inextricably linked to local activities, or what can be referred to as the "backyard" approach.

The "backyard" at the Walt Disney World Resort includes guests and staff at the various theme parks, audiences that are non-traditional to zoos and aquariums, as well as the complex set of diverse businesses that make up The Walt Disney Company®. These non-traditional audiences present an opportunity to bring new people into the conservation arena. In working to reach these new audiences, it has become apparent that the same principles that seem to apply to conservation projects in the developing world apply equally well to catalyzing conservation in the United States. Foremost, these include the necessity to understand the local culture before embarking on a conservation project, and the realization that change does not happen overnight.

CONSERVATION AND THE WALT DISNEY COMPANY®

Perhaps surprisingly to some, involvement in conservation and education is part of the legacy for The Walt Disney Company®. The "True Life

Adventures" series seen on television beginning in the 1950s may have been the first exposure for many people to exotic places and the mysterious lives of wild animals. More than 40 years ago, Walt Disney himself was quoted as saying "... *conservation isn't just the business of a few people. It's a matter that concerns all of us.*" At the Walt Disney World Resort near Orlando, Florida, this legacy extends to the Resort's beginning in 1971, when approximately one-third of the original 30 000 acres was set aside in perpetuity as conservation lands (http://www.disneywildlife.org).

With the opening of Disney's Animal Kingdom Theme Park® in 1998, a specific business unit, the Disney's Animal Programs division, was created to corral all animal care and conservation efforts under one umbrella. Disney's Animal Programs oversees all areas within the Walt Disney World Resort where animals are present: Disney's Animal Kingdom Theme Park®, The Seas with Nemo and Friends at Epcot, Typhoon Lagoon Water Park, the Tri-Circle D Ranch at Fort Wilderness and Disney's Animal Kingdom Lodge. The division also manages Environmental Initiatives, a group focused on encouraging employees (Cast Members) to apply green thinking to their daily work ethic.

Disney's Animal Kingdom believes that conservation is key to both its mission and business. The park's mission is "to make a difference for the future of wild animals and wild places." The approach at the park, and at Walt Disney World, is to target conservation efforts to have impact at the global, neighborhood, and backyard levels. It is useful to visualize conservation efforts as a dartboard, with the backyard as the bull's-eye, the next ring representing the regional neighborhood, and the final outer ring representing the world. "Hitting the bull's-eye," and making an impact locally, in the backyard, will have perhaps the greatest impact at all levels, creating a ripple effect.

THE OUTER RING: CONSERVATION EFFORTS GLOBALLY

Disney is attempting a holistic approach to conservation, combining three critical success factors whenever possible: scientific studies, educational efforts, and ensuring sustainability (including both economic alternatives and opportunities to influence decision-makers). Global conservation is targeted in two ways: through contributions made by the Disney Wildlife Conservation Fund (DWCF), and through support and staff involvement under the Disney's Animal Programs Cast Conservation Program umbrella.

The Disney Wildlife Conservation Fund (DWCF) was established to help identify and support critical needs within the conservation community.

(http://www.disneywildlife.com). Since Disney's Animal Kingdom opened, the DWCF has supported more than $10 million in conservation projects around the world. Almost half of these dollars come directly from guests through the "Add-a-Dollar" program, where guests have the opportunity to add a dollar (or more) to their purchases at many merchandise locations. These dollars are then matched by the office of Worldwide Outreach for The Walt Disney Company, with additional dollars coming from the proceeds of many of the Disney's Animal Programs' fee-based educational programs. To date, the DWCF has funded more than 500 projects conducted by over 100 organizations working in 68 countries, encompassing marine environments and on 5 continents around the world.

The second program is Disney's Animal Programs Cast Conservation Program, which focuses specifically on staff involvement in conservation and concentrates staff where their particular expertise can make a difference. These programs are intended to be holistic in nature and to include all of the above-mentioned critical success factors. Although this program is relatively new, since its inception over 20 projects have been supported, ranging from a comprehensive program supporting the conservation of cotton top tamarins in Colombia, to local efforts to help threatened butterflies (e.g., Miller *et al.* 2004).

THE INNER RING: CONSERVATION EFFORTS IN THE LOCAL NEIGHBORHOOD

At the Walt Disney World Resort, the "neighborhood" is the entire state of Florida, a biologically diverse state in the US facing numerous conservation challenges. Disney's neighborhood programs focus on a number of listed animal species, including sea turtles and manatees, as well as on ecosystem restoration and sustainable development.

For more than 15 years the staff at The Living Seas have been assisting with the rehabilitation and release of injured and cold-stunned sea turtles. This has now expanded to a program including satellite tracking of released turtles, community education efforts, and sea turtle nest monitoring on a section of beach near Disney's Vero Beach Resort, all in partnership with the University of Florida, the United States Department of Fish and Wildlife and Archie Carr National Wildlife Refuge System (e.g., Gill 2005, Miller *et al.* 2005; Savage *et al.* 2005).

Similarly, marine mammal experts from The Living Seas at Epcot have been caring for sick and injured manatees since 1985, as part of the US Department of Fish and Wildlife's manatee recovery program (Fellner *et al.*

2006). Disney serves as an active member of the Manatee Rehabilitation Partnership, a cooperative group of non-profit, private, state, and federal entities with a goal of tracking and monitoring the health and survival of rehabilitated and released manatees (Wells 2006).

Staff's efforts for sea turtle and manatee conservation are well supplemented by DWCF support of these species (http://www.disneywildlife.org). As an example, since 1999 the DWCF has contributed over $250 000 for manatee research and conservation, including support for the Wildlife Trust's manatee research in Belize, and funding of public-awareness waterway signs with the Save the Manatee Club, and web-based turtle tracking with the Caribbean Conservation Corporation. The DWCF has also supported the production of teacher education guides for several Caribbean countries where manatees are endangered. The Living Seas at Epcot also sponsors and distributes these guides and activity books.

From a sustainable development perspective, efforts include The Disney Wilderness Preserve which is 45 min south of Walt Disney World Resort. Owned and managed by The Nature Conservancy, these 12 500 acres of uplands and wetlands are a "teaching hospital" for the environment. Most of the property, previously known as The Walker Ranch, was purchased by Walt Disney World in a then unique program for off-site mitigation to allow future expansion of Walt Disney World Resort. The result is a 10-year partnership with The Nature Conservancy to restore the land and use it as a campus for land management training. The property is of particular importance as it includes the watershed that serves as the headwaters of the Everglades (see Flicker 1994; http://www.disneywildlife.com).

THE BULL'S-EYE: DISNEY'S BACKYARD APPROACH

Catalyzing conservation really starts in one's own backyard, and it starts with people. The Disney backyard includes both the Walt Disney World Resort and its parent company The Walt Disney Company®, which owns businesses worldwide. This represents a very non-traditional audience for a zoo or aquarium, but one that has tremendous potential for developing conservation partners, if effectively nurtured and inspired. Catalyzing conservation thus begins with creating a culture for conservation within organizations through people who can be influenced through daily work.

There is an awareness of the need to be cautious when practising conservation in other countries. However, the same principles that apply to working in other cultures also apply to conservation locally. The cardinal principle is to learn the culture before embarking on the project. In a business environment, this translates to the need to learn and appreciate the business. The culture at Disney is steeped in tradition and a legacy focused on delivering magical experiences to guests. Any success in catalyzing conservation at The Walt Disney Company® requires merging with this culture.

The second key principle is that change does not happen overnight. Change takes time and happens in small steps. It is important to remember that everyone is in a different place regarding taking action for conservation. Conservation can be viewed as a continuum in this way; on one end are those who have no awareness regarding conservation; at the other end are those actively taking steps for conservation. Any conservation organization must meet people where they are on this metaphorical continuum, rather than where they wish them to be. Further, rather than assuming that people will leap from one end of the continuum to the other, they generally need to be helped to take small steps at their own pace. Each small step in the right direction should be celebrated and reinforced. This approach has been used to good effect within Disney to promote conservation.

The final critical step to creating a conservation culture in the Disney backyard has been in determining which experiences are most impactful, and then creating those experiences for target audiences. In Disney, there are three core, non-traditional audiences: guests (who are not necessarily typical visitors to zoos and aquariums), Disney's Animal Kingdom staff, and Walt Disney World Resort and The Walt Disney Company® businesses. Each of these audiences represents an opportunity for catalyzing conservation at many levels.

Guests

Guests to the animal areas at Walt Disney World are not representative of usual zoo and aquarium visitors and members elsewhere. As such, the ability to touch people who may have less awareness of conservation issues than many visitors to traditional zoos and aquariums is exceptional. Disney's education mission is to "create personal experiences that celebrate wildlife, renew everyone's connections to the natural world and inspire conservation action." Working to accomplish this through building partnerships, training and empowering all staff provides a diversity of inspirational experiences for guests.

Fee-based programs for groups are offered, including both school groups and adult guests. These programs are provided at Disney's Animal Kingdom and The Seas with Nemo & Friends Epcot, and include a variety of experiences, from programs focused on critical thinking for teens, to diving (for those Scuba certified) at The Living Seas. In addition, a wide range of experiences designed to promote learning are provided for guests, from the messages woven into the attractions at Disney's Animal Kingdom, to up-close and personal animal demonstrations, to interactive experiences targeted to young audiences at both Disney's Animal Kingdom and The Living Seas.

There is the potential of reaching millions of people with conservation messages at Walt Disney World. The numbers of conservation messages that are actually delivered by Cast members at Disney's Animal Kingdom are tracked; in 2006 alone over 3 926 549 messages were delivered, meaning that conversations included specific discussions of conservation of wildlife or conservation action. While it is pleasing to deliver many conservation messages, assessing the impact of these messages is more critical. Data indicate that visiting Conservation Station, a part of the park featuring animal care and conservation efforts as its focus, results in increased understanding regarding conservation, as well as more limited impact on long-term conservation-related attitudes and behavior (Dierking *et al.* 2004).

Staff

The entire staff at Disney's Animal Kingdom and The Seas, as well as the staff at all zoos and aquariums, should be viewed as catalysts for conservation and the core of the backyard approach. They are truly the epicenter of the bull's-eye. Why is our staff non-traditional? In many zoos and aquariums, there may already exist a conservation culture by virtue of the fact that the majority of the employees are part of the animal care, veterinary, education, and scientific teams. At Disney's Animal Kingdom, however, this same animal division comprises only about 10% of the total staff. Therefore, a conservation culture had to be cultivated. The message that conservation is everyone's business, that every single role has the potential to make a contribution to conservation, from custodians to food servers, to our maintenance workers, was crucial. Each person has the ability to engage in conservation practices such as recycling and using double-sided copies; and most staff have the opportunity to communicate conservation messages to guests.

In working to create a conservation culture, the cardinal rules of conservation were again considered. As an example, as Disney's Animal Kingdom prepared to open, learning and understanding the Disney culture, quickly, was important in delivering conservation-aware staff. Elements that were particularly well incorporated into the culture were identified. One example was "The Seven Guidelines to Guest Service" – a tool created to assist staff in working with guests. The culture at Disney is completely focused on providing great guest service, and all staff know the Seven Guidelines. In order to help staff focus on what they could do to inspire conservation action, the "Seven Guidelines for Conservation Action" were created. In this form, it was easier for staff to understand and value conservation. These Guidelines are: (1) seek out information about conservation issues; (2) spread the word to others about the value of wildlife; (3) look for and purchase products that are friendly to the environment; (4) create habitats for wildlife in your backyard; (5) reduce, reuse, recycle, and replenish; (6) choose your pets wisely; and (7) support conservation organizations through contributions and volunteerism.

In addition to equipping staff with the right tools, inspiring them directly and personally about conservation has been a priority area. Through a series of orientation and training classes, as well as experiences involving direct interaction with the animal programs staff and up-close and personal experiences with animals, staff have been inspired to care about conservation.

To assess success in this area, a series of qualitative focus groups was conducted. The purpose of these focus groups was to determine whether or not the staff's conservation-related knowledge, attitudes, and behavior were impacted by working at Disney's Animal Kingdom, and which training methods were most effective at inspiring conservation action. A cross-section of Cast members from all disciplines and all levels were sampled and two primary discoveries were made:

1. Working at Disney's Animal Kingdom appears to have had a positive impact on Cast members in their personal lives and their professional lives. In their personal lives, they have become more aware of conservation issues and many of them have changed their behavior in some way to benefit conservation.
2. The most effective training methods are not the classroom experiences where they are taught about conservation. Rather the most effective experiences are those where staff have the opportunity to make a personal connection – with living animals and with animal staff (Groff *et al.* 2005).

It is also apparent that staff have applied this to their experiences with guests. Staff across all the Disney parks are expected to create special experiences that are called "Magical Moments." At Disney's Animal Kingdom, the majority of these experiences feature a key conservation message. Every discipline within the park has created games, conversations, and a variety of memorable experiences with conservation messages for the guests. It has been particularly inspiring to see the very creative and inspirational experiences created by staff in the non-animal departments of the park.

Additionally, staff actively participate in volunteer efforts that support conservation; in 2002 staff participated in over 120 team-building activities that were conservation-related, such as the removal of invasive plants on nearby conservation lands. Not only do these experiences provide direct conservation action, but they also serve as reinforcing experiences for staff and Disney's community partnerships.

Walt Disney World Resort and The Walt Disney Company® businesses

Beyond the parks, Disney's backyard can be defined as the *circle of influence within the parent company*. Working in partnership with several Disney businesses (films, television, and publishing) to implement new conservation approaches has been fruitful in mitigating animal-related issues and in providing counsel when new ideas are discussed. In approaching our sister businesses, the necessity to learn their cultures before embarking on sharing the conservation mission, and to understand that change does not happen overnight, were again recognized.

To increase the likelihood of conservation-related decision-making, care has been taken to help share conservation efforts and experiences with many executives (chefs, accountants, merchandise buyers). They have visited the beaches to see a sea turtle nest, and they might even have a nest "named" after them, a sort of an "adopt a nest" concept. Increased levels of receptiveness follow each such experience. An example of this is the Add-a-Dollar program. This started at Disney's Animal Kingdom and The Seas but has now grown to include Disney's Vero Beach and Hilton Head Island Resorts, as well as Disney's Wilderness Lodge, the Fort Wilderness Campground and now Disney Cruise Line. This has only been possible through the support of executives who previously had little conservation awareness.

Another opportunity was apparent within the more than 500 restaurants at the Resort, many of which serve seafood. One of the first issues tackled was the national "Give Swordfish a Break" campaign several years ago. This succeeded in removing swordfish from the menu for two main reasons: it was a national campaign adopted by other well-known chefs (their peers) and animal experiences provided for many of the Disney chefs had a personal impact on them. It was literally an immersion experience: many of the chefs who were serving swordfish were invited to participate in the "Dolphins in Depth" program where they donned wetsuits and got in the water with a dolphin trainer during a husbandry training session. Close encounters with dolphins had a profound effect on these chefs and opened them up to learning about the perils facing swordfish and many other ocean inhabitants. One chef explained that he would not be able to look himself in the mirror if he continued to serve swordfish. It wasn't the facts alone that convinced him, it was first the emotional connection.

When considering Disney television and film businesses, the connection is more distant (literally and figuratively, since those businesses are based in California). Although more distant, the process is very similar. The same cardinal rules of conservation are applied, including sensitivity to their creative culture and the importance of building relationships. Of particular importance has been introducing Disney's Animal Programs staff to these different businesses as a resource – providing ideas for animal-related content and assisting producers in ensuring animal facts are correct. This has been successful at assisting these partners in identifying potential issues related to animals and conservation that might not have been anticipated. By doing so, trust and credibility have been established. Trust leads to open lines of communication, and hopefully to a rich partnership that will allow Disney's Animal Programs to tap the weight of these talented divisions to further catalyze conservation.

CONCLUSIONS

At Walt Disney World, while there is far to go, positive conservation-related outcomes have become apparent by focusing on the people in the Disney backyard. The Disney backyard may be slightly non-traditional to zoos and aquariums; however, the principles utilized seem to hold true in all environments:

- People are the true catalysts for conservation
- Cultural sensitivity is an essential ingredient to success
- Everyone is at a different place along the conservation action continuum; meeting each person where he/she is on that continuum and starting there is essential
- Change doesn't happen quickly
- Back to basics – utilize the tools that are known to work best – build relationships; provide personal experiences with animal staff and animals – do not underestimate the importance of emotion and affect.

Creating a culture of conservation in the "backyard" has facilitated Disney's ability to engage in conservation programs in the nearby locality and the world. Identifying non-traditional audiences such as non-animal staff and governing boards has provided new opportunities to achieve improved conservation outcomes. All zoos and aquariums must ensure that their "backyard" is aware of their conservation aims and use the full weight of their resources, all their staff, to reach their goals.

References

Dierking, L. D., Adelman, M., Ogden, J., Lehnhardt, K., Miller, L., & Mellen, J. D. (2004). Using a behavior change model to document the impact of visits to Disney's Animal Kingdom: a study investigating intended conservation action. *Curator: The Museum Journal*, 47(3), 33–61.

Fellner, W., Odell, K., Corwin, A., Davis, L., Goonen, C., Larkin, I., & Stamper, M. A. (2006). Response to conditioned stimuli of two released rehabilitated West Indian manatees (*Trichechus latirostris*). *Aquatic Mammals*, 32(1), 66–74.

Flicker, J. (1994). Disney catalyst for off-site mitigation. *Urban Land*, p. 14.

Gill, E. (2005). It's a small world after all: a new program at the Disney Vero Beach Resort is helping to protect our sea turtles. *Vero Beach Magazine*, Summer edition.

Groff, A., Lockhart, D., Ogden, J., & Dierking, L. D. (2005). An exploratory investigation of the effect of working in an environmentally-themed facility on the conservation-related knowledge, attitudes and behavior of staff. *Environmental Education Research*, 11(3), 371–387.

Miller, L., Savage, A., & Giraldo, H. (2004). Quantifying remaining habitat within the historic distribution of the cotton-top tamarin. *American Journal of Primatology*, 64(4), 351–357.

Miller, L., Savage, A., Davis, J., Christman, J., Lehnhardt, J., & Lehnhardt, K. (2005). Walt Disney World's sea turtle conservation program at Disney's Vero Beach Resort. Poster presented at International Sea Turtle Symposium.

Savage, A., Koelsch, J., Evans, D. R., Miller, L., Gordon, A., Heyes, G., Appelson, G., Lehnhardt, K., Leeming, D., & Hashimoto, T. (2005). Neighbors Ensuring Sea Turtle Survival (NESTS): a community program to increase awareness and action for the conservation of sea turtles near the Archie Carr National Wildlife Refuge. Poster presented at International Sea Turtle Symposium.

Wells, M. (2006). Manatee rehabilitation project. *Communiqué*, May, 21–23.

9

Message received? Quantifying the impact of informal conservation education on adults visiting UK zoos

ANDREW BALMFORD, NIGEL LEADER-WILLIAMS, GEORGINA M. MACE, ANDREA MANICA, OLIVIA WALTER, CHRIS WEST, AND ALEXANDRA ZIMMERMANN

INTRODUCTION

Humanity is growing ever more disconnected from wild places and wild creatures (Gadgil 1993, Nabhan and St. Antoine 1993; Nabhan and Trimble 1994, Pyle 1993, 2003, Balmford 1999, Kahn and Kellert 2002). Over 50% of people now live in towns and cities, and their numbers are rising by 160 000 daily (World Resources Institute 2000). With this in mind the world's zoos, with more than 600 million visitors each year (WAZA 2005), have enormous potential to educate and inspire the public about conservation. Much is made of this role, in both reviews and policy statements on the conservation significance of zoos (Tribe and Booth 2003, Miller *et al.* 2004, WAZA 2005). Yet there is evidence that some captive facilities pay only limited attention to conservation education (Dunlap and Kellert 1995, Evans 1997, Mazur and Clark 2000). Against this criticism, very few studies have so far attempted to quantify whether zoo visits change people's conservation-related knowledge, attitudes, or behavior, or whether such impacts vary across zoos.

Most assessment of educational impacts to date has instead been non-quantitative (for reviews, see Kellert and Dunlap 1989, Broad and Weiler 1998, Dierking *et al.* 2002; see Discussion for counter examples). Moreover, rather than examining the effects of a zoo visit on people's overall conservation knowledge or attitudes, research has usually focused either on general natural history knowledge, or else on whether a visit specifically changes people's attitudes to zoos or knowledge of zoos' role in conservation (Dierking *et al.* 2002). Likewise, much more work has looked at the effects of particular (usually new) exhibits than at the impact of a zoo visit in its entirety (Broad and Weiler 1998, Dierking *et al.* 2002). Last, few studies have examined the effects of general, informal education on the adult visitor (Kellert and Dunlap 1989, Broad and Weiler 1998, Mazur and Clark 2000). As a result, we are left with remarkably few clear tests of whether and, if so, how zoos change the knowledge, attitudes or behavior of those 600 million visitors.

Given the limited evidence to date on zoos' conservation education impacts, the Zoo Measures Working Group of the British and Irish Association of Zoos and Aquariums (BIAZA) devised a questionnaire aimed at quantifying the effects of informal education on adult visitors to UK zoos. This focused on measuring various aspects of visitors' knowledge about conservation, their level of concern about conservation relative to other issues, and their ability to suggest practical ways in which they could make a difference to conservation. We gauged the impact of a single zoo visit on these response variables by comparing the answers of visitors arriving at a zoo with those of a roughly equal-sized but non-overlapping set of visitors leaving the same zoo (see also Broad 1996, Giusti 1999, Spotte and Clark 2004). We thereby avoided problems caused by differences between zoo visitors and the general public, differences between visitors to different zoos (see below), and any confounding effects of our post-visit sample having been aware of the survey questions during their visit. To test the power of our questionnaire, we also collected data on several aspects of visitors' backgrounds, as well as their general interest in wildlife, to see whether these predicted variation in our response measures. This chapter presents an overview of our findings from questioning 1340 adult visitors at 6 UK zoos (and, for comparison, one nature reserve) during the latter half of 2003.

METHODS

Our survey

Visitors were surveyed at Bristol (365 people), Chester (64), Colchester (46), London (194), and Paignton (411) zoos, Thrigby Hall Wildlife Gardens (47), giving a total of 1127 zoo visitors. In addition, we also surveyed 213 visitors to the Wildfowl and Wetlands Trust Wetland Centre at Barn Elms, London. Questionnaires were handed out by volunteers (at Bristol, Chester, London, Paignton, and Barn Elms) or zoo staff (at Colchester and Thrigby Hall only), but completed entirely by respondents. Respondents classified themselves as "arriving" (mostly in the entrance queue, and 91.8% within 1 h of entry) or as "departing" (mostly >3 h and 92.3% >2 h after entry). Care was taken to ensure that no respondents completed more than one questionnaire; respondents under 18 were excluded from analysis. On average, participation in the survey took 5–10 min, with volunteers or staff generally collecting 40–100 completed questionnaires each per day, depending on visitor numbers.

The questionnaire

The questions asked fell into four main groups (note that we also asked zoo visitors a series of questions about their attitudes to zoos and zoos' contributions to conservation, but these are not analyzed here).

Background information

We asked visitors to tell us their age (<20, 20–39, 40–59 or ≥60 years); their sex; whether or not they had received tertiary education; whether they were UK nationals; and in order to assess their general interest in conservation and wildlife, whether they were members of any conservation-related charities (ranging from the National Trust to Greenpeace), and whether they had visited a zoo in the past.

Conservation knowledge

Respondents were asked to name any globally threatened species and any threatened British species, and for each, to give one reason why it was threatened. They were also asked to rank habitat loss, pollution, overhunting or overharvesting, introduced species, and climate change in terms of their relative importance as threatening processes, both worldwide and in Britain; and to rank three major habitat types (tropical forests, ice caps, and freshwater) in terms of their global threat status. Answers were assessed as correct or not using information from the IUCN Red List (IUCN 2003) and the UK Biodiversity Action Plan (Joint Nature Conservation Committee 2004), and with "species" interpreted loosely (so that "whale" and "elephant" were scored as correct answers to the question on globally threatened species, for example). Respondents were then awarded marks out of 20, with equal weight given to globally threatened "species," British threatened "species," global threats, British threats, and threatened habitats. As well as being treated as a response variable, this conservation knowledge score was also analyzed as a potential predictor of our remaining response variables.

Concern about conservation

We quantified respondents' relative concern about conservation through three questions about how they would allocate £1000 of locally raised money among competing charities. When asked to give it all away to charity, how much would they assign to conservation versus health, domestic social concerns, international aid, and animal welfare? If they had to give all the money to conservation, how much would they assign to international versus national or local conservation projects? And last, if they had to give it all to conservation, how much would they allocate to habitat- versus species-oriented conservation? Hypothetical willingness-to-pay surveys are often criticized as shedding little light on real-world patterns of spending (e.g., see discussion in Kramer and Mercer 1997). However, it is worth noting that, in our sample, respondents' overall contribution to conservation versus other good causes was around 50% higher among those who were already members of conservation charities than among other visitors (means ± SEs of £167.39 ± £5.70 vs. £229.93 ± £6.62; Mann-Whitney test: $W = 79\,985$, $N = 1127$ zoo visitors, $P < 0.001$). Moreover, our questions explicitly incorporated some of the tradeoffs that people face in donating to good causes. We therefore suggest that they do capture meaningful information on people's relative concern about conservation.

Ability to name useful activities

While it was not possible during such a large, short-term survey to directly assess visitors' conservation-related behavior (see Adelman *et al.* 2000), we were able to ask visitors to name something which they could do to help conserve species or habitats. In scoring their answers, we scored as 0 all passive or very general suggestions (such as "learn more" or "be green"), but scored as 1 active suggestions such as "join a conservation organization," "recycle," or "use public transport." This scoring system is simple, but does reveal considerable variation in people's potential to engage in conservation-related behavior (see below).

Analysis

Following some general analyses of respondents' answers and of how these varied across the seven locations we surveyed, we conducted progressively more sophisticated tests of the effects of a visit. First, for each of our five response variables (conservation knowledge, spend on conservation, spend on habitat, international spend, and ability to name a useful activity) and for each of our six zoos, and for comparison the one nature reserve, we used simple univariate statistics to see whether visitors' answers differed between arrival and departure (non-parametric statistics were used to account for the non-normality of the data). We next built generalized linear models (GLMs) for each response variable and each site to test for such differences, this time controlling for other significant predictors of our response variables; these models used a quasi-binomial error structure to account for the nature of our response variables, which could all be expressed as a given score out of a hypothetical maximum value. Finally, after considering the potential limitations of our sample sizes at any single site, we built generalized linear mixed models (GLMMs with a quasi-binomial error structure; Venables and Ripley 2002), which used the data for all zoos combined to identify statistically significant predictors of each of our response measures; note that for this final, pooled set of analyses we excluded non-zoo data.

RESULTS

Overview of responses

The visitors we surveyed differed widely in their scores for the different response variables. The mean conservation knowledge score among all zoo visitors was 9.26 ± 0.23 out of 20. The most common answers when respondents were asked to name globally threatened species were, perhaps predictably, all large mammals, with one interesting exception: the cod (Table 9.1). Fewer zoo visitors could correctly name a threatened British species (51.4% vs 85.4%), but in this case the species they did name were more varied, with the top 10 including several birds, as well as cod and newt (Table 9.1). Although respondents were free to name any species, plants were named in only 0.1% of all (global or UK) answers.

Out of £1000, zoo visitors' mean hypothetical allocation to conservation was £180.32 ± £16.82, substantially higher than the means of £111 and £149 obtained in two separate 2003 surveys which asked the same question not of zoo visitors but of the general British public (A. Beaumont, S. McInnes, R. van Millingen, and K. Vinnicombe, pers. comm.); our sampled visitors are thus generally more concerned about conservation than the public as a whole (see also Giusti 1999, Adelman *et al.* 2000). Mean spend on international rather than national or local conservation was £246.23 ± £18.35 out of £1000, while the mean allocation to habitat-based rather than species-based conservation was £538.99 ± £14.53 out of £1000.

Table 9.1
The 10 most common answers when 1127 zoo visitors were asked to name globally threatened species and threatened British species, and the percentage of correct responses naming each "species." Note that badgers are not threatened in Britain

Globally threatened species	%	Threatened British species	%
Tiger	28.17	Red squirrel	25.79
Giant panda	20.68	Otter	8.50
Rhino	13.39	Badger	6.58
Gorilla	8.09	Bat	5.08
Lion	4.90	Water vole	4.12
Elephant	4.70	Sparrow	4.12
Whale	4.00	Eagle	3.43
Cod	2.00	Cod	3.02
Orangutan	1.70	Newt	3.16
Chimpanzee	0.80	Barn owl	2.88

On average 45.7% ± 2.8% of zoo visitors named an active way in which they could make a practical contribution to conservation. The most common suggestion – giving money to a conservation organization – accounted for 52.8% of responses. Other common responses were recycling (14.4%), environmentally sensitive shopping (9.1%), raising awareness among children or adults (8.7%), wildlife-friendly gardening (8.7%), and giving time to conservation organizations (8.3%). Notably, reducing car use, reducing domestic energy use, and voting strategically or writing to politicians each accounted for less than 2% of responses, suggesting that few respondents see these activities as having a direct bearing on conservation.

Not surprisingly, we found that all of our response variables were intercorrelated with each other, although not very strongly (Table 9.2). Respondents with higher scores for conservation knowledge typically allocated more money to conservation generally and to international and habitat conservation in particular. Those who assigned more money as a whole to conservation usually wanted a greater share of it to be spent abroad and on habitats rather than single species. Those visitors able to suggest how they could make a difference to conservation on average had 16% higher knowledge scores and allocated 8%, 41%, and 7% more money to conservation as a whole, to international conservation, and to habitat conservation respectively, than did those unable to suggest practical activities (Table 9.2).

One last striking general feature of our dataset was that participants' answers differed widely across locations (Figure 9.1). Scores for conservation knowledge were particularly high among visitors to Thrigby, and lowest at Chester and Paignton zoos (but visitors at Barn Elms reserve

Table 9.2

*Relationships among our response variables. The left-hand part of the table gives Spearman rank correlation coefficients between our continuous response variables (N = 1127, zoo visitors; *** $P < 0.001$*; $P < 0.05$). The right-hand part of the table shows how the mean scores for these variables vary with respondents' ability to name a useful conservation action they could undertake (P values are for Mann-Whitney tests; N = 611 and 516 respectively)*

		Correlation with allocation to conservation (/£1000)			Comparison of those unable vs able to name useful action		
		Overall	International	Habitat	Unable mean	Able mean	*P* value
Conservation knowledge (/20)		0.07*	0.16***	0.12***	9.38	10.84	<0.001
Allocation to conservation (/£1000)	Overall	–	0.07*	0.07*	173.76	187.64	<0.001
	International	–	–	0.20***	205.68	290.59	<0.001
	Habitat	–	–	–	520.48	559.02	0.019

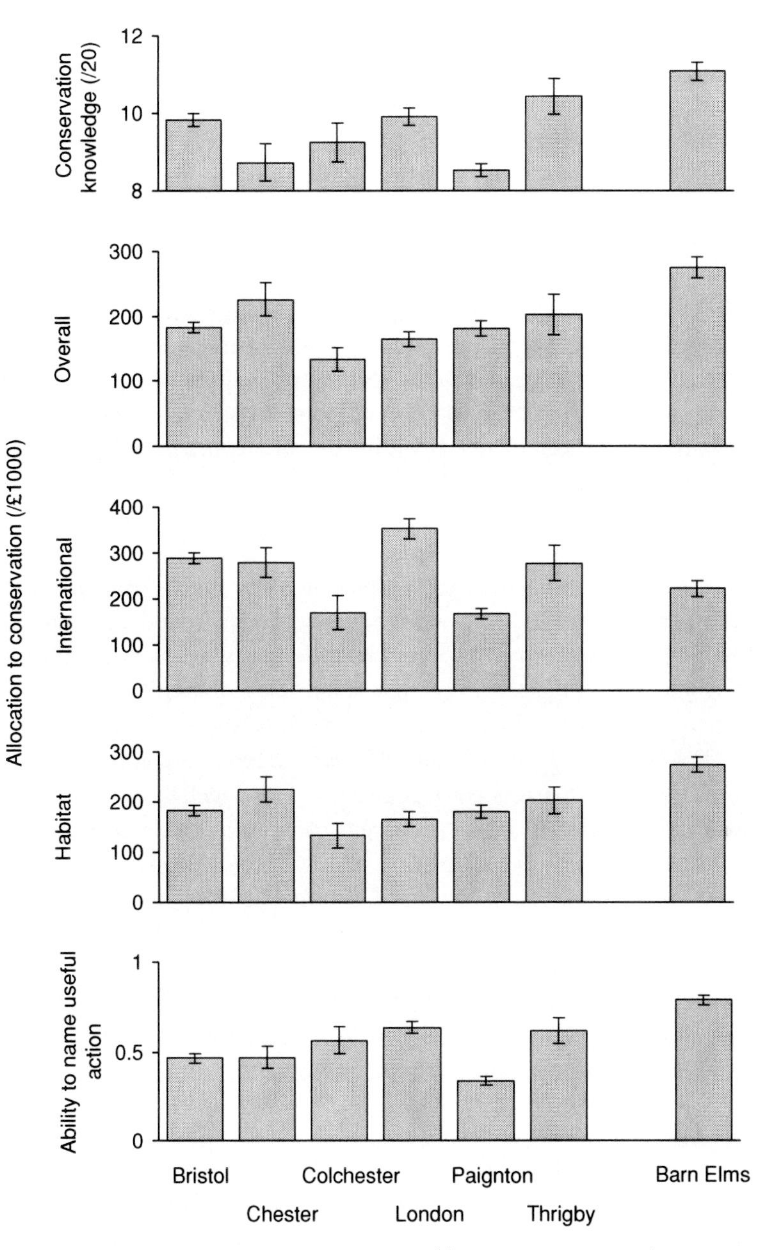

Figure 9.1 Variation in our response variables across sites. Bars show means ± SEs. In each case there was highly significant variation in scores across sites (Kruskal–Wallis tests: all $P < 0.001$)

outperformed visitors to any of the zoos). Hypothetical allocations to conservation were highest at Barn Elms, but also high at Chester. The allocation of conservation money internationally was greatest at London Zoo, where the proportion of visitors surveyed who were not UK nationals was substantially higher (at 14.0%) than elsewhere (mean across the five other zoos: 1.4%; $\chi^2 = 73.78$, df = 1, $P < 0.001$). The relative allocation to habitat rather than species conservation was quite similar across the six zoos, but higher at Barn Elms. Respondents' ability to name useful activities they could do was highest at London, and lowest at Paignton (with respondents at Barn Elms scoring much higher than visitors to any of the zoos). These clear differences in the profiles of visitors to different institutions underline the necessity, in comparing the impacts of different zoos, to rely not on simple exit surveys, but instead to look at within-zoo differences between arriving and departing visitors.

Visit effects – univariate comparisons

Despite looking at five response variables at seven sites, simple univariate comparisons revealed just one difference between respondents entering and leaving our study sites. We found no differences, at any site, between arriving and departing knowledge scores, hypothetical spend on conservation, spend on habitat, or international conservation spend (Mann–Whitney tests, all NS). The only significant difference to emerge was in terms of respondents' ability to name practical contributions which they could make, for which Paignton visitors scored roughly twice as highly on departure as on arrival (0.48 ± 0.03 vs 0.28 ± 0.03, $W = 17\,336.5$, $N = 411$, $P < 0.001$). However, this result was not observed elsewhere (NS for other six sites). Moreover, taken overall, Paignton respondents had by far the lowest average scores for this response variable (Figure 9.1). Both this and the recorded increase in these scores on departure could have been an artifact of Paignton visitors apparently being in a greater hurry to complete the questionnaires on entry than on exit (M. Ebbage-Taylor, pers. comm.).

Generalized linear models

GLMs built to predict variation in each of our response variables at each of our sites showed that this failure to identify significant differences in responses between arriving and departing visitors was not simply because our data were too noisy, or because of confounding effects of other variables. We were able to find significant predictors of one or more of our response variables at every site, and of all variables at Barn Elms and at the best sampled zoo, Paignton (Table 9.3). Moreover, these predictors had generally

Table 9.3
Significant terms in the generalized linear models for each of our response variables at each of our sites. Cells show only those terms significant at P < 0.1. ***P < 0.001; **P < 0.01; *P < 0.05; *italics* = P < 0.1

Site	*N*	Conservation knowledge (/20)	Allocation to conservation (/£1000)			Ability to name useful action (0 or 1)
			Overall	International	Habitat	
Bristol	365	Sex* Charity member***	Charity member** *Conservation knowledge*	*Conservation knowledge*	Sex** Charity member** *Conservation knowledge*	*Conservation knowledge*
Chester	64	*Charity member*	*Charity member*	–	–	*Conservation knowledge*
Colchester	46	*Charity member*	–	Conservation knowledge* Visited zoo before*	–	–
London	194	Tertiary education*	–	*Conservation knowledge*	–	–
Paignton	411	Age** Tertiary education** Charity member** Visited zoo before*	Age* *Charity member* Conservation knowledge*	Tertiary education*** Conservation knowledge*	Age* *Sex*	Conservation knowledge*** Visited zoo before** Arrival vs departure***
Thrigby	47	Charity member**	Charity member**	–	Tertiary education*	Conservation knowledge*
Barn Elms	213	Age* Charity member**	Charity member* *Visited zoo before*	Tertiary education*	Sex* Conservation knowledge*	Conservation knowledge*** *Visited zoo before*

consistent effects across sites and across response variables. For example, respondents who had received tertiary education had higher knowledge scores (at two sites), and assigned more money to international conservation (two sites), and habitat conservation (one site) than other respondents, independently of the effects of other significant predictors. Likewise, there were independent positive effects of charity membership on knowledge score (four sites), and on money assigned to conservation in general (three sites) and habitat-based conservation in particular (one site). Individuals with higher knowledge scores allocated more to general conservation (one site), international conservation (two sites), and habitat conservation (one site), and were more likely to name a useful activity (three sites). Last, we found some positive associations between having visited a zoo before and knowledge score, allocation to international conservation, and ability to name a useful action (one site each) – but these could reflect a general interest in wildlife, rather than being an effect of site visits *per se*.

The only strong test of the effect of visits which these data allow is the comparison of arriving and departing visitors. As with the univariate analyses, the only significant effect detected in the GLMs was on the ability of Paignton visitors to name useful conservation-related activities. Thus despite controlling statistically for the effects of many potentially confounding variables, we could find no strong and consistent signal, within individual sites, of informal education of adults during a single visit.

However, this could still conceivably arise because of undersampling. Some indication of the sensitivity of our findings to sample size is given by a plot of number of significant effects detected for a site (summed across the GLMs of all five response variables) against the number of people interviewed there (Figure 9.2): detecting five or more significant effects required a sample of at least 300 people. With this result in mind, we therefore built generalized linear mixed models (GLMMs), which pooled data for all zoos into a single set of analyses (these analyses exclude Barn Elms data because we were most interested in the effects of visiting a zoo).

Generalized linear mixed models

The minimal GLMMs we derived echoed the GLMs (Table 9.4). Combining data across all six zoos, we found positive effects of tertiary education on knowledge and allocation to international education; of charity membership on knowledge, general conservation allocation, and ability to name a useful activity; of conservation knowledge on each of the other four response

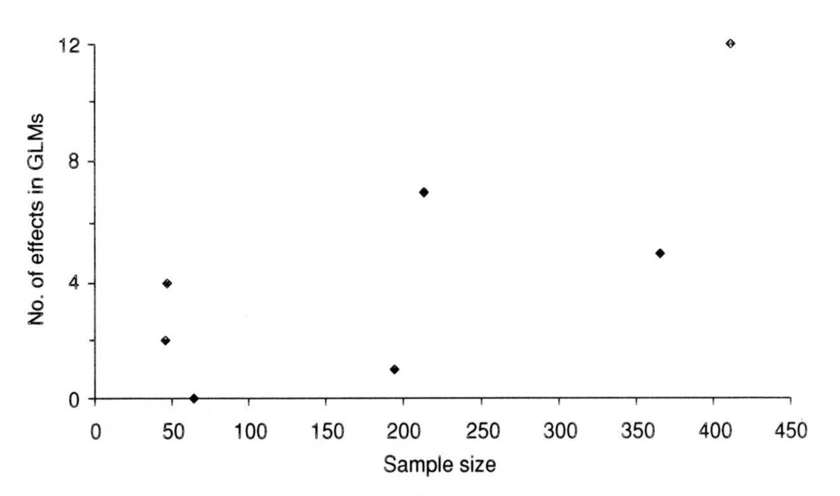

Figure 9.2 The total number of significant effects detected in the five GLMs for a site (from Table 9.3) plotted against the number of respondents interviewed at that site ($r_s = 0.74$, $P = 0.054$)

variables; and of a previous zoo visit on knowledge and allocation to habitat conservation. However, whether visitors were interviewed on arrival or departure entered only one GLMM: that for ability to name a conservation-related activity (which was due entirely to the potentially questionable effect detected at Paignton Zoo). No interaction terms were significant in any GLMM either, so it was not the case that genuine single-visit effects were masked because they held only for first-time zoo visitors, or for less knowledgeable visitors, for example.

Moreover, for these analyses our failure to detect a visit's effect on four out of five response variables was not due to a lack of statistical power. We tested the power of our GLMMs by Monte Carlo simulations, which involved generating artificial differences between entering and exiting visitors and seeing how big these needed to be in order to enter each of our minimal GLMMs (at $P < 0.05$). Arrival vs departure would have entered the models in 90% of 1000 iterations (i.e., with power set to 0.9) even if it affected knowledge score by as little as 0.39 (out of 20), or allocation to overall conservation, international conservation, or habitat conservation, respectively, by as little as £5.26, £6.09 or £13.41 (out of £1000). We can thus conclude that the effect of a single visit on these response variables is at most 5%, 4%, 4%, and 3% of their respective intercept values (Table 9.4).

Table 9.4

The minimal generalized linear mixed models for each of our response variables. Cells show coefficients for all predictors significant at $P < 0.05$ *(together with 95% confidence limits); italicized terms are significant only at* $P < 0.1$ *and are not considered further.* $^{***}P < 0.001$; $^{**}P < 0.01$; $^{*}P < 0.05$; $(^{*}) = P < 0.1$; *NS = not significant. The bottom row gives the smallest effect size for arriving vs departing visitors which entered each final GLMM (at* $P < 0.05$*) in 90% of 1000 simulations, expressed as a percentage of the intercept value*

Response variable:	Conservation	Allocation to conservation (/£1000)			Ability to name useful
Predictor:	knowledge (/20)	Overall	International	Habitat	action (0 or 1)
Intercept	7.86 (7.45;8.26)	131.57 (117.39;147.18)	152.25 (129.65;178.00)	447.13 (418.58;476.02)	0.22 (0.18;0.27)
Age (four categories)	0.71^{**} (0.23;1.17)	NS	NS	NS	NS
Sex (male vs female)	0.42^{*} (0.00;0.84)	NS	NS	38.18^{*} (2.00;77.53)	NS
Tertiary education (yes vs no)	0.79^{***} (0.34;1.23)	NS	54.32^{***} (27.52;86.82)	*23.29 (*)* *(−6.28;51.41)*	NS
Charity member (yes vs no)	1.36^{***} (0.84;1.88)	40.14^{***} (15.65;69.21)	NS	NS	0.06^{**} (0.00;0.12)
Conservation knowledge (/20)	–	3.10^{*} (0.84;6.83)	6.18^{***} (2.61;9.74)	5.36^{*} (1.30;9.46)	0.02^{***} (0.01;0.03)
Visited zoo before (yes vs no)	0.55^{*} (0.01;1.14)	NS	NS	41.32^{**} (13.43;67.45)	NS
Arrival vs departure	NS	NS	NS	NS	0.07^{**} (0.03;0.13)
Minimum effect size detectable	5%	4%	4%	3%	–

DISCUSSION

We found very little evidence, in the zoos we sampled, of any measurable effect of a single informal visit on adults' conservation knowledge, concern, or ability to do something useful. We also found no effect of a single visit to Barn Elms reserve, though this could be partly explained by the generally high score of all visitors here, for nearly all response variables (Figure 9.1). The only effect of a single visit that we did detect was restricted to one response variable at just one zoo, and this might have been an artifact of differences in the time which arriving and departing visitors spent completing our questionnaire (M. Ebbage-Taylor, pers. comm.).

These results are at first rather surprising. However, we do not believe that they are due to problems in the questions we asked or in the use of volunteers to administer the questionnaires, or to any shortage of statistical power. To our knowledge this is the largest survey of its kind conducted to date. Moreover, despite the survey's inevitable limitations, it was able to detect many other, intuitively sensible predictors of our response variables (Tables 9.3 and 9.4). Last, formal power analysis suggests that the overall effects of a single visit, pooled across zoos, must be slight or non-existent to have gone undetected given our sample size and analytical framework.

Our findings are less surprising when seen in the context of the handful of other studies to have pursued these questions. For instance, Kellert and Dunlap (1989), comparing arriving and departing visitors at three US zoos, found that a visit increased wildlife interest at one zoo but increased "dominionistic" attitudes to conservation at another, and, if anything, led to a general decrease in wildlife knowledge. Working at Jersey Zoo, Broad (1996) found departing visitors knew better than arriving visitors which of the species exhibited there were threatened, but in follow-up phone calls 7–15 months later, 80% reported that the visit had not influenced them in any way. L. Lach (pers. comm.) reported that a single visit to Franklin Park Zoo, Boston increased pro-conservation attitudes but did not change the likelihood of visitors donating money to conservation. Adelman *et al.* (2000) found that visitors leaving the National Aquarium in Baltimore had a more focused and broader understanding of conservation, but were no more concerned or likely to change their behavior than on arrival. Last, surveys conducted at the American Museum of Natural History's state-of-the-art Biodiversity Hall could detect only weak impacts on visitors' knowledge or ability to do something practical about conservation (Giusti 1999).

A potential counterargument to these observations is that informal education does have effects, but that these either take time or repeated visits to manifest themselves. However, measuring long-term effects of individual visits is difficult, as any messages received at a zoo will become increasingly overlaid with other, more recent messages. Moreover, the few studies to have conducted follow-up interviews have found the immediate effects of a zoo visit generally wane over time (L. Lach, pers. comm.; Adelman *et al.* 2000). Measuring the effect of repeated zoo visits is even harder, as the frequency of zoo visiting is likely to co-vary strongly with a general interest in wildlife, making it impossible to establish causality. Last, it is worth noting recent evidence that concern and knowledge about another environmental issue, climate change, can be altered, more or less immediately, by a single, dramatic experience – in this case, seeing the disaster film *The day after tomorrow* (Balmford *et al.* 2004, which uses the same survey design as this chapter).

Our results, and others like them, clearly do not mean that zoos do not or cannot educate or inspire their public about conservation. After all, expecting adults (who may already be disproportionately concerned about conservation issues) to further absorb conservation messages informally during a general zoo visit may be unrealistic, especially as they are often preoccupied with looking after young children. It may instead be the case that educational effects are more likely where adults are exposed to intense, focused experiences, via keeper talks or animal shows (Yerke and Burns 1991, cited in Kreger and Mench 1995; Ollason 1993, Kreger and Mench 1995, Broad 1996, Miller *et al.* 2004), or through visits to exhibits which have been explicitly designed to convey conservation messages, such as New York's Congo (Chapter 5), or Zurich's Masoala Rainforest (Chapter 14). Likewise, our data say nothing about the impacts of informal or indeed formal zoo education on two, arguably more important, zoo audiences – children, and either adults or children in developing countries (where most biodiversity occurs, yet where rapid urbanization means that the public are becoming increasingly isolated from wild nature).

However, in all these cases, we see a clear need for more data of the kind collected here, involving rigorous quantitative comparisons of large numbers of visitors before and after their zoo experience. To date such data remain patchy, yet without them we are unable to make a convincing case for the conservation education role of zoos, or to identify best practice and so improve zoo performance in this crucially important enterprise.

ACKNOWLEDGEMENTS

We are extremely grateful to staff at Bristol, Chester, Colchester, London, Paignton, Thrigby, and Barn Elms for hosting and helping us; to Lizzie

Coleman, May Ebbage-Taylor, Ben Feetham, Kat Harrington, Nadine McCarthy, Emilia O'Carroll, Amanda Simeoni, and Michelle Simeoni for collecting data; and to the British and Irish Association of Zoos and Aquariums (BIAZA) and Miranda Stevenson for both inspiration and co-ordination. We thank Lori Lach for access to unpublished data, and the organizers of the Catalysts for Conservation symposium for inviting us to present our work, which was supported by the Zoological Society of London.

References

Adelman, L. M., Falk, J. H., & James, S. (2000). Impact of National Aquarium in Baltimore on visitors' conservation attitudes, behaviour and knowledge. *Curator*, **43**, 33–61.

Balmford, A. (1999). (Less and less) great expectations. *Oryx*, **33**, 87–88.

Balmford, A., Manica, A., Airey, L., Birkin, L., Oliver, A., & Schleicher, J. (2004). Climate change, Hollywood and public opinion. *Science*, **305**, 1713.

Broad, G. (1996). Visitor profile and evaluation of informal education at Jersey Zoo. *Dodo*, **32**, 166–192.

Broad, S. & Weiler, B. (1998). Captive animals and interpretation – a tale of two tiger exhibits. *Journal of Tourism Studies*, **9**, 14–27.

Dierking, L. D., Burtynk, K., Buchner, K. S., & Falk, J. H. (2002). *Visitor Learning in Zoos and Aquaria*. Annapolis: Institute for Learning Innovation.

Dunlap, J. & Kellert, S. T. (1995). Zoos and zoological parks. In *Encyclopedia of Bioethics*, Revised Edition, Volume 1, ed. W. T. Reich. New York: Simon and Schuster, pp. 184–187.

Evans, K. L. (1997). Aquaria and marine environmental education. *Aquaria Sciences and Conservation*, **1**, 239–250.

Gadgil, M. (1993). Of life and artefacts. In *The Biophilia Hypothesis*, eds. S. R. Kellert & E. O. Wilson. Washington: Island Press, pp. 365–377.

Giusti, E. (1999). *A Study of Visitor Responses to the Hall of Biodiversity*. New York: American Museum of Natural History.

IUCN (2003). *2003 IUCN Red List of Threatened Species*. Available at www.redlist.org [accessed 13 September 2004].

Joint Nature Conservation Committee (2004). *UK Biodiversity Action Plan*. Available at http://www.ukbap.org.uk/ (accessed 13 September 2004).

Kahn, P. H. & Kellert, S. R., eds. (2002). *Children and Nature. Psychological, Sociocultural, and Evolutionary Investigations*. Cambridge, MA: MIT Press.

Kellert, S. R. & Dunlap, J. (1989). *Informal Learning at the Zoo: A Study of Attitude and Knowledge Impacts*. Philadelphia, PA: Zoological Society of Philadelphia.

Kramer, R. A. & Mercer, D. E. (1997). Valuing a global environmental good: US residents' willingness to pay to protect tropical rain forests. *Land Economics*, **73**, 196–210.

Kreger, M. D. & Mench, J. A. (1995). Visitor–animal interactions at the zoo. *Anthrozöos*, **8**, 143–158.

Mazur, N. A. & Clark, T. W. (2000). Zoos and conservation: policy making and organizational challenges. *Yale Forestry and Environmental Sciences Bulletin*, **105**, 185–201.

Miller, B., Conway, W., Reading, R. P., Wemmer, C., Wildt, D., Kleiman, D., Monfort, S., Rabinowitz, A., Armstrong, B., & Hutchins, M. (2004). Evaluating the conservation mission of zoos, aquariums, botanical gardens, and natural history museums, *Conservation Biology*, **18**, 86–93.

Nabhan, G. P. & St. Antoine, S. (1993). The loss of floral and faunal story: the extinctions of experience. In *The Biophilia Hypothesis*, eds. S. R. Kellert and E. O. Wilson. Washington: Island Press, pp. 229–250.

Nabhan, G. P. & Trimble, S. (1994). *The Geography of Childhood. Why Children Need Wild Places*. Boston, MA: Beacon Press.

Ollason, R. J. (1993). Getting the message across. *Journal of the International Association of Zoo Educators*, **29**, 186–190.

Pyle, R. M. (1993). *Thunder Tree: Lessons from a Secondhand Landscape*. New York: Houghton Mifflin.

Pyle, R. M. (2003). Nature matrix: reconnecting people and nature. *Oryx*, **37**, 206–214.

Spotte, S. & Clark, P. (2004). A knowledge-based survey of adult aquarium visitors. *Human Dimensions of Wildlife*, **9**, 143–151.

Tribe, A. & Booth, R. (2003). Assessing the role of zoos in wildlife conservation. *Human Dimensions of Wildlife*, **8**, 65–74.

Venables, W. N. & Ripley, B. D. (2002). *Modern Applied Statistics in S*. London: Springer.

WAZA (2005). *The World Zoo and Aquarium Conservation Strategy: Building a Future for Wildlife*. Liebefeld-Bern: WAZA.

World Resources Institute (2000). *World Resources 2000–2001*. Oxford: Elsevier Science.

PART III

Establishing connections between zoos and the wild

10

Animal ambassadors: an analysis of the effectiveness and conservation impact of *ex situ* breeding efforts

ANNE BAKER

The analogy of zoos to conservation "arks" is not a new one (Durrell 1976, Conway 1980). As the concept evolved, demographic and genetic parameters that govern the management of *ex situ* populations were established (Soulé *et al.* 1986). The development of computer models such as GENES (Lacy 1997a), DEMOG (Foose and Ballou 1996), ZRBook (Princée 1990), and PM2000 (Pollak *et al.* 2002) provided a scientific foundation to our population management programs. Throughout the early 1990s, *ex situ* breeding was heralded as zoos' contribution to conservation.

Over the past decade we have expanded that perspective and now recognize that *ex situ* breeding is one component of a broad array of conservation activities to which zoos and aquariums can and should contribute (Hutchins and Conway 1995). At the same time, our notion of what constitutes a "managed" population has expanded, with the recognition that eventually most "natural" populations may need to be managed at some level (Hutchins 1995, Miller and Hobbs 2002). Concepts such as metapopulation management have suggested that animals and their descendants may move back and forth between *in situ* and *ex situ* management as part of a metapopulation model. Indeed this is happening today with species such as the golden lion tamarin, whooping crane, and Puerto Rican crested toad.

There are four questions that are central to an examination of the relationship between managed *ex situ* breeding programs and conservation.

- How can and do managed population *ex situ* breeding programs contribute to *in situ* conservation?
- How effective are our *ex situ* breeding programs at maintaining viable populations?
- How do we and should we select species for our *ex situ* breeding programs?
- How can our *ex situ* programs best contribute, and what are the obstacles we need to overcome?

The answers to these questions are complex and it is not my intent to explore them fully within this chapter. My goal is to stimulate thought and further discussion.

HOW CAN ANIMALS IN *EX SITU* BREEDING PROGRAMS CONTRIBUTE TO CONSERVATION?

The answer to this question draws mainly on examples from the United States, not because these efforts are unique to the United States, but simply because these are the efforts with which I am most familiar.

Genetic and demographic reservoir

Managed *ex situ* populations can serve as a genetic and demographic reservoir for wild populations. Well-managed captive populations have the advantage that they can retain a higher level of genetic diversity than much larger wild populations. Where only one or a few small populations exist in the wild, these populations may be extremely susceptible to disease outbreaks or other catastrophic events. Captive populations can serve as a buffer against loss of these wild populations to such events. May (1986) has suggested that the distemper epizootic in the last remaining wild population of black-footed ferrets would not have been so devastating had a captive population of black-footed ferrets existed.

Population for recovery and reintroduction

Particularly where the reasons for wild population decline are temporary (e.g., over-hunting, disease outbreak, climatic events such as floods or hurricanes, etc.) rather than extensive habitat loss, *ex situ* populations can serve to augment or re-establish populations *in situ*. Successful reintroductions of American bison, golden lion tamarins, Jamaican rock iguanas, peregrine falcons, and the American burying beetle are just a few examples. Of 120 reintroduction attempts identified in the 1993 World Zoo Conservation Strategy, 15 had established self-sustaining populations (IUCN/CBSG 1993).

Basic research

Ex situ populations provide opportunities for research not easily afforded by wild populations, and information gathered through this research can be used in conserving *in situ* populations. Work by Ralls *et al.* (1988) on the effects of inbreeding on zoo populations has had a major influence on the management of *in situ* populations, including the determination of appropriate reserve size and the need for translocation or genetic augmentation. *Ex situ* research has allowed us to identify and characterize diseases and to develop vaccines used to combat wildlife epidemics (Karesh and Cook 1995), and to better understand endocrine function as it relates to reproduction (Hodges 1996, Wildt 1996).

Development of technologies relevant to *in situ* conservation

Closely linked to basic research is technology development (Seal 1988). Zoo populations have been used to perfect radio and satellite monitoring techniques for wild populations (Nobbe 1994, Johnson 1994). Within the realm of wildlife medicine, techniques for chemical immobilization and the diagnosis and treatment of wildlife diseases have been developed using *ex situ* populations (Hutchins *et al.* 1991). Similarly non-invasive hormonal assays (Lasley and Kirkpatrick 1991), head starting techniques, fostering and cross-fostering techniques (Kuehler *et al.* 2000), reproductive techniques that can be used to improve the potential for gene exchange among small isolated populations (Holt 1995), and contraceptive techniques (Asa 1992) are but a few of the technologies that have been developed utilizing *ex situ* populations.

Conservation education

The living animal collections found in zoos and aquariums are becoming mankind's primary contact with living diversity (Conway 1995). It is through this contact that we can engender the caring attitude that is critical to developing a conservation ethic. Zoos and aquariums in the United States alone reached over 142 million visitors in 2003 (Carr, pers. comm., 2004). We have a tremendous potential to reach people who otherwise might not be exposed to conservation issues and to utilize a living animal connection to create links between these visitors and local and global conservation issues.

Animals as fundraising ambassadors

We want to believe that animals in our collections can serve as fundraising ambassadors for their wild counterparts, and there are models that demonstrate that this can occur. The Bronx Zoo's Congo exhibit gives visitors an opportunity to vote with their dollars to support field conservation, and as of the close of 2003 had raised over $2 million, all of which has gone directly into field conservation (Lattis, pers. comm., 2004). Christie's chapter (Chapter 17) documents the success of the European Association of Zoos and Aquaria (EAZA) Tiger Campaign fundraising effort. An often-voiced criticism is that money used to support zoos and aquariums detracts from other conservation organizations and activities. While this may be true in some instances, zoos and aquariums have the ability to tap into sources of support that are not open to other conservation organizations. Many of our zoos and aquariums receive significant municipal funding, unlikely to be available to other conservation organizations.

Professional training

Training programs such as those supported by National Zoo's Conservation Research Center (Wemmer *et al.* 1993), and the Brookfield Zoo in conjunction with Chicago's Field Museum and Shedd Aquarium (Jackson 2001) have trained hundreds of zoo and field biologists and wildlife agency staff. Our facilities offer opportunities for thousands of graduate students each year to be involved with research on exotic species, and some of this research provides information useful in field conservation efforts.

Thus far I have focused on how *ex situ* breeding programs and the animal populations they support can assist conservation efforts. We should also examine what zoos and aquariums do as institutions.

In situ research and conservation

The Wildlife Conservation Society, Durrell Wildlife Conservation Trust, Frankfurt Zoological Society, and Chicago Zoological Society, among others, have long been recognized as significant contributors to *in situ* research and conservation. The number of zoos and aquariums participating in such activities is extensive. As an example, with 70% of facilities accredited by the Association of Zoos and Aquariums (AZA) reporting, the total number of *in situ* research and conservation projects identified in 2002 came to 1240 in 95 different countries, at a cost of more than $6.4 million dollars (Hutchins *et al.* 2002).

Habitat protection and habitat restoration

Minnesota Zoo's Adopt a Park program in Indonesia (Tilson 1995), Brookfield Zoo's long-term efforts at Brookfield Conservation Park in Australia (Parker 1992), Zurich Zoo's support for Masoala (Chapter 14), and the work of the Madagascar Fauna Group (a consortium of zoos and other organizations) are just a few examples of the efforts of zoos and aquariums to restore and protect vital animal habitat.

Coalition building

Both EAZA and AZA served as catalysts for Bushmeat Crisis Task Forces comprising zoos and NGOs. AZA's Butterfly Conservation Initiative includes over 50 AZA zoos, the United States Department of Fish and Wildlife, the National Wildlife Federation, the Xerces Society, and the Environmental Defense Fund. Similar coalitions have been formed in other regions.

Catalysts for local conservation initiatives

Many smaller zoos focus on local conservation issues. The Rosamond Gifford Zoo in Syracuse, New York works closely with the New York State Department of Environmental Conservation and the United States Department of Fish and Wildlife on efforts to preserve endangered New York state endemics such as the Chittenango ovate amber snail. In partnership with Idaho Fish and Game and the United States Department of Fish and Wildlife Service, Zoo Boise in Idaho is working to preserve the endangered Idaho ground squirrel. From the above list of activities it is possible to develop a contribution matrix (Table 10.1).

Many of the activities within this matrix are common to a number of conservation organizations. There are two areas of conservation activity to which zoos and aquariums are especially well qualified to make contributions and for which managed *ex situ* populations are essential. These are the provision of a genetic and demographic reservoir and as the source of animals for reintroduction efforts for species formerly extinct in the wild. Conway (1988, p. 268) points out that "technology is not a panacea for the disease of extinction. It is a palliative – a topical treatment with which to buy time, to preserve options . . ." With the human population rapidly increasing, the increasing urbanization of our landscape, and a growing level of affluence of the average human, wildlife is increasingly at risk. Now more than ever it would seem to be an especially bad time to decrease options.

Table 10.1
How zoos and aquariums contribute to conservation

	Activity		
	Breeding	Animal collection	Zoos & aquariums as institutions
Genetic reservoir	×		
Reintroduction	×		
Research	×	×	×
Technology	×	×	×
Conservation education		×	×
Fundraising ambassadors		×	×
Professional training		×	×
Field research			×
In situ conservation			×
Habitat protection			×
Habitat restoration			×
Coalition building			×
Local conservation catalysts			×

Other chapters in this volume address most of the areas mentioned above in far greater depth. However, the management of our zoo and aquarium populations receives little attention (see Chapter 15). Our ability to sustain populations that are viable over the long term, and suitable for reintroduction should the need arise, depends on good management of our *ex situ* populations. This brings us to the next question.

HOW EFFECTIVE ARE OUR *EX SITU* BREEDING PROGRAMS AT MAINTAINING VIABLE POPULATIONS?

The management of *ex situ* populations is nearly unique to zoos and aquariums, therefore it is important that we manage these populations well. Within North America, individual programs annually look at whether recommendations have been followed, and what this means for the population of that particular species, but *to date there has been no systematic effort to evaluate the effectiveness of our efforts as a whole.* Measures that we might use to assess effectiveness include: the proportion of our collection that is captive born or hatched, the demographic health of our captive populations (as measured by the number of animals/program, the age structure of the population, and whether or not demographic goals of a stable, increasing, or decreasing population are being met), and the genetic health of our captive population (as measured by gene diversity or founder genome equivalents, average inbreeding coefficients). Taken together demographic and genetic assessments can give us some idea of population viability over the long term.

Turning first to the proportion of our collection that is captive born or hatched, information from the International Species Information System (ISIS), October 2003, indicates that: of 125 584 mammals, 79% are captive born; of 162 321 birds, 63% are captive hatched; of 56 820 reptiles, 41% are captive hatched; and of 22 507 amphibians, 27% are captive hatched. While the situation for mammals is reasonably good, especially considering that many of our mammals are long-lived mammals that were wild caught 20–30 years ago, the situation for reptiles and amphibians is less encouraging, with over 33 500 reptiles and 16 000 amphibians harvested from the wild. Although specimens are obtained in accordance with all capture and import requirements and regulations, the numbers indicate the lack of coordinated captive breeding programs for these taxa.

Assessments of demographic and genetic health can provide insight into the question of how viable *ex situ* populations are over the long term. The AZA's 2002 Annual Report on Conservation and Science (Hutchins *et al.* 2002) contains information from 146 of the 243 programs that are cooperatively managed [Species Survival Plans (SSPs) and Population Management Plans (PMPs)]. Of the 146 programs that reported, 89 provided sufficient information to analyze long-term viability. For the purpose of this analysis, populations were said to have good viability, or a high probability of survival over the long term, if the current gene diversity was greater than 90%, the target population size was 200 or greater, and the population was stable or growing. Populations were said to have low viability over the long term if the current gene diversity was less than 90%, the target population size was less than 200, or the population was declining. Using these criteria, ***fewer than 50% of the populations that were assessed were likely to be viable over the long term*** (Table 10.2). The prospects for those that did not report are unlikely to be substantially better, and may be worse.

Some might argue that the criteria used to assess population viability are too stringent. However, the criteria are not arbitrary. Those for maintenance of genetic diversity come from past experience of animal breeders maintaining domestic livestock (Frankel and Soulé 1981), studies of the level at which we see damage from inbreeding depression (Ralls *et al.* 1988, Lacy *et al.* 1993), and calculations of the diversity needed to maintain the long-term evolutionary potential of species (Lande and Barrowclough 1987, Lande 1995, Lacy 1997b).

Table 10.2
Population viability of 89 SSP/PMP[a] populations

	No Reporting	High[b]	Low[c]
Mammals	69	26 (38%)	43
Birds	17	4 (23%)	13
Reptiles	8	2 (25%)	6
Amphibians	1	1	

[a] SSP, Species Survival Plans; PMP, Population Management Plans.
[b] High: >90% gene diversity, target pop. = 200, pop. growing or stable.
[c] Low: <90% gene diversity, target pop. < 200, or pop. declining.

Earnhardt *et al.* (2001) examined 17 PMP and 46 SSP populations, with similar conclusions. On average, the PMP populations were projected to lose 10% or more of founding gene diversity within only 2 years, and the SSP populations were projected to lose 10% or more of founding gene diversity within 40 years. The authors conclude that target population sizes are not scientifically assessed and allocated to captive breeding programs.

HOW DO WE AND HOW SHOULD WE SELECT SPECIES FOR OUR *EX SITU* BREEDING PROGRAMS?

Answering this question makes the assumption that managed *ex situ* breeding programs will continue to exist. If zoos and aquariums are to continue to exist, there are two alternatives to maintain their animal collections. The first is (sustainable) harvest of animals from the wild; the second is to manage self-sustaining *ex situ* populations. Although some have suggested that eliminating *ex situ* breeding programs would lower cost, it also would eliminate our ability to serve as genetic and demographic reservoirs for future reintroduction efforts, the only conservation activity in Table 10.1 that is carried out predominantly by zoos and aquariums. Further, it is doubtful that eliminating breeding would significantly reduce costs in most instances, since for most species the costs of breeding are minimal compared to the cost of animal maintenance. To illustrate this point I looked at breeding and maintenance costs for one of the most expensive land mammals, elephants. At the Rosamond Gifford Zoo in Syracuse, New York there is a herd of one male and five female Asian elephants. Over the past 12 years a calf has been born, on average, every 2 years. The cost of maintaining the herd, including staff, food, facilities maintenance, utilities, and staff training, is approximately $240 000 annually. Incremental costs associated with breeding have averaged less than $3000 per birth. Compare this to a recent

Table 10.3
Number of managed ex situ *breeding programs for AZA & EAZA*

	SSP[a]	PMP[a]	EEP[b]	ESB[b]
Mammals	110	133	105	72
Birds	25	105	37	58
Reptiles	9	59	7	10
Amphibians	2	5		
Fish	1+			
Invertebrates	1+		2	

[a] SSP, Species Survival Plan, and PMP, Population Management Plan, both are programs of AZA.
[b] EEP, European Endangered Species Programs, and ESB, European studbook.

import of elephants to the United States in which legal fees alone were in excess of $400 000 (Myers, pers. comm., 2003). Once in US facilities these animals still needed to be housed, fed, and cared for.

A retrospective look at the selection of species for *ex situ* breeding programs shows that, prior to the early 1990s, our programs reflected a historical bias, with a focus on large mammals, and a tendency not to focus on threatened or endangered species. With the advent of Regional Collection Plans (RCPs) in the early 1990s there was a shift to choosing species based on a combination of factors related to conservation, education, research, and exhibitry, including, but not limited to: conservation status in the wild, recommendations from other conservation bodies such as the World Conservation Union's Species Survival Commission, existence of a viable *ex situ* population, reintroduction potential, scientific research potential, husbandry expertise, conservation education value, exhibit value and taxonomic uniqueness. Unfortunately, many Taxon Advisory Groups have not yet found a good way to prioritize these criteria, and frequently the criteria are used to justify rather than critically evaluate existing programs. In part because of historical bias, and in part because of a continued focus on the charismatic mega-vertebrates, our managed *ex situ* programs continue to reflect a mammal bias (Table 10.3).

Some would argue that this mammal bias is appropriate, suggesting that it is the large mammals that can best act as flagship species, providing a focus for conveying conservation issues to the public and rallying their support (Western 1986). This strategic perspective bases species selection on the ability of our collections to fund-raise and build public awareness. In a fascinating review of the use of flagship species to raise money for conservation, Leader-Williams and Dublin (2000) found that the more simple and the more charismatic the message, the more money could be raised. Inhabitants of North Yorkshire were willing to pay more to secure the future of the Eurasian otter than the equally threatened water vole. What was surprising was that the otter alone elicited more support than the otter and the water vole combined.

Others suggest an ecological approach, selecting keystone or umbrella species that play a vital role in ecosystem maintenance (Roberge and Angelstam 2004). This again might lead us to select the charismatic megavertebrates such as large predators, although more and more we are realizing the importance of the roles of smaller species. Still others (Balmford *et al.* 1996, Balmford 2000) argue for selection based on a reintroduction perspective, arguing for selecting species that breed rapidly and reliably in captivity, for which the cost of *ex situ* maintenance is low, and for which there is a high probability of a successful reintroduction effort over the short term.

If an understanding and appreciation for the rich diversity of species is one of our educational goals, no single approach will suffice, and our collections must reflect a balance of all of the strategies identified above. The species we select should match the conservation goals we have set, and these will vary among institutions and over time.

HOW CAN OUR *EX SITU* PROGRAMS BEST CONTRIBUTE AND WHAT ARE THE OBSTACLES WE NEED TO OVERCOME?

To answer this question we need to look at what makes us unique – what sets us apart from other conservation organizations – and that is the maintenance of living animal collections. Returning to our contribution matrix, we can distinguish those activities that are directly dependent on our living collections. It is within the realm of these activities that we are positioned to make unique contributions to conservation. Well-managed *ex situ* breeding programs are at the core of maintaining our zoo and aquarium populations, and of ensuring that we can continue to contribute in the areas of recovery and reintroduction, research, technology, conservation education, and fund-raising. There are challenges to maintaining effective *ex situ* breeding programs, and these must be addressed if we are to be successful into the future.

At the top of the list of challenges is space. Ten years ago many of us were optimistic that our capacity would grow. For a number of reasons this is not happening. As we renovate old exhibits, many of which were small and did a poor job of meeting animal needs, we are replacing them with larger exhibits that we believe are better in terms of animal welfare. Often these larger exhibits must fit within the same building footprint as the older exhibits, resulting in a loss of carrying capacity. As an example, Lincoln Park Zoo in Chicago has undergone extensive renovation over the past 12 years. Between 1988 and 1997 the total number of exhibits at Lincoln Park Zoo has decreased from 139 to 72, a decrease of 48%. Over the same time period the number of species exhibited decreased by 37%, and the number of individual animals by 41% (Thompson, pers. comm., 2004). This is good news in that the larger exhibits can generally be said to improve both animal welfare and the visitor experience. It's bad news in that it significantly reduces the number of rooms on the Ark. In planning future breeding programs we need to recognize that this trend probably will continue and that there will be fewer, not more, spaces in our zoos and aquariums for large animals.

Smith and Allard (1999) looked at space projections of three AZA Regional Collection Plans (large felids, small felids, and prosimians). Despite recommendations that population size increase for six out of the eight large felids, over a 3-year period in all but one instance space actually decreased for these species. The results for large felids are mirrored by those for small felids, and to a lesser extent by those from prosimians. There is no doubt that space constraints are severely limiting many of our *ex situ* breeding programs, particularly for the larger mammals and birds.

Genetic and demographic problems are another challenge to our *ex situ* breeding programs. Many populations suffer from an insufficient founder base, unequal representation of founders, animals of unknown provenance, and linkage of common and rare alleles, to name just a few of the genetic hurdles. Demographic problems include unstable age pyramids, shortened captive interbirth intervals, imbalanced sex ratios, and high numbers of non-reproductive individuals. Of the 89 SSPs reporting in the 2002 AZA Annual Report on Conservation and Science (Hutchins *et al.* 2002), 25% reported current gene diversity less than 90%, and 67% reported current population size less than 100. Low gene diversity and population size severely limit the ability to maintain a viable population over the long term.

The AZA RCP for Old World Monkeys (Lyon and Pate 2000) illustrates the source of some of the problems. By failing to make choices, it so thinly distributes space resources that it dooms its efforts for all of the guenon species. The plan recommends ten species of guenons for SSP management. Of those ten recommended species, recent genetic and demographic data were available for five (Table 10.4).

Table 10.4
Guenon species recommended for SSP management in the AZA Old World Monkey RCP

	Target population	Current population	Current FGE[a]	FGE projected @ 100 years
Allenopithecus nigroviridis	75	33	3.25	1.6
Cercopithecus diana	125	53	12.75	2.7
C. neglectus	125	77	11.58	2.0
C. mitis Stuhlmanni	75	15	2.37	0.8
C. petaurista	75	11	2.78	1.3

[a] FGE, founder genome equivalent is the number of unrelated wild-caught animals that would contain the gene diversity equivalent to that present in the living population of captive born animals. An FGE of 2.37 means the captive population has the same level of gene diversity as would a sample of 2.37 unrelated animals in the wild.

Applying the criteria established earlier (see Table 10.1) to assess the likelihood of long-term population viability, *none of the populations has much chance of surviving for an extended period of time.* None of the ten recommended species currently has a population above 80 individuals. No target population size is above 125 and only one population is even close to its target population size at present. Founder genome equivalents (FGE) are a measure of gene diversity and while two of the populations have marginally acceptable FGEs at present, looking at FGEs 100 years out no population would have the genetic equivalent of more than three wild caught animals. Were difficult choices to be made, and the number of taxa targeted for managed programs reduced, it is possible that viable populations of the two taxa with the highest current gene diversity could be maintained over the long term.

Institutional needs, preferences, and history place constraints on our breeding efforts. Converting a facility built for lions into a facility for breeding reptiles and amphibians is no small task. Many zoo managers believe that without the large, charismatic mega-vertebrates they would not get visitors through the gates. This assumption has yet to be seriously tested, and research in this area is sorely needed. Balmford *et al.* (1996) suggest that much zoo space could be directed toward smaller species with no loss of visitor satisfaction.

Regulatory constraints imposed by permitting processes and quarantine restrictions limit not only the movement of animals, but the movement of samples as well. We will need to work closely with appropriate government agencies to ensure that they understand the importance of our *ex situ* breeding and research efforts to the future of wild populations.

Finally, a lack of clear conservation-oriented goals hampers many of our *ex situ* breeding programs. In the nearly 25 years since our cooperative *ex situ* breeding programs began we have made great strides in developing the tools that we need to use to manage these populations. In 1996 the AZA, in collaboration with Lincoln Park and Brookfield Zoos, established the AZA Population Management Center to facilitate analysis of our population data and communication of management plans. Unfortunately, our vision for the roles these managed populations might play has not kept pace with our management tools. In too many instances we have lost sight of the forest for the trees.

Returning to the 2002 AZA Annual Report on Conservation and Science (Hutchins *et al.* 2002), of the 99 programs that identified goals, 53% defined their program goals in terms of target population size, gene diversity and anticipated growth rate, but gave no mention of the ultimate reasons behind their population management objectives. Fewer than 50% linked their programs to *in situ* conservation in any way. While there were fewer of them, programs for amphibians and reptiles had much stronger *in situ* conservation links than those for mammals. Clearly there is still much work to be done.

In summary, managed *ex situ* breeding programs can make important and unique contributions to *in situ* species conservation. However, many of our programs have not yet identified conservation goals beyond maintenance of a captive population, and are not yet managed in ways that will ensure that they will be able to contribute. As we continue in our efforts to support wild populations we must attend to four critical areas. Our Regional Collection Plans must be realistic. They must rationally evaluate a program's potential and make the difficult choices that our limited resources require. There must be good collaboration among regions in managing *ex situ* breeding programs. Clear conservation goals must be articulated for every cooperatively managed *ex situ* breeding program. And finally, institutional exhibit plans should closely link to institutional conservation plans. If all of these occur we will have taken a giant step forward in assuring that our *ex situ* breeding programs truly do support animal populations in the wild. If they don't, the basis for arguing for the continued relevance of zoo and aquarium animal populations to conservation will be considerably diminished.

References

Asa, C. S. (1992). Contraceptive development and its application to captive and free ranging wildlife. In *AAZPA/CAZPA Conference Proceedings*, pp. 71–75.

Balmford, A. (2000). Priorities for captive breeding – which mammals should board the ark? In *Priorities for the Conservation of Mammalian Diversity*, eds. A. Entwistle & N. Dunstone. Cambridge: Cambridge University Press, pp. 291–307.

Balmford, A., Mace, G. M., & Leader-Williams, N. (1996). Designing the ark: setting priorities for captive breeding. *Conservation Biology*, **10**(3), 719–727.

Conway, W. (1980). An overview of captive propagation. In *Conservation Biology: An Evolutionary-Ecological Perspective*, eds. M. E. Soulé & B. A. Wilcox. Sunderland, MA: Sinauer Associates, Inc., pp. 199–208.

Conway, W. (1988). Can technology aid species preservation? In *Biodiversity*, ed. E. O. Wilson. Washington, D.C.: National Academy Press, pp. 263–268.

Conway, W. (1995). Zoo conservation and ethical paradoxes. In *Ethics on the Ark: Zoos, Animal Welfare and Wildlife Conservation*, eds. B. G. Norton, M. Hutchins, E. F. Stevens, & T. L. Maple. Washington, D.C.: Smithsonian Institution Press, pp. 1–9.

Durrell, G. (1976). *The Stationary Ark*. New York: Simon & Schuster.

Earnhardt, J. M., Thompson, S. D., & Marhevsky, E. A. (2001). Interactions of target population size, population parameters, and program management of viability of captive populations. *Zoo Biology*, **20**(3), 169–183.

Foose, T. & Ballou, J. (1996). Demographic and genetic management of captive populations. In *Wild Mammals in Captivity: Principles and Techniques*, eds. D. Kleiman, M. Allen, K. Thompson, & S. Lumpkin. Chicago, IL: University of Chicago Press, pp. 263–283.

Frankel, O. H. & Soulé, M. (1981). *Conservation and Evolution*. Cambridge: Cambridge University Press.

Hodges, J. K. (1996). Determining and manipulating female reproductive parameters. In *Wild Mammals in Captivity: Principles and Techniques*, eds. D. Kleiman, M. Allen, K. Thompson, & S. Lumpkin. Chicago, IL: University of Chicago Press, pp. 418–428.

Holt, W. V. (1995). Germplasm cryopreservation and its potential role in wildlife conservation. In *Research and Captive Propagation*, eds. U. Ganslosser, J. K. Hodges, & W. Kaumanns. Furth: Filander-Verlag, pp. 71–80.

Hutchins, M. (1995). What do "wild" and "captive" mean for large ungulates and carnivores now and into the twenty-first century? In *Wildlife Conservation, Zoos and Animal Protection: A Strategic Analysis*, ed. A. N. Rowan. Proceedings of a workshop held at the White Oak Conservation Center, 21–24 April 1994, pp. 1–18.

Hutchins, M. & Conway, W. (1995). Beyond Noah's ark: the evolving role of modern zoological parks and aquariums in field conservation. *International Zoo Yearbook*, **34**, 117–130.

Hutchins, M., Foose, T., & Seal, U. S. (1991). The role of veterinary medicine in endangered species conservation. *Journal of Zoo Wildlife Medicine*, **22**, 277–281.

Hutchins, M., Smith, B., Allen, R., Souza, M., & Lankard, J. (2002). *AZA Annual Report on Conservation and Science*. Washington, D.C.: AZA.

IUCN/CBSG (IUCN/SSC). (1993). *The World Zoo Conservation Strategy: The Role of Zoos and Aquaria of the World in Global Conservation*. Brookfield, IL: Chicago Zoological Society.

Jackson, W. M. (2001). Conservation training consortium: a consortium of the Field Museum, Chicago Zoological Society, University of Illinois, Chicago, John G. Shedd Aquarium, and University of Chicago. In *American Zoo and Aquarium Association Field Conservation Research Guide*, eds. W. G. Conway, M. Hutchins, M. Souza, Y. Kapentanakos, & E. Paul. Atlanta, GA: Wildlife Conservation Society, New York & Zoo Atlanta, pp. 236–239.

Johnson, R. R. (1994). Model programs for reproduction and management: *ex-situ* and *in-situ* conservation of toads of the family Buffonidae. In *Captive Management and Conservation of Amphibians and Reptiles*, eds. J. B. Murphy, K. Alder, & J. T. Collins. Ithaca, NY: Society for the Study of Amphibians and Reptiles, pp. 243–254.

Karesh, W. B. & Cook, R. A. (1995). Application of veterinary medicine to *in-situ* conservation efforts. *Oryx*, **29**, 653–658.

Kuehler, C., Lieberman, A., Oesterle, P., Powers, T., Kuhn, M., Juhn, J., Nelson, J., Snetsinger, T., Herrmann, C., Harrity, P., Tweed, E., Fancy, S., Woodworth, B., & Tefler, T. (2000). Development of restoration techniques for Hawaiian thrushes: collection of wild eggs, artificial incubation, hand-rearing, captive-breeding, and re-introduction to the wild. *Zoo Biology*, **19**, 263–277.

Lacy, R. C. (1997a). GENES v.11.2 (software). Brookfield, IL: Chicago Zoological Society.

Lacy, R. C. (1997b). Importance of genetic variation to the viability of mammalian populations. *Journal of Mammology*, **78**, 320–325.

Lacy, R. C., Petric, A., & Warneke, M. (1993). Inbreeding and outbreeding in captive populations of wild animal species. In *The Natural History of Inbreeding and Outbreeding: Theoretical and Empirical Perspectives*, ed. N. W. Thornhill. Chicago, IL: University of Chicago Press, pp. 352–374.

Lande, R. (1995). Breeding plans for small populations based on the dynamics of quantitative genetics. In *Population Mangement for Survival & Recovery: Analytical Methods and Strategies in Small Population Conservation*, eds. J. D. Ballou, M. Gilpin, & T. J. Foose. New York: Columbia University Press, pp. 318–340.

Lande, R. & Barrowclough, G. F. (1987). Effective population size, genetic variation, and their use in population management. In *Viable Populations for Conservation*, ed. M. E. Soulé. Cambridge: Cambridge University Press, pp. 87–123.

Lasley, B. L. & Kirkpatrick, J. F. (1991). Monitoring ovarian function in captive and free-ranging wildlife by means of urinary and fecal steroids. *Journal of Zoo Wildlife Medicine*, **22**, 23–31.

Leader-Williams, N. & Dublin, H. T. (2000). Charismatic megafauna as "flagship species." In *Priorities for the Conservation of Mammalian Diversity*, eds. A. Entwistle & N. Dunston. Cambridge: Cambridge University Press, pp. 53–81.

Lyon, E. H. & Pate, D., eds. (1999). *AZA North American Regional Collection Plan for Old World Monkeys*. Chicago, IL: Lincoln Park Zoo.

Lyon, E. H. & Pate, D. E. (2000). *North American Regional Collection Plan for Old World Monkeys*, Second Edition. Silver Spring, MD: American Zoo and Aquarium Association.
May, R. (1986). The cautionary tale of the black-footed ferret. *Nature*, **320**, 13–14.
Miller, J. R. & Hobbs, R. J. (2002). Conservation where people live and work. *Conservation Biology*, **16**(2), 330–337.
Nobbe, G. (1992). Going into orbit. *Wildlife Conservation*, **95**(5), 62–64.
Parker, P. (1992). The Chicago Zoological society's affair with Australian conservation. *Bison*, **6**, 12–19.
Pollak, J. P., Lacy, R. C., & Ballou, J. D. (2002). *Population Management 2000*, version 1.205. Brookfield, IL: Chicago Zoological Society.
Princée, F. P. G. (1990). Zooresearch Studbook Management Program version 1.04, software and update manual. Amsterdam: National Foundation for Research in Zoological Gardens.
Ralls, K., Ballou, J. D., & Templeton, A. (1988). Estimates of lethal equivalents and cost of inbreeding in mammals. *Conservation Biology*, **2**(2), 185–193.
Roberge, J.-M. & Angelstam, P. (2004). Usefulness of the umbrella species concept as a conservation tool. *Conservation Biology*, **18**(10), 76–85.
Seal, U. S. (1988). Intensive technology in the care of *ex-situ* populations of vanishing species. In *Biodiversity*, ed. E. O. Wilson. Washington, D.C.: National Academy Press, pp. 269–295.
Smith, B. & Allard, R. (1999). Regional Collection Plan: lifeboat or dinghy? *American Zoo and Aquarium Association Annual Conference Proceedings*. Minneapolis, MN: Minnesota Zoo, pp. 75–79.
Soulé, M., Gilpin, M., Conway, W., & Foose, T. (1986). The millennium ark: how long a voyage, how many staterooms, how many passengers? *Zoo Biology*, **5**(2), 101–113.
Tilson, R. L. (1995). In support of nature: the Minnesota Zoo's adopt-a-park program. In *The Ark Evolving: Zoos and Aquariums in Transition*, ed. C. Wemmer. Front Royal, VA: National Zoological Park, Conservation and Research Center, Smithsonian Institution.
Wemmer, C., Rudran, R., Dallmeier, F., & Wilson, D. E. (1993). Training developing county nationals is the critical ingredient to conserving global biodiversity. *BioScience*, **43**, 762–767.
Western, D. (1986). The role of captive populations in global conservation. In *Primates: The Road to Self-Sustaining Populations*, ed. K. Benirshchke. New York: Springer-Verlag, pp. 13–20.
Wildt, D. (1996). Male reproduction, assessment, management, and control of fertility. In *Wild Mammals in Captivity: Principles and Techniques*, eds. D. Kleiman, M. Allen, K. Thompson, & S. Lumpkin. Chicago, IL: University of Chicago Press, pp. 429–450.

11

Reintroductions from zoos: a conservation guiding light or a shooting star?

MARK R. STANLEY PRICE AND JOHN E. FA

INTRODUCTION

The release of either wild-caught or captive-born animals into an area in which they have either declined or disappeared, as a result of human pressures (e.g., overhunting) or from natural causes (e.g., epidemics), is an important tool for conservation. A reintroduction project can involve translocation of wild-caught individuals from other areas of natural habitat, it can involve release of naive, captive-born animals, or it can be attempted with equally naive, wild-caught animals captured as either infants or young juveniles and raised in captivity.

Zoos have unhesitatingly espoused the rationale for captive breeding of endangered species as an eventual return of the species back to the wild. Outside the community of zoo professionals, however, views on reintroduction as a conservation strategy are somewhat more reserved. In fact, some would even say that reintroductions are ill-conceived, wasteful of conservation money, and little more than romanticized schemes (Lindburg 1992). Arguments are presented for zoos to divert funds spent on reintroductions to *in situ* conservation efforts.

Advocates of captive breeding counter this by pointing to well-known cases such as the Arabian oryx (*Oryx leucoryx*) and the California condor (*Gymnogyps californianus*). They dispute, furthermore, that reliance on any single strategy such as conserving reduced wild populations actually increases the risk of extinction to an unacceptable level. In response to the charge that reintroduction costs could be deferred to *in situ* conservation, it is held that those who contribute to zoo programs may not feel the same compassion for the inhabitants of a distant patch of jungle. The two approaches are probably not in competition for the same funds.

The appreciation of the role of zoos in conservation has evolved substantially since the 1960s. The first World Zoo Conservation Strategy [IUDZG/CBSG 1993] stated explicitly, "The time has come for the zoo community as a whole to come forward and clarify its commitment to conservation." The strategy also reproduced the well-known trajectory that demonstrates how zoos have evolved from being 20th century living museums to the anticipated environmental resource centers of the 21st century, with accompanying changes in their themes, subjects, concerns, and exhibits. While much of this has been aspirational, the new World Zoo and Aquarium Conservation Strategy (WZACS) is more ambitious, yet more practical (WAZA 2005). It states that the conservation of any species, defined as the "securing of long-term populations . . . in natural ecosystems and habitats wherever possible," should increasingly become the main focus for any zoological institution with commitment to conservation in the wild as the primary goal. WZACS identifies the reintroduction of animals bred in zoos as a useful conservation tool, but cautions over the prospects for reintroductions because of the complexities of returning captive-bred animals to the wild. Is this more modest expectation for reintroductions based on experience and realism?

In this chapter, we submit that reintroductions offered zoos a lifeline at a critical time 20–30 years ago when most were then, in effect, providers of public recreation, even entertainment, based on their animal inhabitants. Thus, diversity of species, the presence of large flagship animals, as well as animal charisma, color, or dramatic behavior were significant determinants of collections. On this basis, zoos were menageries, and could not be expected to evolve swiftly to proclaim conservation as a central role; to maintain this would have been spin rather than substance. Further, many zoos might find unacceptable the consequent trade-off between commercial activity and the essentially charitable role of conservation. At the same time, the zoo movement was sensitive to criticism that zoos were collectively holding animals in inadequate conditions (for welfare and natural behavior) for little purpose beyond public entertainment (e.g., Anon 1994).

The 1980s saw two trends that were mutually dependent and which supported the concept of zoos as significant conservation bodies. The first trend was the burst of theoretical work on the concept of minimal levels of genetic and demographic diversity required for viable captive populations to survive for at least 200 years (e.g., Soulé *et al.* 1986). Such theoretical foundations have spawned diverse organizational structures for the zoo world to contribute to the cooperative, intensive management of endangered species (de Boer 1993). Such Species Survival Plans (SSPs), or their equivalents, were complemented by structures such as Conservation Assessment and Management Plans, that enabled objective setting of priorities for intensive management (including captive breeding) within the broader conservation needs of threatened taxa (Seal *et al.* 1993). Further tools for assessing and improving survival prospects of small populations in the wild, such as the process workshops for Population and Habitat Viability Analysis, developed progressively in parallel (Seal *et al.* 1993).

While theory advanced and methods and organizations came into existence, several high-profile species were being saved from extinction and, for each, the release of captive-bred individuals was either being done or was envisaged as a key component of recovery strategies. These included species such as the black-footed ferret (*Mustela nigripes*), for which removal from the wild for captive breeding started in 1985 (Thorne and Oakleaf 1991); the last wild Californian condors, taken into captivity, started to breed successfully in 1988 (Toone and Wallace 1994); the Arabian oryx was released in Oman in 1982 using stock mainly from the USA (Stanley Price 1989); the golden lion tamarin (*Leontopithecus rosalia*) was the subject of great inter-zoo collaboration and successful releases back into Brazil, starting from 1984 (Kleiman *et al.* 1991). Each of these species was kept and bred successfully in captivity, and became iconic examples, as well as opening the opportunity for zoos to begin to rationalize and promote the potential to restore the world's fauna by supplying animals from well-managed captive populations in the world's zoos. These captive-breeding successes or actual reintroductions also struck a chord with the public, for which "taking animals back home" is a simple but powerful message, and with considerable emotional impact. Further, it has suggestions that humans are atoning for their past abuses of wildlife and the environment that were causing extinctions.

Thus, zoos developed the theoretical basis and practice of keeping captive populations while enhancing their conservation credentials as sources of animals to restore animal populations and communities. Given the further challenges for zoos in the new WZACS, it is timely to ask whether reintroductions have been a major part of zoos' conservation efforts. Have they been a guiding light or have the success stories been merely a shooting star for an era, now over? If the latter, what should zoos be doing for the future?

CONSERVATION AND CAPTIVE BREEDING

In 1987 IUCN (The World Conservation Union) produced a policy statement on captive breeding (IUCN 1987). This was the first official recognition of the important role that captive breeding and zoos could play in species conservation (see also McNeely *et al.* 1990, IUCN 1992). Reinforcing the Convention on Biological Diversity (CBD), the foremost international treaty adopted at the Earth Summit in Rio de Janeiro in 1992, with the goal of conservation of biological diversity (or biodiversity), sustainable use of its components and fair and equitable sharing of benefits arising from genetic resources, the World Conservation Union published technical guidelines on the management of *ex situ* populations for conservation. These proclaimed that "Moreover, *ex situ* conservation should only be considered as a tool to ensure the survival of the wild population" (Secretariat of the Convention on Biological Diversity 2005). Despite these clear statements on purpose for out-of-the-wild populations for conservation of their species, various authors (e.g., Rahbek 1993, Snyder *et al.* 1996) have pointed out that while zoos play an important role in maintaining threatened species, they can be criticized for not carrying this out effectively.

ASSESSING THE CONTRIBUTION OF ZOOS TO REINTRODUCTIONS

The role of zoos in a few signal successful reintroductions is evident, and these are well documented. But, the issue is whether zoo-originated reintroductions have made a significant quantitative contribution to conservation, or have the potential to do so. The following list defines four measures by which zoo populations might be judged for their reintroduction potential:

1. The re-establishment of animals in the wild is usually in response to either total or local extinction; animals may also be released to reinforce small populations. Thus, reintroduction is most likely to address a situation of relative or absolute rarity. If zoos are to meet these demands for animals, they should be keeping rare species. Do they?
2. No source population, captive or wild, should be endangered by demands for individuals for release (IUCN 1998). Therefore, source populations for reintroduction must be self-sustaining, at least as large as the minimum size for this criterion. Are zoo populations large enough?

3. Reintroduction projects vary greatly in the numbers of animals released and the number of releases. One attribute of successful wild-to-wild translocations is that they involve the release of many individuals and over several to many years (Griffith *et al.* 1989). Models of swift fox releases also showed that an initial population of 100 individuals supplemented by 20 individuals/year for 10 years had the lowest probability of extinction of all models tested (Ginsberg 1994). Thus, an adequate supply of animals must be available; are zoo populations large enough to meet such demands?
4. Many reintroductions have now been carried out. Given the variety in situations and types of animals released, have zoos been the major provider?

We now attempt to answer each of these questions. We first use data contained in the reference section of the *International Zoo Yearbook* (IZY). The IZY publishes a census of zoos and aquariums of the world in alternate years. These are listings of zoos, aquariums, and other institutions with significant animal collections. The information includes institution name, address, staff names, institution size, total staff numbers, visitor attendance, number of species and specimens of mammals, birds, reptiles, fishes, amphibians, invertebrates, and total collection, institution specialty and governmental affiliation. The data are arranged in paragraph format by country and are based on answers supplied to a questionnaire provided by the IZY. Responding institutions provide all answers to the questionnaire but these responses are not subject to editorial verification or inspection. Four decade periods spanning a 30-year period (1960–1990) were chosen for study (Olney 1970, 1980, Olney and Fiskin 1990, Olney *et al.* 2000). Thirty years was regarded as long enough for significant changes in zoos and shifts in philosophies to be detected. Additionally, the choice of a 30-year period ensured the inclusion of only the more established zoos. In any case, information prior to 1960 was available on considerably fewer zoos in the format reported in the IZY. Zoos within three main regions (USA & Canada, Europe, and Japan) of the world were included in the analyses. Europe and the USA & Canada had the highest number of zoos reporting in the 1990 sample. A cohort of zoos was chosen within the main regions. For inclusion in the cohort, a zoo had to report to the IZY in each decade period in order to avoid fluctuation in calculated statistics through zoos appearing in one sample and not in another. The European and Middle East sample comprised 87 zoos in 21 countries, while the samples for North

America covered 52 zoos in the USA and Canada, with only 9 institutions from Japan.

In addition, as a regional example, we present collection data from South American zoos, with the caution that this is for illustrative purposes and not because this region's performance was expected or known to differ radically from that of any other region. The South American sample includes 188 zoos in 6 countries, containing 49 665 specimens of 1211 species. Within these, mammals, birds, and reptiles predominated, with relatively few amphibians, fish or invertebrates. Most birds (72%) and reptiles (74%) were native to the continent, while fewer than half the mammals (43%) were native.

Information from the zoo world is unable to answer the question of the importance of zoos compared to other sources as providers of animals for release. The Reintroduction Specialist Group of IUCN/SSC promotes responsible reintroductions and tracks projects around the world, and its data are used as indicated.

THE PERFORMANCE OF ZOOS IN REINTRODUCTIONS

Table 11.1 shows the average specimens per species for all world zoos taken from IZY. For mammals, mean specimens per species is 7.6 ± 0.7 ($n = 736$, median 4.7), followed by birds (7.5 ± 0.4; $n = 734$, median 5.0), but fell for amphibians to 6.2 ± 0.5 ($n = 352$, median 3.3) and for reptiles to 5.4 ± 0.3 ($n = 590$, median 3.7).

Thus population sizes in 2000 were small. If zoos were aiming to keep populations, under collaborative management arrangements, larger than minimum critical sizes, one might have seen increases in population sizes with developments in the theory of zoo populations. Average specimens per species in all zoo regions was low, and there were no significant changes across the four decades studied (Figure 11.1).

The average number of specimens per species does not answer questions about the adequacy of total population size for species. If many zoos, each of which holds a few animals of a species, collaborate fully in a breeding program, the total population size might exceed minimum numbers. To explore this we use the South American zoo data. Figure 11.2 separates native and non-native species in the South American collections. But, the picture is similar and clear in that most species are found in only one collection; very few species are found in more than 10% of all zoos. When total population sizes for different species groupings are examined, only 8% of all species have combined populations larger than 100 specimens

Table 11.1
Specimens per species for main taxonomic groups of animals held in zoos within the main world regions

		Australasia	Africa	Europe/Middle East	South America	North America	C America/Caribbean	Far East
Mammals	Mean	9.0	8.8	9.7	4.9	6.0	6.0	6.6
	SE	1.6	3.1	1.8	0.5	0.8	0.7	1.1
	Median	5.8	4.3	5.5	4.7	4.2	5.3	4.4
	n	21	21	248	21	163	9	253
Birds	Mean	6.0	6.8	5.8	4.9	5.2	9.8	11.0
	SE	0.6	2.1	0.2	0.5	0.7	1.6	1.0
	Median	6.0	4.8	4.9	3.8	3.7	9.7	6.9
	n	21	21	247	21	163	9	252
Reptiles	Mean	4.0	6.2	6.0	8.9	3.9	7.2	5.7
	SE	0.4	5.4	4.8	8.1	3.1	7.2	4.6
	Median	3.9	4.7	4.0	6.9	2.8	7.4	3.5
	n	19	19	205	20	154	9	164
Amphibians	Mean	5.4	14.9	7.6	5.2	4.4	23.8	6.0
	SE	1.5	10.2	1.2	2.1	0.5		0.7
	Median	3.3	7.6	4.0	3.5	2.9	23.8	3.0
	n	12	4	116	6	108	1	105

(Figure 11.3). This suggests that surpluses for release will be minimal and not available for many years.

Our final test of the South American data was to look at the conservation status of the species kept in the collections. Species status was taken from the *IUCN Red List* (Hilton-Taylor 2000). Table 11.2 shows the data for native and non-native species. While the South American zoos held twice as many native species as non-native species, and four times the number of individuals, 90% of the native species and 88% individuals were from species that were non-threatened. Species that were critically endangered in South America were represented by only 10 species and 49 individuals.

Our final analysis looks at the source of animals for reintroductions. As an example we use the data relating to all reintroductions of amphibians and reptiles for which the Reintroduction Specialist Group (RSG) has been

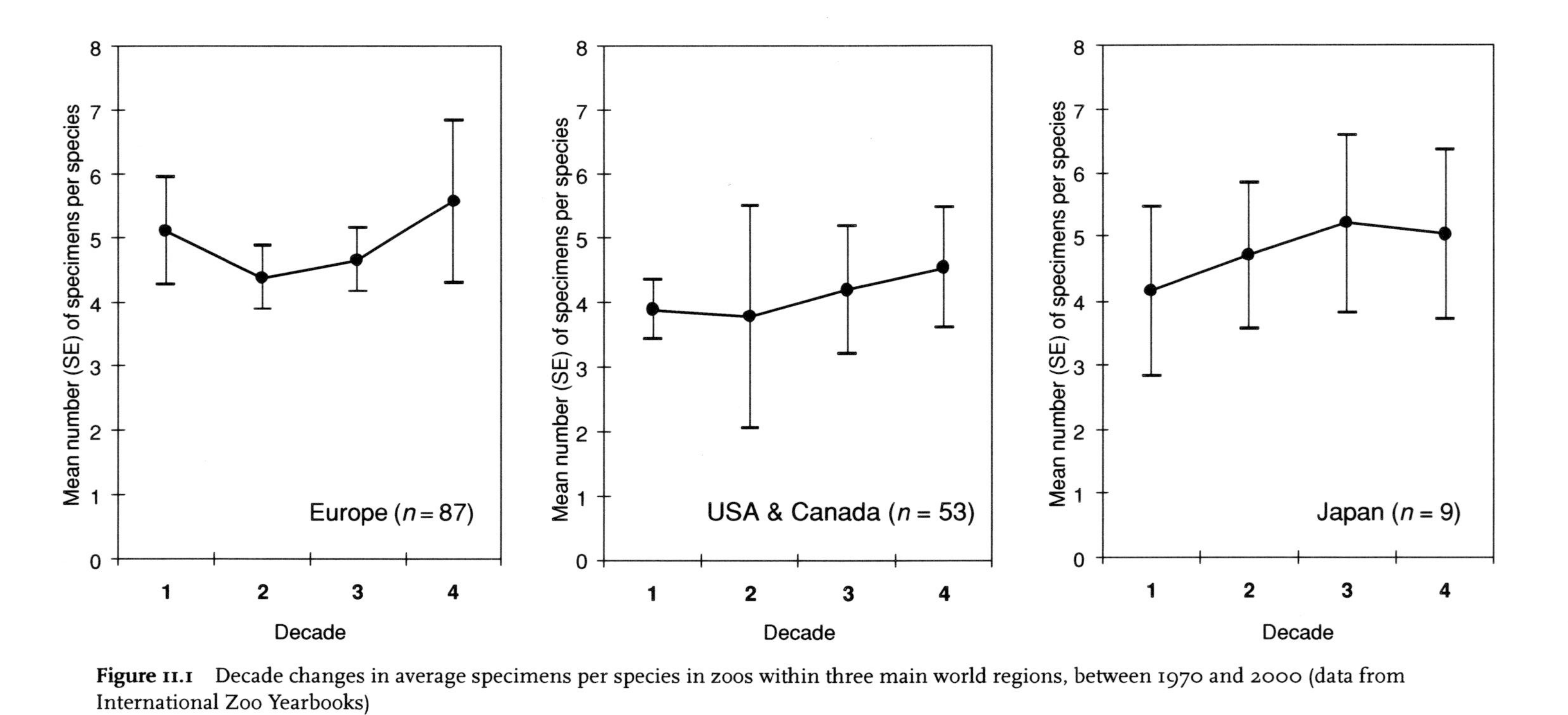

Figure 11.1 Decade changes in average specimens per species in zoos within three main world regions, between 1970 and 2000 (data from International Zoo Yearbooks)

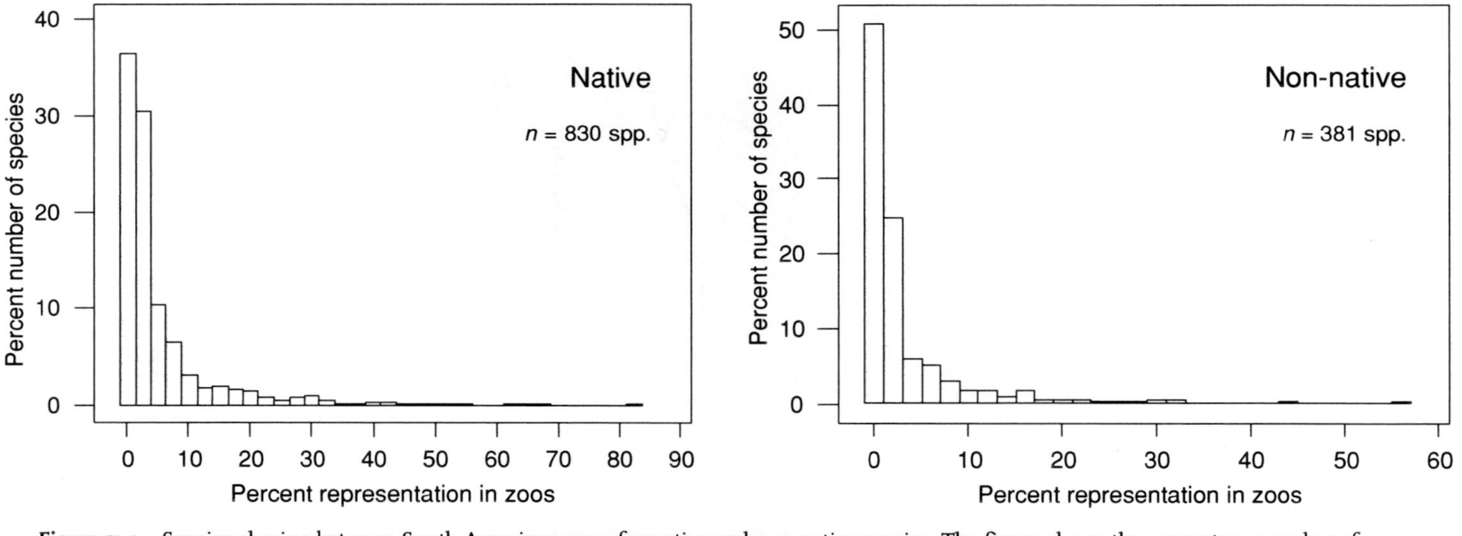

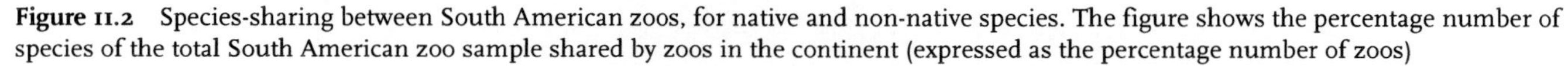

Figure 11.2 Species-sharing between South American zoos, for native and non-native species. The figure shows the percentage number of species of the total South American zoo sample shared by zoos in the continent (expressed as the percentage number of zoos)

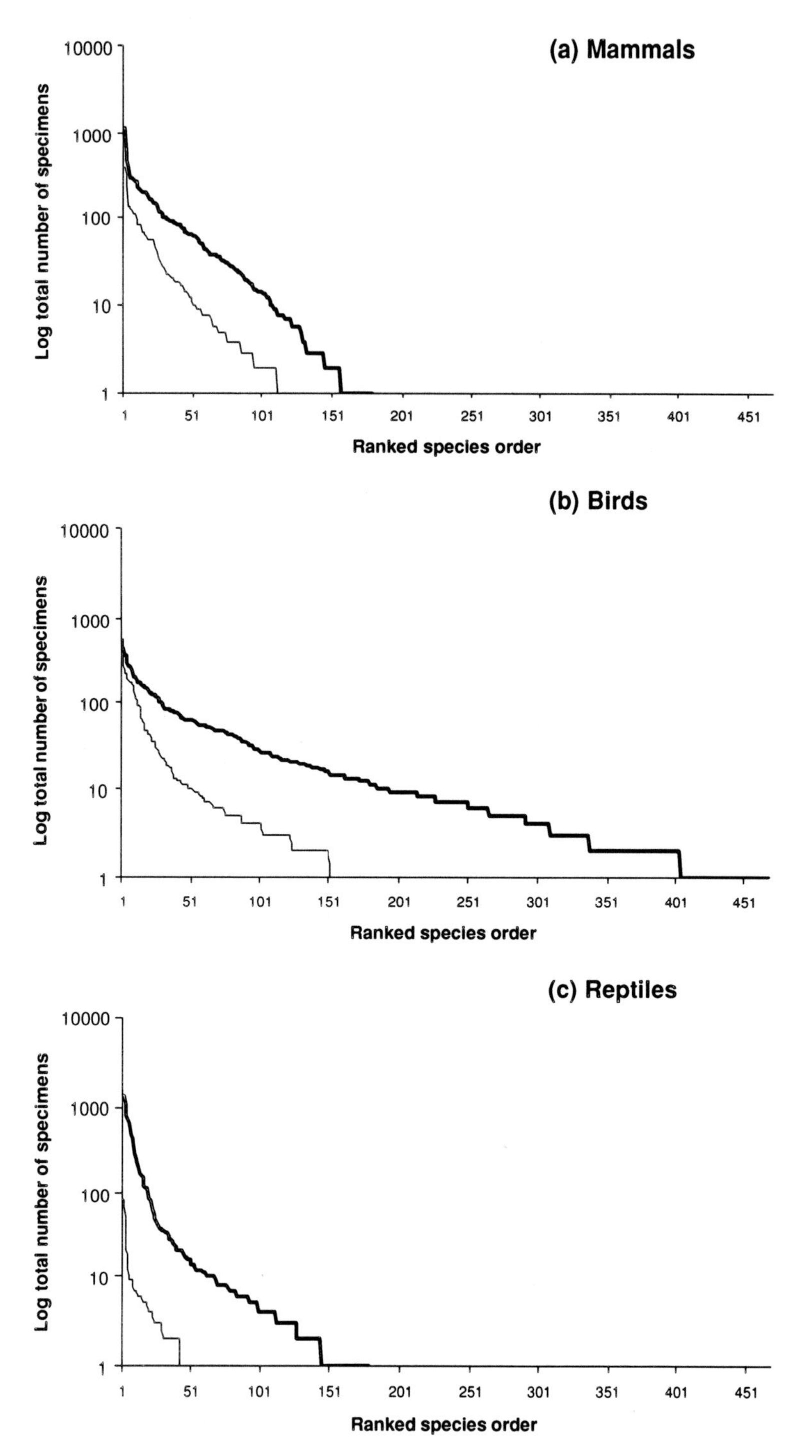

Figure 11.3 Total number of specimens of native and non-native species of mammals, birds, and reptiles found within South American zoos, ranked in descending order. Native species indicated by thicker line and non-native species indicated by thinner lines

Table 11.2

Numbers of threatened non-native and native bird, mammal and reptile species held in South American zoos

	Non-native species					Native species					
	n	Birds	Mammals	Reptiles	Total	*n*	Birds	Mammals	Reptiles	Total	Grand total
Critically endangered	3	2	3		5	10	6	35	8	49	54
Endangered	16	3	137	1	141	20	15	79	30	124	265
Vulnerable	22	42	115	7	164	26	46	135	98	279	443
Lower risk	13	5	72	53	130	25	115	107	21	243	367
Non-threatened	329	655	439	111	1205	754	2572	1706	912	5190	5268

Table 11.3
Origin of specimens of threatened reptiles and amphibians, used for reintroductions, according to threatened status

	Origin of released animals			
IUCN status	*Specialized facility*	*Wild caught*	*Zoo*	Grand total
Critically endangered	6	3	3	12
Endangered	9	0	0	9
Vulnerable	7	9	1	17
Lower risk	3	4	2	9
Non-threatened	11	33	3	47
Grand total	**36**	**49**	**9**	**94**

Source: Unpublished data, IUCN/Reintroduction Specialist Group and Seddon *et al.* (2005).

able to collect adequate data (see also Seddon *et al.* 2005). These groups of animals are small, breed well in captivity, and have a large component of their behavior non-learned; they are generally expected to be easy to reintroduce (Table 11.3). Out of 94 species with data, more than 50% relied on wild-caught individuals that were translocated to a second in-wild destination. Where captive-bred animals were released, only 10% cases came from zoos, and the remainder from specialized facilities of various types. Thus zoos have not been major producers for reintroduction of amphibians and reptiles, although it is notable that zoos did contribute proportionately more for species that were critically endangered.

Many reintroductions are complex operations, requiring diverse skills. The role of providing animals for release is merely one component. Are zoos playing other roles in reintroductions? To explore this, we took on an issue of the RSG newsletter (Soorae 2003) and assessed the organizations involved in the reintroductions described and the institutional affiliations of the authors listed. Table 11.4 shows the data based on a total of 28 papers on reintroductions and 51 authors involved. While only about 20% authors listed a zoo affiliation, and twice as many authors came from conservation organizations, zoos were listed as involved in almost 50% projects. Where zoos were involved in a reintroduction, zoos had provided animals for release in slightly less than half of them. In just over half, the zoos had played other roles. These roles fell into three areas: (1) to provide overseas project administration services; (2) as a provider of funds for the project; or (3) most commonly, as sources of specialized technical expertise, amongst which veterinary assistance was notable. There may be many biases in using

Table 11.4
Analysis of affiliation and origin of reintroduction projects reported in the 2004 Reintroduction Specialist Group newsletter (n = *28 papers, 51 authors)*

Author affiliation	10 from zoos 19 conservation organization 11 government 9 university 2 other/?
Zoo involvement in project?	Yes in 13 No in 15
Where zoos were involved	6/13 zoos provided animals for release 7/13 zoos played other roles

this author information, but it does suggest that zoos are significant contributors to reintroductions, but in ways that use their particular resources and skills rather than providing animals for release.

In conclusion, our analysis suggests that zoos are not keeping the species whose level of endangerment would suggest they might be suitable candidates for reintroduction. Zoos also keep small numbers of animals both individually and collectively. Managing these as a single population requires more active management and transfers between collections. Further, the low numbers suggest that a constant supply of animals for release through multiple years might be hard to meet. Perhaps it is for such reasons that many animals bred for release are produced in dedicated facilities. Overall, while zoos have been responsible for some major successes in reintroductions, they appear neither to have been the sources of animals nor even to have the animal collections to meet the reintroduction priorities of the future.

THE FUTURE FOR REINTRODUCTIONS

Many of the species with successfully established new populations have been medium-to-large mammals or birds. Such species tend to be most vulnerable to extinction, and many reintroductions are efforts to restore one such species into otherwise intact communities. But, we ask whether the pressures on biodiversity will continue this trend, or whether the extinction threats have changed quantitatively and/or qualitatively. If this is so, then the prospects for reintroductions may be different, and zoos should revise their potential roles in conserving biodiversity. There are a number of trends at work today which are directly relevant to future reintroduction, as outlined below.

Disease

The threat of extinction facing many amphibian species from the chytrid fungus is great (Weldon *et al.* 2004). But, our ignorance of the disease, its epidemiology, and ecology is profound. Diseases, such as West Nile virus, are spreading, appearing at new sites, and crossing between species, and there is concern that environmental perturbations are causing new infectious diseases to emerge (Daszak *et al.* 2001). Until these diseases are researched and understood and removed as a threat, reintroduction is premature (IUCN 1998). Further, disease may be endemic in captive populations [which include clinically healthy but infected individuals (Lloyd and Flanagan 1991)], may spread between species in captivity (e.g., Soares *et al.* 2004), and may be passed unknowingly by infected animals between collections (J. Lopez, pers. comm., 2005). Consequently, release from captive populations may never be without risks, especially if releases are used to reinforce small, surviving wild populations.

Invasive species and restoration ecology

Invasive species are one of conservation's greatest problems (Baillie *et al.* 2004). Where areas are invaded by multiple plant and animal species, the effects on native species are complex and interactive. We know little about the dynamics of multiply invaded communities. While the science of restoration ecology is growing fast, the prospects for successful reintroduction into such highly modified communities are greatly reduced. It is a large jump from restoring a single, dominant species to an otherwise intact community to the increasingly prevalent situation of multiple invading species and complex consequences and interactions.

Unsustainable harvesting

Commonly known as the bushmeat issue, the world is now seeing biodiversity reduction over huge areas as more people desperately try to obtain animal protein (Fa *et al.* 2003). This is potentially the greatest short-term extinction factor for whole communities. Reintroduction in these circumstances is not valid until the pressure for animal protein has been alleviated through access to alternative sources. But, even if that day comes, the complexity of entire animal community restoration is currently beyond current human experience (see Maina and Howe 2000).

Global climate change

Over-shadowing all the above factors lies the increasing prospect of significant climate change (Thomas *et al.* 2004). Its impacts on wild populations of plants and animals can rarely be predicted with confidence, but we know that changes in habitat composition or distribution will occur. Animals will respond by emigration and/or changes in their numbers, both increasing and decreasing, and there will be local extinctions. Yet it is the species that are extinct through habitat loss that are least amenable to reintroduction (Balmford *et al.* 1996). It is a corollary, and yet a paradox, that climate change will result in new habitats, communities and with new distribution patterns for species. The concept of natural habitat or historical range may become so anachronistic that the release of animals into semi-wild conditions or into modified habitats may become not just acceptable but the norm, and preferable to keeping animals only in conventional zoo assurance populations.

These phenomena are already happening, and are all likely to increase in severity. Our knowledge base to counter or cope with them is inadequate. Conservation will need all the resources at its disposal to maintain diversity or even attain the CBD's goal of reducing the rate of biodiversity loss by 2010 (Secretariat of the Convention on Biological Diversity 2005). The threats suggest that single species-extinctions and single species restorations will be overtaken by threats which either cannot be countered by reintroduction or will require complex, multi-species restoration efforts for which the current experience base is scanty. Against such scenarios, we ask how zoos can adapt to future conservation needs to show they are committed to being effective agents for conservation.

HOW CAN ZOOS ADAPT TO MEET FUTURE CONSERVATION NEEDS?

Export expertise rather than repatriate animals

While it is evident that the world's biodiversity is being destroyed rapidly (Balmford *et al.* 2003), only for amphibians and birds do accurate and comprehensive data allow precise quantification (Baillie *et al.* 2004). Species declining fast in the wild or on the edge of extinction require intense efforts to save them (Jones 2004). It is inevitable that in light of the prospects for many species, in the face of the trends above, the conservation world will find active interventions on behalf of species increasingly common. The areas or situations in which biodiversity can continue to thrive through being left alone under conditions of relatively pristine nature are increasingly rare.

Yet zoos have been maintaining small populations of rare species for many years, under conditions which are so unnatural that they require great skill, knowledge, and understanding to be successful. These are attributes increasingly needed in the field. Management of small populations in the wild is increasingly sophisticated as techniques that were recently confined to zoo situations are employed in the wild. Where there may be complex manipulations of eggs and birds in the wild or between the wild and captive populations, it is imperative that there is veterinary advice and supervision of conservation efforts to restore wild populations (Greenwood 1996). The distinction between the classic *in situ* and *ex situ* work is breaking down to be replaced by a broad zone of interventions in which the tools and technologies are common to each.

As zoos have found that they have rarely repatriated the products of captive breeding back to home countries for release, they have been exporting knowledge and expertise instead. Thus, Bristol Zoo provides expertise on a continuing basis to Cameroon's primate rescue centre (http://www.bristolzoo.org.uk/). But we would argue that making skills and experience from zoos available to the conservation of wild populations should become more intense. Conventional conservation organizations do not routinely have the skills of handling individual animals, understanding their needs, or the rigorous attention to detail and schedule, all of which effective zoo work requires.

Zoos are individually involved in ever-increasing conservation activities. Yet, the challenge is how to make these contributions add up for maximum and massive conservation impacts. We suggest that zoos could make a major contribution, based on their areas of strength, namely in driving development of a theoretical framework for the recovery and management of small populations, and, second, in the development of the field of translocation biology. The latter covers the range of wild-to-wild transfers to the release of captive-bred animals. The study of translocation biology would be based on (1) accumulated experience, (2) the growing range of opportunities, (3) greater understanding of the developmental psychology of individual animals, (4) the increasing scope for different rearing, learning, and pre-release experiences, and (5) more sophisticated management and monitoring of performance in the wild. With the benefits of greater university links with zoos (Chapter 16), zoos could become the driver in this sector of conservation biology.

Evolve the image and messages from zoos

The most common perception by the visiting public of the conservation role of zoos is "they keep animals that can go back to the wild one day." We contend that this has not been the strength of zoos, and is unlikely ever to be so. Given the record, it would behove the zoo world to disseminate a more subtle and realistic image to the public. The risk of not doing so is that zoos' record in providing animals for reintroduction is challenged, and their legitimacy as conservation organizations then discredited despite their many other contributions to conserving wild populations.

There is enormous scope for zoos to convey multiple messages about animals, their conservation, and society through coherent strategies of education, communications, marketing, and public relations. The sheer number of visitors to zoos gives them a unique headstart for impact.

And, as the WZACS acknowledges, the conservation role of zoos is fully compatible with the need to provide recreational opportunity to visitors. The annual EAZA campaigns demonstrate the collective power of zoos to engage public interest and financial support for a single, time-limited, high-profile cause (Stanley Price *et al.* 2004). Zoos face a serious challenge in transforming the stereotyped and outdated images and associations of the day out at the zoo, which reinforce the menagerie image of small numbers of many species all easily visible.

Evolve animal collections, management, and research

The heyday of confidence in the value of zoo populations as a resource for ultimate return to the wild stimulated research on minimum population sizes, maintaining genetic diversity, and demographic security (Soulé *et al.* 1986). The conclusions have guided collaborative breeding programs since then. But, we ask whether this needs to be revisited, based on subsequent experience with the extent of reintroductions, and on the greater knowledge of the effects of genetic stochasticity, especially the effects of inbreeding on small populations in the wild (Brook *et al.* 2002). Although some scientists have been skeptical about the occurrence of inbreeding depression in wild populations, compelling evidence for it now exists. Of 157 valid data sets across 34 taxa reviewed by Crnokrak and Roff (1999), 90% showed differences indicating that inbreeding was deleterious to reproductive fitness (Frankham 2000).

There have also been advances in molecular genetics techniques, with significant insights into the status of wild populations and conservation units. These should enable a more subtle, taxon-specific approach for zoo populations, each with specific objectives. This would also help optimize the use of limited zoo space (Conway 1986, Balmford *et al.* 1996).

Much zoo-based research is aimed at improving the living conditions and welfare of zoo animals. The more challenging component is to do research on aspects that can directly benefit conservation in the wild. Clearly, there are areas in which the artificial yet controlled conditions of the zoo are advantageous compared to the natural but uncontrolled conditions of the wild. Each has its advantages and opportunities. If zoos can develop their expertise in the interventionist management of small wild populations, many opportunities for zoo-based research will open up. For example, the Durrell Wildlife Conservation Trust holds a population of free-flying Livingstone's fruit bats. Critically endangered in their native Comoros Islands (Mickleburgh *et al.* 2002), it is imperative to study their nocturnal movements, for which wildlife telemetry (such as radio or satellite tracking) will be necessary. Catching bats to attach radios will be difficult; yet the value of each capture from the wild will be increased by prior use of captive bats to develop and test the harness for the radios. Here Durrell Wildlife's captive population in Jersey is providing specific opportunities, the results of which can be applied to the direct benefit of their wild conspecifics. And, the visiting public can observe and learn from seeing harnessed bats on display.

There is a further aspect if zoos wish to realign their activities with conservation. Many zoos have significant conservation programs, yet the composition of their collection has little overlap with the targets of their in-the-wild conservation efforts. While the collection plan may owe much to historical chance and opportunities, and to commercial considerations, we suggest that under the new realism of zoos as conservation bodies, the public will increasingly expect the animals held in zoos to be those conserved by the zoo in the wild. While closely related species or others with similar habits or occupying the same region can be used to make a conservation message about a species not in the collection, it is harder to demonstrate a zoo's conservation work without the relevant species. If a species is in a zoo for purposes of applied research, and not as part of a collaborative effort to maintain a minimum viable number, then very few individuals may be needed. These individuals can then play multiple roles as exemplars for their species in the wild and as subjects for applied research, which will benefit their wild populations.

If zoos can orient their collections to their conservation interests, then we see the power of their conservation messaging increasing simply through greater direct relevance. It will also allow zoos to be more differentiated with fewer species common to many zoos; the charge of being "stamp-collecting" menageries will have less substance. Zoos could also more legitimately claim to be carrying out non-commercial, charitable activities with the public benefit of environmental conservation. Movement in this direction will increase the conservation contribution and image of zoos, and need not diminish the recreational resource that zoos must usually provide to maintain their visitor numbers.

CONCLUSIONS

The zoo world embraces institutions of many shapes and sizes, occupying many, very diverse cultures. Hence, generalizations are difficult and open to charges that they are wrong or do not apply. Nonetheless, we feel it useful to explore at this time one of the accepted wisdoms of the zoo world, and its associated public, namely that zoos hold animals to send back to the wild.

The information for the analyses presented here comes from diverse sources, and is necessarily illustrative rather than comprehensive. But, the conclusions are clear: zoos are not keeping the priority species for reintroduction, nor are species kept in adequate numbers as sources for reintroduction. But, zoos and their staff are significantly involved in reintroductions, in ways that are not merely providing the animals for release.

There have been some successful reintroductions of zoo-bred animals. But, despite the high profile of successes, reintroduction is not a common conservation technique (Stanley Price *et al.* 2004). Reintroduction was a guiding light of convenience in the 1970s and 1980s as zoos strove to demonstrate their conservation credentials in their evolution from zoological parks. Yet it has proved to be a shooting star instead, providing an eye-catching attraction but not long-term illumination for conservation. Conservation itself has changed. The concept of restoring extinct populations is shifting to the conservation imperative of active management of small and declining populations. Integrated management between truly wild and zoo populations is increasing, and we see a major niche and opportunity for the zoo world to take. Grasping the opportunity would demonstrate sincere conservation commitment, with concomitant opportunities for image, messaging to the public, and evolution of species collections and their management objectives.

Given the crisis that biodiversity now faces, and the ambitious goals set for the next few years (Secretariat of the Convention on Biological Diversity 2005), there is powerful reason why the zoo world, with 10 000 institutions worldwide and hundreds of millions of users, must be a significant player in meeting biodiversity objectives. If zoos are perceived as conservation bodies then they will have met the grand aspiration of Miller *et al.* (2004) that "collection-based institutions . . . change their prime function from visitor recreation to the fundamental community service function of conservation." The evolution that we propose is offered as a liberating opportunity to the zoo world. And this evolution is fully consistent with the new WZACS.

ACKNOWLEDGEMENTS

We thank Karen Clark for assistance in collecting and analyzing the IZYB data, and to Moacir Tinoco, Bruno Carpinetti, Itala Yepez Jean-Marc Touzet and Diana Sarmiento for their help with gathering the South American zoo data, especially for Brazilian, Argentinean, and Andean zoos respectively. We are most grateful to the numerous directors of South American zoos who have supplied us with their animal inventory information.

References

Anon (1994). *The Zoo Inquiry*. London: World Society for the Protection of Animals; and Horsham: Born Free Foundation.

Baillie, J. E. M., Hilton-Taylor, C., & Stuart, S. (2004). *2004 IUCN Red List of Threatened Species: A Global Species Assessment*. Gland: IUCN.

Balmford, A., Mace, G. M., & Leader Williams, N. (1996). Designing the ark: setting priorities for captive breeding. *Conservation Biology*, **10**, 719–727.

Balmford, A., Gaston, K. J., Blyth, S., James, A., & Kapos, V. (2003). Global variation in terrestrial conservation costs, conservation benefits, and unmet conservation needs. *Proceedings of the National Academy of Sciences*, **100**, 1046–1050.

Boer, L. E. M. de (1993). Development of coordinated genetic and demographic breeding programmes. In *Creative Conservation, Interactive Management of Wild and Captive Animals*, eds. P. J. S. Olney, G. M. Mace, & A. T. C. Feistner. London: Chapman & Hall, pp. 304–311.

Brook, B. W., Tonkyn, D. W., O'Grady, J. J., & Frankham, R. (2002). Contribution of inbreeding to extinction risk in threatened species. *Conservation Ecology*, **6**, 16. [online] URL: http://www.consecol.org/vol6/iss1/art16/.

Conway, W. G. (1986). The practical difficulties and financial implications of endangered species breeding programmes. *International Zoo Yearbook*, **24/25**, 210–219.

Crnokrak, P. & Roff, D. A. (1999). Inbreeding depression in the wild. *Heredity*, **83**, 260–270.

Daszak, P., Cunningham, A. A., & Hyatt, A. D. (2001). Anthropogenic environmental change and the emergence of infectious diseases in wildlife. *Acta Tropica*, **78**, 103–116.

Fa, J. E., Currie, D., & Meeuwig, J. (2003). Bushmeat and food security in the Congo Basin: linkages between wildlife and people's future. *Environmental Conservation*, **30**, 71–78.

Frankham, R. (2000). Genetics and conservation: commentary on Elgar and Clode. *Australian Biologist*, **13**, 45–54.

Ginsberg, J. R. (1994). Captive breeding, reintroduction and the conservation of canids. In *Creative Conservation, Interactive Management of Wild and Captive Animals*, eds. P. J. S. Olney, G. M. Mace, & A. T. C. Feistner. London: Chapman & Hall, pp. 365–383.

Greenwood, A. G. (1996). Veterinary support for *in situ* avian conservation programmes. *Bird Conservation International*, **6**, 285–292.

Griffith, B., Scott, J. M., Carpenter, J. W., & Reed, C. (1989). Translocation as a species conservation tool: status and strategy. *Science*, **245**, 477–480.

Hilton-Taylor, C. (2000). *2000 IUCN Red List of Threatened Species*. Gland: IUCN.

IUCN (1987). *IUCN Policy Statement on Captive Breeding*. Gland: IUCN, Species Survival Commission, Captive Breeding Specialist Group.

IUCN (1992). *Global Biodiversity Strategy: Guidelines for Action to Save, Study and Use Earth's Biotic Wealth Sustainably and Equitably*. Gland: IUCN.

IUCN (1998). *Guidelines for Re-introductions*. Prepared by the IUCN/SSC Re-introduction Specialist Group. Gland: IUCN, pp. 1–10.

IUDZG/CBSG (IUCN/SSC) (1993). *The World Zoo Conservation Strategy: The Role of the Zoos and Aquaria of the World in Global Conservation*. Executive Summary. Brookfield, IL: Chicago Zoological Society.

Jones, C. J. (2004). Conservation management of endangered birds. In *Bird Ecology and Conservation: A Handbook of Techniques*, eds. W. J. Sutherland, I. Newton, & R. Green. Oxford: Oxford University Press, pp. 269–301.

Kleiman, D. G., Beck, B. B., Dietz, J. M., & Dietz, L. A. (1991). Costs of a reintroduction and criteria for success: accounting and accountability in the Golden Lion Tamarin Conservation Program. In *Beyond Captive Breeding: Re-introducing Endangered Mammals to the Wild. Proceedings of Symposium of the Zoological Society of London*, **62**, 125–142.

Lindburg, D. G. (1992). Are wildlife reintroductions worth the cost? *Zoo Biology*, **11**, 1–2.

Lloyd, M. L. & Flanagan, J. (1991). Recent developments in ophidian paramyxovirus research and recommendations on control. In *Proceedings of the American Association of Zoo Veterinarians Annual Meeting 1991*, Calgary, Canada.

Maina, G. G. & Howe, H. F. (2000). Inherent rarity in community restoration. *Conservation Biology*, **14**, 1335–1340.

McNeely, J. A., Miller, K. R., Reid, W. V., Mittermeier, R. A., & Werner, T. B. (1990). *Conserving the World's Biological Diversity*. Gland: IUCN.

Mickleburgh, S. P., Hutson, A. M., & Racey, P. (2002). A review of the global conservation status of bats. *Oryx*, **36**, 18–34.

Miller, B., Conway, W., Reading, R. P., Wemmer, C., Wildt, D., Kleiman, D., Monfort, S., Rabinowitz, A., Armstrong, B., & Hutchins, M. (2004). Evaluating the conservation mission of zoos, aquariums, botanical gardens, and natural history museums. *Conservation Biology*, **18**, 86–93.

Olney, P. J. S., ed. (1970). *International Zoo Yearbook*, Volume 10. London: Zoological Society of London.
Olney, P. J. S., ed. (1980). *International Zoo Yearbook*, Volume 20. London: Zoological Society of London.
Olney, P. J. S. & Fisken, F. A., eds. (1990). *International Zoo Yearbook*, Volume 29. London: Zoological Society of London.
Olney, P. J. S., Fisken, F. A., & Stanley, H. F. (2000). *International Zoo Yearbook*, Volume 37. London: Zoological Society of London.
Rahbek, C. (1993). Captive breeding – a useful tool in the preservation of biodiversity? *Biodiversity and Conservation*, **2**, 426–437.
Seal, U. S., Foose, T. J., & Ellis, S. (1993). Conservation Assessment and Management Plans (CAMPs) and Global Captive Action Plans (CAPs). In *Creative Conservation: Interactive Management of Wild and Captive Animals*, eds P. J. S. Olney, G. M. Mace, & A. T. C. Feistner. London: Chapman & Hall, pp. 312–325.
Secretariat of the Convention on Biological Diversity (2005). *Handbook of the Convention on Biological Diversity Including its Cartagena Protocol on Biosafety*, Third Edition.
Seddon, P. J., Soorae, P. S., & Launay, F. (2005). Taxonomic bias in reintroduction projects. *Animal Conservation*, **8**, 51–58.
Snyder, N. F. R., Derrickson, S. R., Beissinger, S. R., Wiley, J. W., Smith, T. B., Toone, W. D., & Miller, B. (1996). Limitations of captive breeding in endangered species recovery. *Conservation Biology*, **10**, 338–348.
Soares, J. F., Chalker, V. J., Erles, K., Holtby, S., Waters, M., & McArthur, S. (2004). Prevalance of *Mycoplasma agassizii* and Chelonian herpesvirus in captive tortoises (*Testudo* sp.) in the United Kingdom. *Journal of Zoo and Wildlife Medicine*, **35**, 25–33.
Soorae, P. S., ed. (2003). *Re-introduction News*. Newsletter of the IUCN/SSC Re-introduction Specialist Group, Abu Dhabi, UAE. No. 23: 48 pp.
Soulé, M. E., Gilpin, M., Conway, W., & Foose, T. (1986). The millennium ark: how long a voyage, how many staterooms, how many passengers? *Zoo Biology*, **5**, 101–113.
Stanley Price, M. R. (1989). *Animal Re-introductions: The Arabian Oryx in Oman*. Cambridge: Cambridge University Press.
Stanley Price, M. R., Maunder, M., & Soorae, P. S. (2004). Ex situ support to the conservation of wild populations and habitats: lessons from zoos and opportunities for botanic gardens. In Ex Situ *Plant Conservation: Supporting Species Survival in the Wild*. Washington, D.C.: Island Press, pp. 84–110.
Thomas, C. D., Cameron, A., Green, R. E., Bakkenes, M., Beaumont, L. J., Collingham, Y. C., Erasmus, B. F. N., Ferreira De Siqueira, M., Grainger, A., Hannah, L., Hughes, L., Huntley, B., Van Jaarsveld, A. S., Midgley, G. F., Miles, L., Ortega-Huerta, M. A., Townsend Peterson, A., Phillips, O. L., & Williams, S. E. (2004). Extinction risk from climate change. *Nature*, **427**, 145–148.
Thorne, E. T. & Oakleaf, B. (1991). Species rescue for captive breeding: black-footed ferrets as an example. In *Beyond Captive Breeding: Re-introducing Endangered Mammals to the Wild*, ed. J. H. W. Gipps. Oxford: Clarendon Press, pp. 241–261.
Toone, W. D. & Wallace, M. P. (1994). The extinction in the wild and reintroduction of the California condor. In *Creative Conservation; Interactive*

Management of Wild and Capitve Animals, eds. P. J. S. Olney, G. M. Mace, & A. T. C. Feistner. New York: Chapman and Hall, pp. 411–419.
WAZA (2005). *The World Zoo and Aquarium Conservation Strategy: Building a Future for Wildlife.* Liebefeld-Bern: WAZA.
Weldon, C., du Preez, L. H., Hyatt, A. D., Muller, R., & Speare, R. (2004). Origin of the amphibian chytrid fungus. Emerging Infectious Diseases [serial on the Internet]. 2004 Dec [date cited]. Available from http://www.cdc.gov/ncidod/EID/vol10no12/03-0804.htm.

12

Research by zoos

DAN WHARTON

WHY ZOO RESEARCH?

Zoos represent unique venues for scientific research and conservation action (Kleiman 1992, Hutchins *et al.* 1996, Wharton 1995, Stoinski *et al.* 1998, Fraser 2004, WAZA 2005). The existence of institutions and professionals that are dedicated to the effective keeping of large and small vertebrates from all over the globe has always, to some degree, elicited powerful visions of the unique role that zoos could play in the world of science and conservation. Indeed, zoos have played significant roles in both worlds (Conway 1992, Stoinski *et al.* 1998, Conway *et al.* 2001, WAZA 2005) but the reality has some distance to go to live up to the full vision (Mazur and Clark 2001, Miller *et al.* 2004).

Formal zoo-based research has increasingly informed successful zoo animal keeping for more than 25 years by investigating numerous aspects of reproduction, nutrition, behavior, genetics, health, and other details of animal husbandry (Hutchins *et al.* 1996, Stoinski *et al.* 1998, WAZA 2005). Zoo-based research and accumulated knowledge are also important to wildlife management and conservation as implied by the citing of numerous zoo research articles in basic works on wildlife biology (Hutchins *et al.* 1996; also see Nowak 1999). Many zoos now seek more direct links to field conservation by conducting research outside the zoo. It is suggested, although not widely acknowledged, that the strongest and most effective links between zoos and aquariums and field programs will be based on a unified vision of successful wildlife conservation and the unique capacities that zoo biologists and field biologists bring to the process (WAZA 2005). Already *ex situ* scientific research is asking the questions that inform conservation processes while zoos are developing the communication infrastructure for pursuing (and efficiently sharing) the answers. For example, the zoo priority for scientifically managing small but stable populations of rare and endangered species has become the obvious point of departure for laboratory investigation of relevant questions in organismal biology, including genetics, nutrition, ethology, reproduction, preventative medicine and pathology, all essential to individual animal health and reproduction (WAZA 2005).

Perhaps of even greater importance, zoos are ever-more organized around the cultural institution imperative to engage the public by informing and educating it (Fraser 2004). As the gap widens between the needs of wildlife conservation and the resources to fulfil those needs, it is clear that someone needs to package the message a lot better than it has been to date. This process, in itself, is an area in need of scientific investigation although much less obvious to zoo managers than the biological intricacies of wildlife conservation. Creative capacity building in both wildlife research and care plus conservation results-oriented communication could pre-adapt zoos to become highly effective conservation organizations in the 21st century (Conway 2003, WAZA 2005).

A BRIEF HISTORY OF ZOO RESEARCH

Although the "amusement factor" has played a significant role in the development of support for zoos (Kisling 2001), the link to natural history science has always been robust (Hutchins *et al.* 1996). Exotic zoo specimens not only make their species intimately known to millions of people who might not otherwise experience them, but their living presence in urban centers makes them available to scientists for behavioral studies and, potentially, for anatomical studies and/or museum display post-mortem. This dual purpose is most prominently recognizable in the early publications of the *Proceedings of the Zoological Society of London*. Meanwhile, the Menagerie du Jardin des Plantes in Paris (established in 1794, it is the oldest public institution exhibiting live animals) is still considered a part of Paris's Museum National d'Histoire Naturelle (Chavot and Bomsel 2001). The New York Zoological Society was founded in 1895, and by 1903 young curator William Beebe was already launching expeditions to study wildlife in nature (Bridges 1974, Goddard 1995). As he put it in his first publication in *Zoologica*: "We should keep them all in mind . . . [that] the collecting of thousands of skins will be of no service nor will the study of those now in our museums be of any direct use. We must have careful and minute tabulation of the ecological conditions under which the phenomena under discussion appear" (Bridges 1974). So was launched the concept of zoos as a force for studying wildlife in natural habitats.

By the end of the 19th century, the zoo as a cultural amenity and scientific resource was further augmented by an ever-deepening wildlife crisis. Rare and endangered species driven to near or total extinction could sometimes be found in good condition in the world's zoos. Although not all of these cases resulted in long-term success stories (e.g., passenger pigeon, quagga, and Carolina parakeet), the examples of the European bison, the Père David deer, and the Mongolian wild horse were impressive enough to place zoos firmly in the world of wildlife conservation (Wharton 1995) even though zoos and human society in general were still relatively oblivious to the downward spiral in which much of the natural world was headed (Dowie 1997).

The struggle since that time has been to make the conservation accomplishments of zoos less a matter of luck and more a matter of intent (Wharton 2006). The application of population biology and some reproductive technologies to zoo breeding programs has revolutionized zoos' potential to manage captive populations for the long-term (Wharton 1995, Wildt 2003, WAZA 2005). Genetically and demographically healthy populations of captive wildlife represent all the traditional value of zoo specimens – recreation, science, education, and conservation – plus the modern value of being the auxiliary populations that keep zoos in business with diminished dependence on wild-caught specimens. The relevance of scientific research to zoo keeping has become ever more obvious with each passing decade as zoo-specific questions have arisen in all aspects of animal management and zoo-based conservation (Ralls and Ballou 1986, Kleiman 1992, Beck *et al.* 1994, Asa and Porton 2005, WAZA 2005).

PUBLISH OR PERISH: THE STRENGTHS AND LIMITATIONS OF CURRENT ZOO RESEARCH

Publication of high-quality research by zoo scientists (and other scientists using the zoo as a research venue) has never been especially problematic (Hutchins *et al.* 1996). Many science and museum journals deem the scientific insight gained in zoos as unique and well qualified for publication. The *Proceedings of the Zoological Society of London*, *Der Zoologischer Garten*, and the *International Zoo Yearbook* are, or have been, zoo-specific publications to a great degree, often presenting varying proportions of empirical research along with trade-useful technical reports. Peer review of scientific work is historically, and even currently, most common in mainstream science journals outside the zoological garden circles (Hutchins *et al.* 1996, Stoinski *et al.* 1998).

By 1982, research in zoos was expanding and so well embraced by the profession that it was deemed essential that it should have its own peer-reviewed science journal, *Zoo Biology*, dedicated to presenting the scientific findings of zoo biologists. By 1996, over 11% of zoo research publications listed in the annual AZA Annual Report on Conservation and Science were published in *Zoo Biology* (Hutchins *et al.* 1996). *Zoo Biology* was also established in order to underscore the importance of zoo research to the larger world of biology and conservation. In documenting the research activities of 120 zoos, Finlay and Maple (1986) found that wildlife conservation was among the many reasons why zoos and aquariums were expanding their research capacity, although increased interest in animal welfare and the ability to solve animal management problems were also primary.

Zoo research covers such a broad range of topics that much of it remains attractive to mainstream journals such as those devoted to primatology, nutrition, and wildlife medicine (Hutchins *et al.* 1996). One issue regarding zoo research that is particularly relevant to the need for a zoo science journal is centered on both the constraints and opportunities that are specific to zoos. Rare and endangered animal species, by definition, often present very few opportunities to learn much about them. Daily management priorities and animal well-being also take precedence over strictly controlled research methodologies; therefore, some creative logistical and statistical strategies are often employed to make the most of the limited access to the rare and endangered. Appropriate temperament for the zoo research setting (Kleiman 1996) and good training in research techniques will yield scientifically defensible conclusions despite some of the limitations that the zoo setting imposes (Crockett 1996). Both the editorial board of *Zoo Biology* and its extensive list of reviewers are sensitive to the overall need to report on all increments of scientific knowledge among thousands of different species currently maintained in zoos.

Finlay and Maple (1986) reported that approximately 70% of zoos were conducting some form of research in the 1980s. By 1998, this number increased to 88%, including increases in zoos with relatively small attendance figures (Stoinski *et al.* 1998). This same 1998 study cited conservation studies by 59% of respondents, exceeded by reproduction and behavior studies coming in at 75% and 85% respectively. Although a somewhat different question, the majority of respondents supported domestic or international field research projects, with only 21% supporting neither.

The 1998 study by Stoinski *et al.* also reported that the percentage of zoos expanding programs of zoo research had actually decreased since Finlay and Maple's study of 1986 – for all but two of the largest zoos surveyed. This could suggest that the discretionary limit for pursuing science is approaching, although "discretionary limits" is perhaps just another way of saying that the justification for pursuing programs of research is not well developed. Mazur and Clark (2001) observed that "the views of senior management currently dominate zoo policy and practices, in what may be described an ideological tug-of-war between the business-oriented concerns of senior management and the animal/education-based concerns of operational staff." Mazur and Clark specifically advocate the use of the policy sciences by zoos and policy analyses, noting that "the policy challenge for zoos is to ensure that organizational designs, cultures, and operations promote ecological values . . ." It might very well be that the best way to promote research and conservation in zoos is to turn a scientific eye to the organizational cultures of the zoos themselves. Cost/benefit analyses with the relatively long view in mind could be successful in assigning monetary value to scientific research well beyond its straightforward capacity to attract some research-oriented contracts, contributions, and gifts. Examined the other way around, how solid will support remain for cultural institutions charged with the interpretation of organismal biology, animal health and welfare, and the wildlife conservation crisis itself if they have insufficient, intrinsic scientific capacity built within? As pointed out by Benirschke (1996), most industries allocate 3% to 5% of their budgets to research and development, a commitment not yet reached by the zoo profession.

ZOO BIOLOGY VERSUS CONSERVATION BIOLOGY

Zoos have been conceived, and operated primarily, as cultural institutions (Kisling 2001, Fraser 2004) and, as such, it is no surprise that scientific investigation is often limited to questions that preserve collections and enhance the visitor experience. Therefore, the cultural institutional role is consistently emphasized first, while programs of any conservation significance are a far-second despite significant achievement in taking on numerous and successful conservation projects (Hutchins *et al.* 1996, Conway *et al.* 2001). How can zoos improve upon this for a more solid and comprehensive role in wildlife conservation? Can zoos be organized to have serious conservation significance as the first order of business? Can zoos support the science that a strong conservation agenda would require? Would zoos as conservation organizations make a better cultural institution for the communities that zoos serve?

It could be argued that there are four legs on which a zoo, as a legitimate conservation organization, can easily stand without invoking the "discretionary limits" of science and conservation within the cultural institution model. All four have been pursued to some degree; but never as a primary institutional priority (see Conway 2003):

1. **Captive or conservation breeding of endangered species.** Already demonstrated to be an effective conservation strategy for many of the most endangered species, zoos could easily do more and fulfil their local commitment to display some of the most compelling natural history stories on the planet. Currently, conservation breeding is easily displaced by other exhibition priorities wherever logistical resistance to obtaining, managing and/or displaying the endangered counterpart exists. The cultural institution paradigm supports this approach while the conservation organization would make it a priority to address the resistance.
2. **Scientific investigation of species biology, natural history, and new field technology in direct response to field needs.** While requiring good communication with a field conservation organization, this kind of science falls well within the range of research priorities that zoos are likely to have for better animal management. The zoo as a conservation organization would find that this kind of scientific relationship will fit with the business model. Not only does this open doors for additional, collaborative funding but the developing partnerships can allow for the "two way street" in which science from the field that is relevant to zoo animal management could become more readily available to the zoo.
3. **Well-developed articulation and dissemination of the wildlife conservation agenda as a form of "conservation marketing."** The cultural institution invests in educational material and educational processes as ends in themselves. The conservation organization uses education as a means to an end and invests in a conservation outcome. Unlike conservation education, conservation marketing is far less concerned with educational processes that cannot demonstrate a clear result. In the age of the Internet, zoos are perhaps the most qualified of all potential conservation organizations to develop and succeed in the promotion of conservation as a top human priority.
4. **Well-developed education of the influential with emphasis on political leaders and philanthropists.** Similar to the above, this presupposes that exhibits and interpretives that can move the influential will also better

serve the casual visitor. Such an approach would also have dedicated programming for ensuring that leaders see the exhibits and get the message either through the zoo's own program or that of a collaborating conservation organization.

Inevitably, all four of the above require scientific staff to continue ongoing development, improvement, and assessment of these programs. The zoo as a conservation organization first and foremost very quickly resolves Mazur and Clark's (2001) "ideological tug-of-war" between the business-oriented concerns of senior management and the zoo's ecological values. Where ecological values, i.e., conservation programming, are the means to a better cultural organization, then the science and the staffing that supports it becomes indispensably linked to the cost of doing business (see Benirschke 1996).

PRIORITIES AND NICHE FOR THE FUTURE

The science of species conservation *in situ* has advanced rapidly in the last decade with numerous new concepts both in strategy and definition for saving species (Fraser and Bernatchez 2001, Redford and Feinsinger 2001, Sanderson *et al.* 2002, Agapow *et al.* 2004, O'Brien *et al.* 2005, Soulé *et al.* 2005). The science of conservation breeding, if relegated to its proper place as a conservation strategy, is as *in situ* an objective as any other kind of wildlife protection (Wharton 2006). In fact, the economics of *in situ* species conservation strongly indicate that the critically endangered should be a lesser priority than species whose relatively early, downward spiral can be corrected with a reasonable monetary investment (Kinnaird and Redford 2005). This further emphasizes the unique role that zoos can play with the critically endangered species. Zoos, with their hands-on role with individual animals, are already invested in meticulous, specialized attention to small populations and the ultra-protection of captivity. Not only is this a division of labor that makes economic sense, it potentially diffuses the philosophical debate inherent in any kind of "economic triage" associated with current-day realities in conservation decision-making.

It is important to note that like the rest of modern society, zoological gardens are great consumers of science and its products. Certain advances in medicine, husbandry, and communication will come to us whether or not we have one single scientist in the profession. For example, the development of a wide range of antibiotics has provided numerous benefits to zoo animal keeping. Similarly, the science of anesthesia in both humans and domestic animals has provided the enormous breakthroughs that make zoo animal medical treatment a matter of routine (Bush 1996). However, there are some applications of mainstream science and some scientific questions in and of themselves that will go unanswered without our zoo scientists. Minimally, it requires scientific training in many cases to study the actual application of new medical and other animal management resources, from the pharmokinetics of drugs given to exotic species to nutrient uptake of well-formulated feeds where efficacy has been studied only in domestic animals. Most uniquely, there is no other profession so invested in behavioral studies as they apply to animal management and welfare. And, of course, the zoo itself as a societal change agent is a matter of study for zoos themselves. We need more scientists and a firm rationale for supporting them as we move into the future.

NEW RESEARCH OPPORTUNITIES IN ZOO CONSERVATION

Many of the obvious areas of research to do with the management of zoos and their varied animal collections are illustrated in assessments produced at the 58th Annual World Association of Zoos and Aquariums (WAZA) Conference in Costa Rica (Table 12.1). The topics in this list represent areas in which greater knowledge can contribute to better animal care. In some cases, zoo research in these areas can contribute directly to wildlife conservation. This list also represents a great resource for mutually advantageous relationships with universities wherein the conservation and institutional development agendas of zoos can provide unique research opportunities for students. Where research agendas are thoughtfully robust and mentoring is well developed, students can help produce high-quality work for peer-reviewed publication. In cases where agendas are not well developed, student research projects will generally fall into the category of training and should not be expected to be of publication quality. These training projects can help cement relationships with universities and fulfil the expectation that zoos will help communities and their educational institutions. The only downside is in the potential for confusing student training alone with advancing a highly strategic research agenda.

If zoos are able to surmount the complexities of internal policies, outdated mission statements, finances, and untested assumptions to make the transition to full-time conservation organization, traditional research agendas will be revisited with new insight and renewed vigor in achieving highly practical objectives. New research agendas will appear, especially as the

Table 12.1
Basic and applied research in zoos: primary disciplines involved in studying priority topics

	Discipline											
Topic	Anatomy/morphology	Behavior	Ecology	Physiology	Systematic biology	Genetics	Nutrition	Population biology	Biogeography	Veterinary medicine	Social science	Education
Husbandry	×	×	×	×			×	×		×		
Welfare	×	×	×	×		×	×			×		
Population management		×	×	×	×	×		×	×	×		
Disease	×	×	×	×		×	×	×	×	×		
Aging	×	×	×	×		×	×	×		×		
Reproduction	×	×	×	×		×	×	×	×	×		
Domestication	×	×	×	×	×	×	×	×				
Visitor research		×									×	×
Nutrition	×	×	×	×	×	×	×	×		×		
Taxonomy	×	×		×	×	×			×			
Disturbed behavior		×	×	×		×	×			×		
Life history	×	×	×	×		×	×	×	×	×		
Marketing											×	×
Identification	×	×		×	×	×						
Biotechnology	×			×		×	×			×		
Biomaterials banking				×	×	×		×		×		
Contraception	×	×		×						×		

58th Annual World Association of Zoos and Aquaria (WAZA) Conference 2003, San Jose, Costa Rica.

sociological questions surrounding conservation objectives become more important to all of the conservation movement. Here are some examples:

Reproduction science to date has opened the door to some interesting possibilities. We have reached the moment when research on the application of reproduction science and cryopreservation of gametes and embryos can help us organize this science around some specific programmatic goals to do with small population management, and perhaps even in wildlife restoration programs. Precise data on the quantitative relationship of cryo-banked gametes and embryos to living offspring and surviving adults set the stage for much more powerful genetic management capability. "Frozen genetics" has a great payback in the zoo's ability to maintain viable populations of wildlife with far fewer animal transfers and with fewer animals (Soulé *et al.* 1986, Dresser 1988, Wildt *et al.* 1997). Not only does this have the potential to make population management less expensive and risky, it also makes it more precise for greater periods of time. It presents the opportunity to manage more species in smaller captive populations. In turn, this provides for the management of more taxa, including subspecies, and overall greater collection diversity for more conservation outreach (Wildt *et al.* 1997).

Although still not completely perfected, cloning has the potential for additional efficiency in frozen genetics (Ryder and Benirschke 1997, Ryder 2002), especially if the process for stimulating certain somatic cells to function like a fertilized egg is fairly universal for all species. If cloning is therefore much less species-specific than is gamete retrieval and/or *in vitro* fertilization, embryo transfer, etc., it would clearly provide a path of lesser resistance in a field where cryopreservation concern is potentially focused on hundreds, if not thousands, of different species. Already, the preservation of tissue from which clonable cells can be obtained is much more singular in process in contrast to the specificity of conditions under which gametes and embryos must be cryopreserved (Wildt *et al.* 1997). On the conservation front, and in anticipation of future perfection of cloning, programs of systematic tissue collection of the rare and endangered should be seen as a "must do" by all zoos (Ryder and Benirschke 1997).

Now turning the focus to the non-biological, another very interesting area of investigation is in the use of tractable animals. Are we correct about the species that appear to enjoy regular, human contact? Is this only domestic animals? Have we overlooked some species that might be better or at least add justifiable variety? This is perhaps an area in which untested assumptions set the parameters for the animals used for human contact. Perhaps it would not matter if we have overestimated the power of tractable animals to galvanize concern for animals (and, specifically, for wildlife). It is entirely possible that we greatly underestimate the role of tractable animals in forging our alliances with the zoo-going public and the support for conservation that this engenders. Serious research on this otherwise "light" topic might be surprising, especially given the anecdotal evidence that most zoo professionals and many other conservation biologists developed their resolve to work with wildlife in response to personal contact with living animals.

Although studies of the effects of educational mechanisms can be extremely difficult (National Research Council 2002), we have some potential opportunities to understand better the effect of zoo presence by studying countries and cities with zoos versus those culturally similar countries and cities without zoos. The popularity of zoos as a venue for education and entertainment provides ample evidence that zoos have a good degree of importance in human cultural life; however, there are numerous variables (e.g., nature television and cinema, popular publications, even pet-owning) that confound our ability to sort out the direct effect of zoos on eliciting meaningful conservation sympathies in its visitors. This is an important and difficult question but should be considered fundamental to future development of zoos with a serious conservation mission.

Practical arguments for the preservation of nature are not difficult to find (Wilson 2003) although it can certainly be argued that the message has not been effectively received. Among the practicalities of preserving an intact nature are perhaps the advantages that nature provides the human mind and imagination. Human mental health is often associated with the degree to which people have access to nature (Kellert and Wilson 1993). Perhaps urban zoos can study the need for nature in human lives, and perhaps become a voice for better urban planning, for inclusion of green spaces and perhaps even appropriate floral and faunal assemblages to create wildlife-friendly communities. Can we engage the social sciences to test whether or not nature and wildlife are an underestimated spiritual goldmine? If so, can we further study the process by which this understanding is translated into meaningful conservation action? In the final analysis, the simple lesson that humankind has yet to learn can be found in the words of a poet: "A culture is no better than its woods" (Auden 1976).

CONCLUSIONS AND RECOMMENDATIONS

Not only can scientific research be firmly reflective of the role of the zoos in conservation, it can help define the role itself. As noted above, the process by which zoos become genuine conservation organizations (with full financial and moral support to be such) will require policy changes and continuing feedback on effectiveness. Meanwhile, the zoo emphasis on conservation and its relationship to the "better cultural institution" will require empirical evidence. The unique role that zoos can play in conservation action also requires study. Is it appropriate for zoos to mimic, to some degree, the role of the field conservation organization? What are the evident flaws in the conservation industry-wide concepts of "conservation marketing?" Can zoos address the gaps that traditional conservation organizations have failed to fill? The enhancement of zoo strengths would appear to show far more promise in accomplishing the conversion of zoos to conservation organizations than partial adoption of strategies used by other kinds of conservation organizations. These assumptions remain largely untested. But we can say that no other kind of conservation organization is designed to reach and teach millions of subscribers in the way that zoos can. No other profession has such a richness of tools and insight for managing individual animals and the zoo populations of which they are a part. The zoo as a science-based conservation organization will become ever more compelling as zoos become much more strategic in finding that powerful balance between conservation significance and the new kind of cultural institution they are pre-adapted to become (Mazur and Clark 2001, Conway 2003, Miller *et al.* 2004).

References

Agapow, P. M., Bininda-Edmonds, O. R. P., Crandall, K. A., Gittleman, J. L., Mace G. M., Marshall, J. C., & Purvis, A. (2004). The impact of species concept on biodiversity studies. *Quarterly Review of Biology*, **79**(2), 161–179.

Asa, C. S. & Porton, I. J., eds. (2005). *Wildlife Contraception*. Baltimore, MD: The Johns Hopkins University Press.

Auden, W. H. (1976). Bucolics: II Woods. In *W. H. Auden Collected Poems*, ed. E. Mendelson. New York: Random House.

Beck, B. B., Rapaport, L. G., Stanley-Price, M. R., & Wilson A. C. (1994). Reintroduction of captive born animals. In *Creative Conservation: Interactive Management of Wild and Captive Animals*, eds. P. J. S. Olney, G. M. Mace, & A. T. C. Feistner. London: Chapman and Hall, pp. 265–286.

Benirschke, K. (1996). The need for multidisciplinary research units in zoos. In *Wild Mammals in Captivity*, eds. D. G. Kleiman, M. E. Allen, K. Thompson, & S. Lumpkin. Chicago, IL: University of Chicago Press, pp. 537–544.

Bridges, W. (1974). *Gathering of Animals*. New York: Harper and Row.

Bush, M. (1996). Methods of capture, handling, and anesthesia. In *Wild Mammals in Captivity*, eds. D. G. Kleiman, M. E. Allen, K. Thompson, & S. Lumpkin. Chicago, IL: University of Chicago Press, pp. 25–40.

Chavot, P. & Bomsel, M. (2001). Menagerie du Jardin des Plantes. In *Encyclopedia of the World's Zoos*, ed. C. Bell. Chicago, IL: Fitzroy Dearborn Publishers, pp. 800–802.

Conway, W. (1992). The conservation park: a new zoo synthesis for a changed world. In *AAZPA Annual Conference Proceedings*, Toronto.

Conway, W. (2003). The role of zoos in the 21st century. *International Zoo Yearbook*, **38**, 7–13.

Conway, W. G., Hutchins, M., Kapetanakos, Y., & Paul, E., eds. (2001). *The AZA Field Conservation Resource Guide*. Georgia: Zoo Atlanta.

Crockett, C. (1996). Data collection in the zoo setting, emphasizing behavior. In *Wild Mammals in Captivity*, eds. D. G. Kleiman, M. E. Allen, K. Thompson, & S. Lumpkin. Chicago, IL: University of Chicago Press, pp. 545–565.

Dowie, M. (1997). *Losing Ground: American Environmentalism at the Close of the Twentieth Century*. Cambridge, MA: The MIT Press.

Dresser, B. L. (1988). Cryobiology, embryo transfer and artificial insemination in ex situ animal conservation programs. In *Biodiversity*, ed. E. O. Wilson. Washington, D.C.: National Academy Press, pp. 296–308.

Finlay, T. W. & Maple, T. L. (1986). A survey of research in American zoos and aquariums. *Zoo Biology*, **5**, 261–268.

Fraser, D. J. & Bernatchez, L. (2001). Adaptive evolutionary conservation: towards a unified concept for defining conservation units. *Molecular Ecology*, **10**, 2741–2752.

Fraser, J. (2004). Museums and civility. *Curator*, **47**(3), 252–256.

Goddard, D., ed. (1995). *Saving Wildlife: A Century of Conservation*. New York: Wildlife Conservation Society.

Hutchins, M., Paul, E., & Bowdoin, J. M. (1996). Contributions of zoo and aquarium research to wildlife conservation and science. In *The Well-Being of Animals in Zoos and Aquarium Sponsored Research*, eds. G. M. Burghardt, J. Bielitzki, J. Boyce, & D. Schaeffer. Greenbelt, MD: Scientists Center for Animal Welfare, pp. 23–39.

Kellert, S. R. & Wilson, E. O., eds. (1993). *The Biophilia Hypothesis*. Washington, D.C.: Island Press.

Kinnaird, M. & Redford, K. (2005). WCS-International and Development of a Species Strategy. Unpublished report, April 2005.

Kisling, V. N., ed. (2001). *Zoo and Aquarium History*. London: CRC Press.

Kleiman, D. G. (1992). Behavioral research in zoos: past, present and future. *Zoo Biology*, **11**, 301–312.

Kleiman, D. G. (1996). Introduction: captive mammal research. In *Wild Mammals in Captivity*, eds. D. G. Kleiman, M. E. Allen, K. Thompson, & S. Lumpkin. Chicago, IL: University of Chicago Press, pp. 529–530.

Mazur, N. A. & Clark, T. W. (2001). Zoos and conservation: policy making and organization challenges. In *Species and Ecosystem Conservation: An Interdisciplinary Approach*, eds. T. W. Clark, M. J. Stevenson, K. Ziegelmayer, & M. B. Rutherford. Bulletin Series, Yale School of Forestry and Environmental Studies. Number 105. New Haven, CT: Yale University, pp. 185–201.

Miller, B., Conway, W., Reading, R. P., Wemmer, C., Wildt, D., Kleiman, D., Monfort, S., Rabinowitz, A., & Hutchins, M. (2004). Evaluating the conservation mission of zoos, aquariums, botanical gardens and natural history museums. *Conservation Biology*, **18**(1), 1–8.

National Research Council (2002). *Scientific Research in Education*. Washington, D.C.: National Academy Press.

Nowak, R. M. (1999). *Walker's Mammals of the World*, Sixth Edition. Baltimore, MD: The Johns Hopkins University Press.

O'Brien, T. G., Kinnaird, M. F., Groom, M., & Coppolilo, P. B. (2005). Ecologically functional populations: revisiting how much is enough. In *Principles of Conservation Biology*, Third Edition, eds. M. Groom, G. K. Meffe, & C. R. Carroll. Sunderland, MA: Sinauer, pp. 435–438.

Ralls, K. & Ballou, J. (1986). Captive breeding programs for populations with a small number of founders. *Trends in Ecology and Evolution*, **1**(1), 19–22.

Redford, K. H. & Feinsinger, P. (2001). The half-empty forest: sustainable use and the ecology of interactions. In *Conservation of Exploited Species*, eds. J. D.

Reynolds, G. M. Mace, K. H. Redford, & J. G. Robinson. Cambridge: Cambridge University Press, pp. 370–399.
Ryder, O. A. (2002). Cloning advances and challenges for conservation. *Trends in Biotechnology*, **20**(6), 231–232.
Ryder, O. A. & Benirschke, K. (1997). The potential use of "cloning" in the conservation effort. *Zoo Biology*, **16**, 295–300.
Sanderson, E. W., Redford, K. H., Vedder, A., Copolillo, P. B., & Ward, S. E. (2002). A conceptual model for conservation planning based on landscape species requirements. *Landscape and Urban Planning*, **58**, 41–56.
Soulé, M., Gilpin, M., Conway, W., & Foose, T. (1986). The millennium ark: how long a voyage, how many staterooms, how many passengers? *Zoo Biology*, **5**, 101–113.
Soulé, M. E., Estes, J. E., Miller, B. J., & Honnold, D. L. (2005). Strongly interacting species: conservation policy, management, and ethics. *Bioscience*, **55**(2), 168–176.
Stoinski, T. S., Lukas, K. E., & Maple, T. L. (1998). A survey of research in North American zoos and aquariums. *Zoo Biology*, **17**, 167–180.
WAZA (2005). *The World Zoo and Aquarium Conservation Strategy: Building a Future for Wildlife*. Liebefeld-Bern: WAZA.
Wharton, D. (1995). Zoo breeding efforts: an ark of survival? *Forum for Applied Research and Public Policy*, Spring.
Wharton, D. (2006). Miracle under fire. In *State of the Wild*, ed. S. Guynup. Washington, D.C.: Island Press, pp. 256–264.
Wildt, D. (2003). The role of reproductive technologies in zoos: past, present and future. *International Zoo Yearbook*, **38**, 111–118.
Wildt, D. E., Rall, W. F., Critser, J. K., Monfort, S. L., & Seal, U. S. (1997). Genome resource banks: living collections for biodiversity conservation. *Bioscience*, **47**(10).
Wilson, E. O. (2003). What is nature worth? In *Annual Editions: Environment*, ed. J. L. Allen. Guilford, CT: McGraw-Hill/Dushkin, pp. 121–130.

13

Conservation medicine

JOHN C. M. LEWIS

INTRODUCTION

In the past there has been a strong tendency to consider conservation science and the medical sciences as distinct disciplines, only occasionally overlapping and even then peripherally. However, a growing realization that the health of any individual organism, population of organisms or even entire ecosystems is largely a function of the general health of the environment has led to the emergence of a new and eclectic discipline – that of conservation medicine. The basic principle that an animal's health is affected by its environment is familiar to both veterinary and medical practitioners. However, conservation medicine represents a radical and profound expansion of this approach to include consideration of the widest range of environmental and ecological factors in attempts to understand changing patterns of health and disease, and how resultant health problems may be predicted and addressed.

Largely due to the activities of man, the global environment is being dramatically altered at an ever accelerating rate and as a result disease patterns are constantly changing. Conservation medicine focuses on the complex interactions among changes in climate, habitats, biodiversity, and land use; environmental pollution; the emergence and re-emergence of pathogens and parasites; the health of animal communities both free-living and domestic; and human health and social behavior. Understanding the impact of these factors on the health of ecosystems is a prime concern for conservation medicine practitioners, as are the development of practical solutions to ecological disease problems and the promotion of ecological health through education and public policy development (Aguirre *et al.* 2002a). Thus, when compared to a traditional medical model of disease, conservation medicine represents a shift from addressing individual organism or species-specific problems to addressing those of populations and ecosystems with the primary goal being overall ecological health – a goal shared with that of all other conservation disciplines.

Conservation medicine is not the exclusive domain of zoo and wildlife veterinarians, although they may have a primary role. To understand the cause(s) of any particular disease in a community of free-living organisms, consideration of a wide range of ecological parameters is essential and therefore conservation medicine must involve an equally wide range of scientific expertise. Conservation biologists, epidemiologists, ecologists, veterinarians, zoologists, animal behaviorists, educators, botanists, medical practitioners, microbiologists, immunologists, geneticists, endocrinologists, toxicologists, and even social and political scientists all have a part to play. Conversely, as it is equally impossible to understand the ecology of any particular ecosystem without consideration of the health and disease processes affecting its inhabitants, other branches of conservation science must embrace principles of conservation medicine if our goal of ecological health is to be achieved.

ECOLOGICAL CHANGE AND DISEASE

There can be little doubt that since the middle of the 20th century unprecedented local and global environmental change has occurred due to the activities of an ever increasing human population. Species extinctions, loss of biodiversity, habitat degradation, global climate change, over-exploitation of limited resources, and widespread chemical pollution are all familiar characteristics of the Earth's recent history. The genetic composition of many species has also been altered through human activity that has reduced population sizes or restricted gene flow between populations. Taken together, these factors have had a profoundly negative effect on the health of terrestrial and marine ecosystems. No attempt is made here to comprehensively document the health implications of global environmental degradation, but consideration of a few examples of how disease patterns alter in response to environmental change serves as a sobering illustration. Although the focus here is on diseases of animals, the principles apply equally to plant disease.

The evolution of disease is not a new phenomenon – ecosystems have always been in a state of flux and as a result the incidence and impact of disease has never remained static. Changes in human populations and behavior have affected disease patterns for thousands of years (Rapport *et al.* 2002). As agricultural settlements began to replace hunter-gatherer societies some 10 000 years ago, infectious and parasitic diseases became more significant due to the increased level of contact between humans and between humans and their livestock. The crowded and unhygienic conditions of early cities some 7000 years ago led to a rise in intestinal and respiratory infections, and the development of major trading routes allowed the wider spread of infectious disease between human populations. In the 17th and 18th centuries, the colonial activities of Europeans spread measles, smallpox, and other fatal infectious diseases to the immunologically naive populations of the new world with devastating consequences.

Newly identified infectious diseases, or those that are expanding in geographical range, incidence or host range are known as emerging infectious diseases (EIDs) (Daszak *et al.* 2000). All pose significant threats to humans, domestic livestock or wild animals, and the majority affecting humans are zoonotic or probably so. Since the mid 1970s the World Health Organization has recognized at least 30 new human diseases in the category including Ebola virus, human immunodeficiency virus/acquired immunodeficiency syndrome (HIV/AIDS), Lyme disease and severe acute respiratory syndrome (SARS), plus several well-known diseases that are changing, such as malaria, dengue fever, and tuberculosis. A consideration of malaria and other vector-borne diseases (VBDs) is instructive. Many VBDs follow a seasonal pattern in their incidence due to the effect of temperature and rainfall on the vectors. Profound changes in disease incidence can therefore be anticipated as a consequence of global temperature rises. Not only may shorter, warmer winters extend a disease transmission season, but also the geographical range of vectors (Matsuoka and Kai 1994, Martin and Lefebvre 1995, Martens *et al.* 1997) and their ability to survive over winter. Furthermore, wetter, warmer weather over wider areas will almost inevitably lead to greater vector abundance. Other changes in the pattern of VBDs may accompany continued deforestation and reduction in the numbers of tropical forest animals – disturbances that may promote vector adaptation to using humans as their main food source instead of animals (Ward *et al.* 1973, Ryan *et al.* 1990). In the absence of radical counteraction, we should expect human malaria, dengue fever, and other VBDs to occur outside of the tropics with ever increasing frequency.

A well-documented example of climate-related disturbance in infectious disease patterns occurred in the Four Corners area in the southwest of the USA. Six years of drought ended with heavy rainfall and snow in the late winter of 1993. The drought had reduced numbers of predators of the native deer mouse and the onset of wet weather allowed rapid growth in mouse food supplies. Consequently the mouse population exploded, which in turn led to a significant increase in the excretion of a hantavirus carried by the mice, and an outbreak of serious disease (Hantavirus pulmonary syndrome) among local people resulted (Duchin *et al.* 1994, Wenzel 1994).

Serious outbreaks of EIDs have also caused significant wildlife mortalities. For example, prior to 1987, morbilliviruses were only known to cause disease in terrestrial mammals – measles, rinderpest, peste-des-petits-ruminants, and canine distemper. However, since then, newly characterized morbilliviruses related to Canine Distemper Virus are known to have been responsible for mass die-offs in Baikal seals (*Phoca sibirica*) in Russia (Grachev *et al.* 1989), in Harbor seals (*Phoca vitulina*) around the coasts of northwest Europe (Kennedy *et al.* 1988, Osterhaus *et al.* 1990), and in cetaceans along the Atlantic coast of the USA (Lipscomb *et al.* 1994, Duignan *et al.* 1996) and around Western European shores (Domingo *et al.* 1990, Kennedy *et al.* 1992, Thompson *et al.* 1992, Van Bressem *et al.* 1993). Although the reasons why these marine viruses were able to suddenly evolve from terrestrial progenitors have not been clearly identified, contamination of the marine ecosystem with toxins and chemicals, loss of habitat, and starvation may all have increased morbidity and mortality during the viral outbreaks. Large, heterogeneous populations have the capacity to recover from such losses, but small, fragmented and genetically impoverished populations may be extremely vulnerable to local extinctions as a result (e.g., Woodroffe 1999).

Increasing contacts between wildlife and humans or their domesticated animals are also likely to have an impact on patterns of wildlife disease. For example, a high percentage of infectious disease affecting wildlife in the national parks of the USA is now thought to have originated in human or domestic animal populations (Aguirre and Starkey 1994). Modern modes of rapid transport and increasing travel frequency have enormous potential to affect the spread and emergence of infectious diseases. Travelers can access virtually every part of the planet within 24–48 h. Given that the incubation period of most infectious diseases is longer than 48 h it is easy to appreciate how quickly and widely such diseases can be spread globally. Although unproven, it is possible that the recent appearance of West Nile virus affecting birds, humans, and other species in the USA resulted from an infected and viraemic person traveling home from the Middle East where the disease is endemic. The increasing demand for ecotourism in remote areas of the planet combined with rapid means of transportation can create the potential for ecological health disasters – especially in the case of species susceptible to human disease agents such as the great apes. Close contact with habituated wild animals is common, travelers often combine short visits to several sites in a single holiday, and the health screening of tourists is minimal. Furthermore, ecotourists increasingly demand more comfortable facilities with all the local infrastructural development that entails. Rigorous risk assessments are required before endorsing ecotourism projects as a solution to revenue generation in developing countries.

Non-infectious diseases are also of major concern to conservation medicine. As an example, the potential impact of chemical pollutants known as endocrine disruptors has only just begun to be realized. A large number of synthetic chemicals (derived from plastic components, pesticide ingredients, organochlorine compounds, etc.) capable of interfering with the endocrine system are already present in all ecosystems across the world and therefore all animals are exposed to a greater or lesser degree. Their effects include developmental abnormalities, failure to thrive or reach sexual maturity, premature death, and behavioral disturbance, and it is easy to appreciate that their significance to the conservation of biodiversity may be far more profound in the coming years than even the most dramatic of infectious diseases (Colborn 2002).

THE ROLE OF ZOOS IN CONSERVATION MEDICINE

In most countries it is very unlikely that zoos will have the financial or professional resources to lead comprehensive studies in conservation medicine. This role is more likely to be undertaken by specialized centers established by university departments, large conservation non-Governmental organizations (NGOs) or a combination of the two. In the USA some veterinary faculties have already created conservation medicine units (e.g., the Consortium for Conservation Medicine at Tufts University School of Veterinary Medicine) and conservation medicine interests have featured strongly in the activities of the Wildlife Conservation Society (WCS) for some years. However, zoos, or perhaps national organizations representing zoos, could adopt a catalytic role in the creation of new centers for conservation medicine where none exist in a country.

All zoos could contribute to conservation medicine in a wide range of relatively simple and practical ways, but a necessary first step is the intellectual shift toward integrating the principles, concerns, and requirements of conservation medicine into all their activities. Included in the planning for any zoo-based conservation program should be an audit of the conservation medicine implications or opportunities. Even if projects are judged to pose no significant disease risk to the animals under study, there may be unique opportunities for monitoring health or establishing baseline data of use to the wider conservation medicine community. The challenge to zoos is to make a contribution to conservation medicine whatever their resources, and to do so from a genuine belief that such activities are fundamental to conservation efforts in all fields.

SUPPORT FOR EXISTING CONSERVATION MEDICINE PROGRAMS

It is open to any zoo to provide financial support for existing broad-based conservation medicine activities, but they often find it easier to target support on projects featuring so-called flagship species. Currently there are few *in situ* medical programs focused on individual species, but perhaps the best example is provided by the Mountain Gorilla Veterinary Project (MGVP). The MGVP initially provided clinical assistance to the highly endangered mountain gorillas (*Gorilla gorilla beringei*) when deemed appropriate, but has since developed beyond addressing the immediate needs of individual gorillas by seeking to monitor health more generally using a range of non-invasive sampling methods (Cranfield *et al.* 2002). Fecal parasite and bacteria surveys track the intestinal pathogens of these apes; fecal cortisol assays in habituated gorilla groups are used to assess the degree of stress imposed by ecotourism activities; and urine samples from habituated gorillas allow the study of endocrine function, urinary tract infections, kidney and liver function, etc. Due to their genetic closeness to humans, the health of the animals is inextricably linked to the health of local people, visiting tourists, and project staff. Surveys of human diseases reported by local hospitals in Western Uganda have revealed that between approximately 70% and 80% result from an infection potentially transmissible to great apes (Cranfield *et al.* 2002) – a finding which serves as a potent reminder of the dangers of ecotourism to susceptible animals and the need for veterinary and public health input into the design of ecotourism projects and the management of wild animal populations. Many other projects that focus on individual species would probably benefit from a similar approach and there is little to prevent zoo directors and zoo vets stimulating such activities.

Although financial support is perhaps the easiest way in which zoos can contribute to conservation medicine activities, it is the least direct and perhaps the most difficult to exploit in terms of public support, and certainly not very motivating for zoo staff. However, there are numerous other opportunities for zoos to contribute in more direct ways.

WILDLIFE DISEASE MONITORING

The monitoring and assessment of wildlife health and disease status are major components of conservation medicine activities and the bedrock upon which any future interventions would be based. Adequate monitoring programs are necessary on a global scale to assess the health of fragile ecosystems, and they need to be conducted over many years to allow discrimination between long-term trends and short-term variations. However, insufficient government funds are available to meet this need in any country (Daszak and Cunningham 2002).

Large zoos with comprehensive veterinary departments and links to *in situ* conservation projects may be in a good position to contribute directly to the monitoring of wildlife health, but only in concert with epidemiologists, virologists, ecologists, toxicologists, etc. Associations need to be developed and/or strengthened between such zoos and university or government departments capable of providing the appropriate scientific expertise. Even where such ambitious programs are beyond the resources of zoos, a substantial contribution can still be made by experienced zoo vets through clinical veterinary input in interpreting clinical signs, selecting appropriate diagnostic tests, and interpreting results in concert with other experts in the conservation sciences.

It is also important to promote regional laboratory expertise in countries with *in situ* conservation programs and again zoos could be instrumental in this process. One of the first and most important elements of the proposed Amur leopard reinforcement program currently being developed by the Zoological Society of London (ZSL), WCS-Russia and regional veterinary agencies in the Russian Far East is to survey the health status of free-living Amur leopards (*Panthera pardus orientalis*), their prey species and the local domestic stock. The survey is being achieved by developing the necessary laboratory expertise and capacity in the area rather than relying on the export of samples to Britain or the USA. Not only would this approach simplify sample processing, it would also establish a more permanent capability in the area to carry on ecosystem health monitoring studies beyond the specific focus on Amur leopards, stimulate a local interest in wildlife health issues, and crucially transfer responsibility for local wildlife disease issues to local professional personnel.

Pathogen pollution is the term applied to the anthropogenic introduction of pathogens to new geographic areas, and it is a very common cause of disease emergence in wild flora and fauna (Daszak *et al.* 2000). Therefore, the screening of animals for potential pathogens prior to their inclusion in reintroduction or reinforcement projects is essential. Screening animals for disease alone is insufficient – it is necessary to focus both on disease and the mere presence of potentially pathogenic organisms that may not cause overt disease in the host species but may represent a threat to other species in a release area. Clearly zoos have a responsibility to carry out comprehensive pathogen screening on animals destined for reintroduction projects, but it should also be standard practice to maintain populations of healthy animals in our zoos that do not harbor significant pathogens, and that requires zoos to maintain vigorous screening programs for as many species as possible. Many management and veterinary guidelines have been drafted for species held in zoos (for example by the Regional Species Breeding Programmes and Taxon Advisory Groups) that include advice on screening, but despite this too many collections still do not know the status of their resident animals.

Zoos may also consider cooperation with wildlife rehabilitation centers as a way of generating valuable information on wildlife health. Being driven primarily by concerns for individual animal welfare (and hence not by a desire to conserve animal populations), rehabilitation centers are often overlooked by zoos. These centers may handle large numbers of wild animals but very often the opportunity to collect biomaterials that would allow health screening is missed or simply beyond their resources. By developing links with rehabilitation centers, relevant biological samples could be collected and assessed by zoos to the mutual benefit of both organizations. An excellent example of the value of rehabilitation to conservation medicine is the post-mortem study of Californian sea lions (*Zalophus californianus*) dying in rehabilitation, which revealed metastatic carcinoma in 18% of sexually mature animals (Gulland *et al.* 1996). A finding such as this in a top predator could highlight serious problems in the near-shore ecosystem. Further exploration of the concept is given by Aguirre *et al.* (2002b) in their description of marine ecosystem health monitoring. Although valuable, disease or health surveys carried out in rehabilitation centers do need very careful interpretation as the animals concerned are not necessarily representative of a free-living population. Crude or direct parallels in disease prevalence between animals in rehabilitation and free-ranging populations should therefore be avoided.

EDUCATION

In recent years zoos have rightly laid considerable emphasis on the value of public education to further the aims of conservation. The development of conservation medicine will also depend heavily on education. Zoos could easily incorporate basic concepts of conservation medicine into their public educational messages to illustrate the significance of changing disease patterns to animals and humans, and the causal role of environmental degradation, travel, ecotourism, etc.

In recent years modern zoos have introduced more naturalistic enclosures which are not completely isolated from the wider environment. Changes in the ecology of local wildlife species which may carry infectious disease agents can easily affect the health and incidence of disease in resident zoo animals, providing ideal material for micro studies in disease monitoring and modeling. The Durrell Wildlife Conservation Trust has adopted this approach for many years in its conservation biology training center (Allchurch 2002). Even where such comprehensive training programs are not part of a zoo's activities, allowing *bona fide* students of conservation sciences to learn applied practical skills can make a contribution toward the overall aims of conservation medicine. For example, basic training for wildlife biologists in post-mortem examination technique and the collection of appropriate samples for pathology, serological surveying, genetic analysis, toxicology studies, etc. can be offered by many zoos with a veterinary capacity. The basic principles of animal restraint, immobilization and biopsy techniques can be taught, and the clinical expertise of experienced zoo vets shared with visiting veterinary and non-veterinary students.

Most countries do not have any undergraduate or graduate wildlife health components in their veterinary training, although the process has started in North America and Europe. In the USA several universities have developed conservation medicine programs, including UC Davies (Wildlife Health Program) and the University of Illinois (Conservation Medicine Center). In Britain an MSc course in Wild Animal Health is available at Royal Veterinary College, and throughout the neotropics the Wildlife Trust with the Consortium for Conservation Medicine at Tufts University School of Veterinary Medicine is providing conservation medicine training and support in zoos. However, the majority of developing countries have nothing comparable to these programs, although their need is arguably as great if not greater. As many zoos are involved in conservation projects in developing countries, the provision of basic or applied veterinary training for local veterinarians and wildlife biologists by experienced zoo clinicians can make a significant contribution to local skill bases and is far more cost-effective than bringing students to developed countries for training. Furthermore, by providing training in country, home-grown solutions to conservation medicine issues are encouraged rather than an inappropriate reliance on highly technical and expensive solutions practised in the northern hemisphere. A logical extension of this process is the encouragement of wildlife medicine programs in the veterinary and medical schools of developing countries. Although clearly a long-term aim, considerable effort and support should be applied to this goal at every opportunity.

RESEARCH

There are numerous ways in which practical, non-invasive, zoo-based research can contribute to conservation medicine, although any research has limited value unless the results are published in peer-reviewed journals and hence available to the wider scientific community. Zoos with established veterinary departments are well placed to conduct research into the diseases of free-living wildlife and to develop or validate novel diagnostic tools for use by others, although the very nature of conservation medicine requires that such activities involve input from a wide variety of scientific disciplines. The majority of zoos however are not in this position but relevant micro research projects can still be undertaken. By pooling resources and data with other collections, equipment and techniques useful to field biologists can be developed under the controlled conditions of zoos. Examples of possible study areas include the development of safer anesthetic techniques for use in the field; the testing of less traumatic and remote darting systems; the design of more effective animal identification methods; and the generation of normal values of relevance to the study of free-living species. The medical training of zoo animals can be of immense value in determining many normal values and could be used more widely in this context. Other relevant areas for zoo-based research include vaccine response studies, novel techniques in contraception, and assisted reproductive technologies that may have a future use in the management of small, fragmented, wild animal populations.

Many other subjects could be added to the list of research projects that could be carried out in zoos, but it is perhaps more important to consider the nature of the relationship between zoos and research rather than the details. Modern zoos can, and often do, carry out research projects. Participating in research from which conservation benefits accrue to species of wild animals is one of the options available to European zoos to implement the conservation measures required by the EC Zoos Directive 1999. In order to meet this requirement however, many British zoos currently rely too heavily on undergraduate students conducting brief and scientifically weak investigations mainly into captive animal behavior. Although there may be intrinsic value in such investigations, there is a danger that the choice of research topics is largely dictated by the needs of undergraduate students and not by the conservation needs of free-living wild animals. European zoos should avoid this convenient solution to their commitments and concentrate more on conducting applied research to real problems, such as the development of techniques, tools, and information necessary for field biologists to conduct better and safer studies of free-living wild animals. In order to do that links between *in situ* conservation projects and zoos need to be strengthened and expanded so that priorities for zoo-based research can be clearly identified. This approach is likely to improve the quality and relevance of zoo-based research, and zoos would be in a better position to explain and justify their research activities to the public. Benefits to conservation medicine activities and the management of wild animal populations would follow.

THE ADOPTION OF ZOOS IN DEVELOPING COUNTRIES

The adoption of zoos in developing countries by wealthier collections in the northern hemisphere is becoming common practice. Real benefits can be derived by both sponsor and sponsored zoos, but only when such associations are long-term and regularly audited. Most zoos in developing countries are severely under-funded and their immediate needs are often basic – improvements to animal enclosures, food type and quality, staff training, simple veterinary care, etc. However, such collections can have enormous potential as a source of data relevant to conservation medicine and sponsoring zoos should at least assess any such potential and factor it into their involvement. All of the activities suggested in this chapter by which zoos may contribute to conservation medicine could be carried out in sponsored collections and there is no reason why conservation-medicine-related activities cannot be conducted in parallel with general improvements to basic zoo operations. However, it is important that sponsors do not underestimate the amount of staff time and money required to sustain their involvement.

CONCLUSION

Support for conservation medicine activities by zoos, university departments, conservation groups, governments, NGOs, etc. goes far beyond just being a good idea. If the principles of conservation medicine are not fully integrated into our thinking about conservation in general, there is a considerable risk that our best efforts to conserve biodiversity on this planet will be seriously jeopardized, if not totally derailed. To maximize the value of our zoos to the global conservation effort it is essential that the principles and needs of conservation medicine are factored into all aspects of their operations.

References

Aguirre, A. A. & Starkey, E. E. (1994). Wildlife disease in U. S. National Parks: historical and coevolutionary perspectives. *Conservation Biology*, **8**, 654–661.

Aguirre, A. A., Ostfeld, R. S., Tabor, G. M., House, C. A., & Pearl, M. C. (2002a). *Conservation Medicine: Ecological Health in Practice*. New York: Oxford University Press.

Aguirre, A. A., O'Hara, T. M., Spraker, T. R, & Jessop, D. A. (2002b). Monitoring the health and conservation of marine mammals, sea turtles, and their ecosystems. In *Conservation Medicine: Ecological Health in Practice*, eds. A. A. Aguirre, R. S. Ostfeld, G. M. Tabor, C. A. House, & M. C. Pearl. New York: Oxford University Press.

Allchurch, A. (2002). Zoological parks in endangered species recovery and conservation. In *Conservation Medicine: Ecological Health in Practice*, eds. A. A. Aguirre, R. S. Ostfeld, G. M. Tabor, C. A. House, & M. C. Pearl. New York: Oxford University Press.

Colborn, T. (2002). Effects of endocrine disruptors on human and wildlife health. In *Conservation Medicine: Ecological Health in Practice*, eds. A. A. Aguirre, R. S. Ostfeld, G. M. Tabor, C. A. House, & M. C. Pearl. New York: Oxford University Press.

Cranfield, M., Gaffikin, L., Sleeman, J., & Rooney, M. (2002). The mountain gorilla and conservation medicine. In *Conservation Medicine: Ecological Health in Practice*, eds. A. A. Aguirre, R. S. Ostfeld, G. M. Tabor, C. A. House, & M. C. Pearl. New York: Oxford University Press.

Daszak, P. & Cunningham, A. A. (2002). Emerging infectious diseases: a key role for conservation medicine. In *Conservation Medicine: Ecological Health in Practice*, eds. A. A. Aguirre, R. S. Ostfeld, G. M. Tabor, C. A. House, & M. C. Pearl. New York: Oxford University Press.

Daszak, P., Cunningham, A. A., & Hyatt, A. D. (2000). Emerging infectious diseases of wildlife – threats to biodiversity and human health. *Science*, **287**, 443–449.

Domingo, M. L. *et al.* (1990). Morbillivirus in dolphins. *Nature*, **336**, 21.

Duchin, J. S., Koster, F. T., Peters, C. J. *et al.* (1994). Hantavirus pulmonary syndrome: a clinical description of 17 patients with a newly recognized disease. *New England Journal of Medicine*, **330**, 949–955.

Duignan, P. J., House, C., Odell, D. K. *et al.* (1996). Morbillivirus infection in bottlenose dolphins: evidence for recurrent epizootics in the western Atlantic and Gulf of Mexico. *Marine Mammal Science*, **12**, 499–515.

Grachev, M. A. *et al.* (1989). Distemper virus in Baikal seals. *Nature*, **338**, 209.

Gulland, F. M. D., Trupkiewicz, J. G., Spraker, T. R., & Lowensteine, L. J. (1996). Metastatic carcinoma of probable transitional cell origin in free-living California sea lions (*Zalophus californianus*): 64 cases (1979–1994). *Journal of Wildlife Disease*, **32**, 250–258.

Kennedy, S., Smyth, J. A., McCullough, S. J., Allan, G. M., McNeilly, F., & McQuaid, S. (1988). Confirmation of cause of recent seal deaths. *Nature*, **335**, 404.

Kennedy, S., Smyth, J. A., Cush, P. F., McAliskey, M., Moffett, D., McNiven, C. M., & Carole, M. (1992). Morbillivirus infection in two common porpoises (*Phocoena phocoena*) from the coast of England and Scotland. *Veterinary Record*, **131**, 286–290.

Lipscomb, T. P., Schulman, F. Y., Moffatt, D., & Kennedy, S. (1994). Morbillivirus disease in Atlantic dolphins (*Tursiops truncatus*) from the 1987–1988 epizootic. *Journal of Wildlife Disease*, **30**, 567–571.

Martens, W. J. M., Jetten, T. H., & Focks, D. (1997). Sensitivity of malaria, schistosomiasis and dengue to global warming. *Climate Change*, **35**, 145–156.

Martin, P. H. & Lefebvre, M. G. (1995). Malaria and climate: sensitivity of malaria potential transmission to climate. *Ambio*, **24**, 200–209.

Matsuoka, Y. & Kai, K. (1994). An estimation of climate change effects on malaria. *Journal of Global Environmental Engineering*, **1**, 1–15.

Osterhaus, A. D., Groen, M. E., J., Spijkers, H. E. M. *et al.* (1990). Mass mortality in seals caused by a newly discovered morbillivirus. *Veterinary Microbiology*, **23**, 343–350.

Rapport, D. J., Howard, J., Lannigan, R., McMurtry, R., Jones, D. L., Anjema, C. M., & Bend, J. R. (2002). Introducing ecosystem health into undergraduate medical education. In *Conservation Medicine: Ecological Health in Practice*, eds. A. A. Aguirre, R. S. Ostfeld, G. M. Tabor, C. A. House, & M. C. Pearl. New York: Oxford University Press.

Ryan, L., Vexenat, A., Marsden, P. D., Lainson, R., & Shaw, J. J. (1990). The importance of rapid diagnosis of new cases of cutaneous leishmaniasis in pinpointing the sandfly vector. *Transactions of the Royal Society of Tropical Medicine and Hygiene*, **84**, 786.

Thompson, P. M., Cornwell, H. J. C., Ross, H. M., & Miller, D. (1992). A serological study of the prevalence of phocine distemper virus in a population of harbour seals in the Moray Firth, N. E. Scotland. *Journal of Wildlife Disease*, **28**, 21–27.

Van Bressem, M. F., Visser, I. K., De Swart, R. L. *et al.* (1993). Dolphin morbillivirus infection in different parts of the Mediterranean Sea. *Archives of Virology*, **129**, 235–242.

Ward, R. D., Shaw, J. J., Lainson, R., & Fraiha, H. (1973). Leishmaniasis in Brazil: VIII. Observations on the phlebotomine fauna of an area highly endemic for cutaneous leishmaniasis in the Sierra dos Carajas, para State. *Transactions of the Royal Society of Tropical Medicine and Hygiene*, **67**, 174–183.

Wenzel, R. P. (1994). A new hantavirus infection in North America. *New England Journal of Medicine*, **330**, 1004–1005.

Woodroffe, R. (1999). Managing disease threats to wild animals. *Animals and Conservation*, **2**, 185–193.

14

The Masoala rainforest: a model partnership in support of *in situ* conservation in Madagascar

MATTHEW HATCHWELL AND ALEX RÜBEL

INTRODUCTION

With the publication of the first *World Zoo Conservation Strategy* (WZCS) in 1993 (IUDZG/CBSG 1993), the international zoo community challenged itself to establish closer links between the activities of zoological parks worldwide and the *in situ* conservation of wildlife. The aim of the first strategy was "to help conserve the Earth's fast-disappearing wildlife and biodiversity in general" by, among other measures, identifying "the conservation areas in which zoos and aquariums can make a contribution and [determining] how these institutions can support and facilitate the processes leading to nature conservation and sustainable use of natural resources" (Wheater 1995). The new *World Zoo and Aquarium Conservation Strategy* (WZACS) of 2005 is considerably bolder, proposing as a common vision for zoos and aquariums around the world that their "major goal . . . will be to integrate all aspects of their work with conservation activities" (conservation being defined as "the securing of long-term populations of species in natural ecosystems and habitats wherever possible") (WAZA 2005).

First through WZCS and more recently through WZACS, zoos around the world have thus resolved to become agents for *in situ* wildlife conservation. In the decade since the publication of the first strategy, they have developed a range of mechanisms for increasing their involvement in conservation activities on the ground (Miller *et al.* 2004). Many of those mechanisms are described in detail elsewhere in this book. One particular development that is relevant to the present case study was the Wildlife

Conservation Society's Congo Forest exhibit at the Bronx Zoo in New York, which opened in 1998 and was the first Bronx Zoo exhibit designed with strong conceptual links into WCS conservation initiatives on the ground. The Congo Forest was remarkable too in that zoo visitors were invited to direct their entry fee to support *in situ* conservation projects in Africa.

Also during the 1990s, in developing countries worldwide, the creation of new protected areas as the most effective tool for wildlife conservation accelerated dramatically. The latest United Nations List of Protected Areas estimates that there are now over 102 000 protected areas in the world, together encompassing over 11% of the planet's land surface (Chape *et al.* 2003).

It has been recognized for many years that protected area managers in developing countries face a different set of technical challenges to their counterparts in more developed nations (MacKinnon *et al.* 1986). In addition, it is now recognized that one of the most significant issues affecting the long-term viability of protected areas in Africa, Asia, and Latin America is that of financial sustainability (James *et al.* 1999, Wilkie *et al.* 2001). This is not generally the case in developed countries, whose national budgets are sufficiently stable for them to be able to guarantee funding for protected area management from year to year. Madagascar now possesses over 44 protected areas, but not one of them can be considered financially secure at this point. All depend for their financing either on the internationally funded 15-year National Environmental Action Plan (NEAP), which is due to come to an end in 2008, or on other forms of international support. Despite the recent creation of a conservation trust fund, it is unlikely that the government of Madagascar will be in a position to finance protected area management at anywhere close to current levels once the NEAP comes to an end (Keck 2001). As a result, protected area managers realize that they must seek new funding sources if they are to maintain management activities at levels sufficient to preserve the integrity of the sites under their responsibility.

It is clear from the above that, on the one hand, zoos in developed countries have been seeking for some time to develop stronger links between their activities and *in situ* conservation, and, on the other, that protected areas in countries such as Madagascar are being forced to explore innovative ways of assuring their own financial sustainability.

MASOALA NATIONAL PARK

From a financial point of view, Masoala National Park in northeastern Madagascar is typical of the country's protected areas over the past 12 years in that it was created and has been run almost entirely on funding provided by the international community, either through the NEAP or through non-Governmental organizations. The Masoala Integrated Conservation and Development Project (ICDP) was established in 1993 and the national park was gazetted in 1997 (Figure 14.1). At 2300 square kilometers, the park is the largest in Madagascar and is managed by the Malagasy national parks service (Association Nationale pour la Gestion des Aires Protégées, or ANGAP) in partnership with the Wildlife Conservation Society (WCS). It contains a wide diversity of ecosystems, from offshore coral reefs to highly threatened littoral forests and lowland tropical rainforests up to high altitude forests at over 1300 m (Hatchwell 1998, Kremen *et al.* 1999). It is suggested that the Antongil Bay watershed, including Masoala National Park, contains 50% of the plant diversity of Madagascar (G. Schatz, pers. comm.), an estimate that appears to be valid too for other well-documented taxa such as mammals and birds. Since Madagascar's endemic plants and vertebrates represent about 3% of the total for the entire planet (Myers *et al.* 2000), the value of the region for biodiversity conservation is obvious: the successful protection of the wildlife of this one site would protect 1.5% of the known plant and vertebrate species on Earth.

ZOO ZÜRICH AND THE MASOALA HALL

Zoo Zürich was established on its present site overlooking the city of Zürich in 1929. In 2001, the year before the opening of the Masoala Hall, it attracted about one million visitors to its 27-ha site. In 1991, the zoo compiled a new strategic plan, in line with the first World Zoo Conservation Strategy which appeared the following year, in which it resolved to develop a spectacular rainforest exhibit as the centerpiece of a new site development plan. In 1993, the zoo decided on the Masoala peninsula as the focus of the new exhibit, entered into negotiations with the city of Zürich for a major extension of the existing site onto neighboring land where the new hall would be built, and began raising funds for construction. In 1994, the city granted permission for the land extension and agreed to cover the associated infrastructural costs, but it was agreed that all new exhibits, including the Masoala Hall, would be paid for with funds raised from private donors. The same year, significant support for the new exhibit was pledged by Dr Hans Vontobel, whose express purpose was to "make a significant gift to the children of the city of Zürich by enabling them to enjoy close encounters with animals and nature" (Rübel *et al.* 2003). Other donors too contributed to the construction of the Masoala Hall primarily as an

Figure 14.1 Map of Malagasy protected areas showing location of Masoala National Park

enhancement of Zoo Zürich and to attract more visitors to the city (see Rübel *et al.* 2003).

In June 2003, the new Masoala Hall at Zoo Zürich opened its doors to the public (Figure 14.2). The exhibit covers an area of about 11 000 m^2 and cost approximately CHF 52 m to build, or about US$41 m at the time of writing.

The development of the exhibit began in earnest in 1993, when contacts were established between Swiss and Malagasy authorities and a feasibility study for the construction of the exhibit in Zürich was conducted. In following years, detailed plans were put together and closer relationships nurtured with Malagasy institutions, as the result of which two formal agreements were signed in 1996 between Zoo Zürich, the Malagasy Ministry of Water and Forests, and CARE International as the lead organization of the Masoala ICDP.

In the years following the agreements signed in 1996, a series of linkages were developed between the evolving exhibit in Zürich and the national park on the ground in Madagascar, some of them as the result of contractual obligations and others of the synergies between the zoo exhibit and the national park that emerged as the collaboration between institutions progressed. Linkages that were introduced during this development phase of the Masoala Hall in Zürich were:

1. Small-scale development projects conducted in communities around Masoala National Park, financed with grants of $10 000 per year for 10 years managed in collaboration with local authorities and the Ministry of Water and Forests. These projects have been very successful in winning support for the Park among local villagers and local authorities, who are closely involved in the decision-making process.
2. Two plant nurseries established adjacent to the Park as community resources and to provide seeds and seedlings for the exhibit. About 15% of the 17 000 plants in the Masoala Hall were grown from seeds or seedlings originating in the nurseries at Antanambao and Andrakaraka (Figure 14.3).
3. Institutional support for the national seed-bank (Silo National des Graines Forestières or SNGF) as part of a contract to manage seed and plant exports from Madagascar to Switzerland. The non-commercial export of plants to Zürich by the ICDP was frequently misconstrued in the Masoala region and contracting SNGF to take over that role solved the problem instantly – at the same time as strengthening a national institution with a potentially important role to play in the sustainable management of the country's remaining forests.

Figure 14.2 The 1.1-ha Masoala Hall incorporates the latest energy-saving technology

Figure 14.3 Many of the plants and other features in the Masoala Hall were sourced directly from Madagascar

4. A study on development needs around Masoala National Park conducted by Swiss researchers in 2002–03 coinciding with the end of the ICDP and constituting a needs assessment for development around Masoala over the next 5 years.
5. Support for scientific research on the ground at Masoala, for example on plant diversity and on the reproductive ecology of chameleons, which could help not only to provide a sustainable source of animals for exhibition in zoos but also, potentially, to reduce the unsustainable collection of wild chameleons for the pet trade.
6. Promoting Masoala National Park within Madagascar and in Europe. Visits to Switzerland for Malagasy officials and Masoala technical staff, a Masoala open day in the capital Antananarivo, and the publication of a book about Masoala with a foreword by the President of Madagascar resulted in a much higher profile for Masoala than would have been the case without the link with Zoo Zürich. In Europe too, publicity and press attention surrounding the opening of the exhibit and the launch of the Masoala book helped to raise the profile of Madagascar and Masoala as international priorities for biodiversity conservation.
7. Finally, the many contacts between Zoo Zürich and the National Park during the development phase of the exhibit resulted in a two-way exchange of information, ideas and materials between exhibit design staff in Zürich and Masoala staff in Madagascar, which was vital in establishing the on-going links that are in place now that the exhibit is open.

A precise monetary value can be assigned to several of these areas of linkage, which together amounted to an investment of approximately $280 000 by Zoo Zürich between 1996 and June 2003 when the exhibit opened (A. Rübel, pers. comm.). Other links are very difficult to quantify in monetary terms, for example the promotion of Masoala in Madagascar and Europe and the exchange of ideas and information between partners. Over the years, these less direct linkages have proved to be particularly important. Promotion of Masoala has helped to make it, within less than a decade, one of the best-known protected areas in Madagascar. Thanks in part to this higher visibility, Masoala is now on-track to be accorded World Heritage Site status.

The exchange of ideas and information between park and zoo has resulted not only in a zoo exhibit that reflects faithfully the broad reality of the National Park in its ecological, human, cultural, and historical contexts, but also in an on-going partnership for *in situ* wildlife conservation. By 2000, CARE International, Zoo Zürich's original partner at Masoala, had handed over responsibility for park management to the Malagasy national parks service (ANGAP) and the Wildlife Conservation Society, who have been co-managing the park ever since. Even before that transition, however, detailed discussions were held about potential ways to maintain the close relationship between the National Park and the new exhibit in Zürich once it opened. These mechanisms include both financial and non-financial support for the National Park itself, as well as more indirect support for conservation in Madagascar through increased tourism to Masoala and other sites.

Mechanisms to ensure financial and non-financial support

1. Voluntary cash contributions made by zoo visitors were initially the main source of direct support for park management and associated activities at Masoala. In the year following the opening of the exhibit (to 30 June 2004), these amounted to CHF 232 700 (about $190 000 at the time), made up from cash contributions, pledges, and a 2% contribution from the turnover of the Masoala gift shop and restaurant. Overall contributions since 2004 have been lower but are still sufficient to support Zoo Zürich's commitment to contribute $100 000 per year to the National Park, which represents 25%–33% of annual operating costs.
2. An association called the Friends of Masoala, or *Freunde Masoalas*, was created by Zoo Zürich in 2004 with the express purpose of creating the $7–8 million trust fund that is needed to endow Masoala National Park in perpetuity. New members are attracted through brochures and computer terminals in the Masoala Hall and through mailings to European visitors to the National Park, whose names and addresses are collected using questionnaires. The Friends are kept in touch with events at Masoala through a regular electronic bulletin compiled by park staff on the ground in Madagascar and translated into German by Zoo Zürich. In 2003, a similar group, the Friends of Galapagos Switzerland, contributed $140 000 to *in situ* conservation in the Galapagos Islands. In the long run, it is hoped that contributions by the Friends of Masoala association will more than make up for any decline in direct visitor contributions (see previous paragraph).
3. A third linkage between Zürich and the National Park is the sale of Malagasy handicrafts at the Masoala gift shop. In the year following the opening of the exhibit, these amounted to approximately $100 000. These sales not only contribute to the percentage cited above, but also provide direct financial revenues back to the producers of handicrafts in communities around Masoala National Park and elsewhere

in Madagascar, with corresponding benefits for conservation as long as local people are aware of the link between their revenues and the existence of the National Park. The products sold in Zürich are now part of the Fair Trade labeling scheme and an estimated 120 families at Masoala benefit from the revenues (E. Steiner, pers. comm.).

4. Independently from direct contributions generated within the Zoo Zürich exhibit, other donors have been inspired by the exhibit to give to Masoala in the form of both in-kind contributions [binoculars, global positioning system (GPS) units, etc.] and financial support for park management. In future, such support – and other funds raised by the *Freunde Masoalas* – will contribute to the proposed Masoala trust fund.
5. On-going support – financial, technical and in-kind – for research projects, continued seed exports by SNGF, interpretation center graphics, books about Masoala published in 2003 and 2005, and so on. These are generally projects developed jointly by Zoo Zürich and Masoala National Park, funded sometimes jointly and sometimes by Zoo Zürich alone. The annual value of this support will vary but is likely to stabilize at the level of approximately $25 000. These additional contributions by Zoo Zürich are more than covered by increased revenues from visitor admissions to the zoo, which rose by 25% in the 12 months following the opening of the Masoala Hall in June 2003 and now appear to have stabilized at approximately 15% above pre-opening levels.
6. A sixth area of linkage is increased Swiss institutional support for and involvement in conservation and development issues in the Masoala region. These include on-the-ground projects following up on the study of development needs in Masoala peripheral zone conducted in 2002–03, and diplomatic representation by the Swiss Embassy and Chargé d'Affaires in Madagascar.
7. A less easily measured linkage is the educational value of the Zürich exhibit not only in terms of conservation but also as a window on another culture: a living outpost of Madagascar at the heart of Europe. The Malagasy delegation at the opening in June 2003 was struck by the fact that the exhibit does not focus just on a narrow range of conservation issues, but presents a broad profile of their country in all its complexity (G. Ramangoson, pers. comm.). As such, the exhibit presents wildlife conservation in something close to its true context, thereby establishing a crucial link in visitors' minds between biodiversity conservation and human livelihoods. Achieving long-term conservation success in a country like Madagascar without understanding that linkage is essentially impossible.

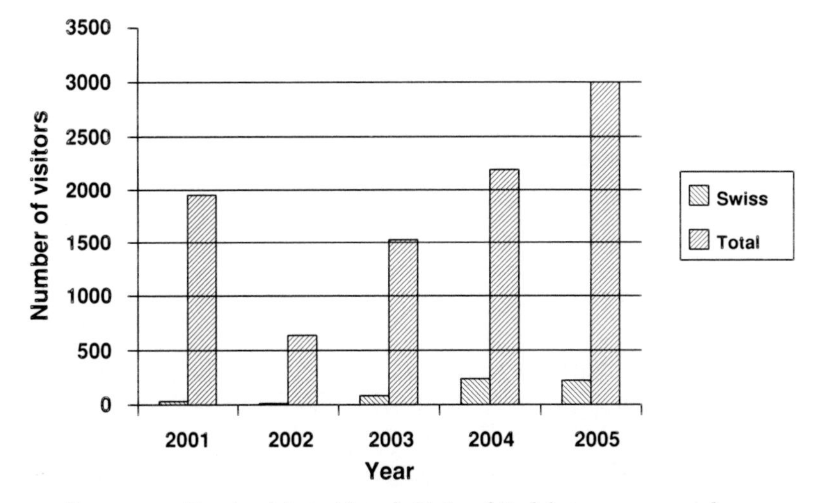

Figure 14.4 Tourist visits to Masoala National Park between 2001 and 2005

8. A final on-going linkage between Zürich and Masoala is in the form of increased tourism. First of all, visitors to Masoala and other protected areas in Madagascar make direct contributions to park management in the form of entry fees. But these are very low: currently less than $10 per person for a 3-day entry permit, which translates into a total of about $375 for the 50 or so additional Swiss nationals who visited Masoala in 2003 relative to previous years. The latest figures available (for 2000) indicate that each visitor to Madagascar that year was worth about $860 to the national economy (INSTAT 2002). While one might expect those earnings to rise with inflation, the steady loss in value of the Malagasy franc means that the dollar value of tourist revenues remained flat for the 1995–2000 period for which data are available, so we have used that same figure as the basis for calculating per-tourist revenues associated with the Masoala Hall.
9. According to visitor data from the National Park, 31 Swiss citizens visited the park in 2001 and 10 in 2002. In 2003, the number of Swiss visitors to Masoala had more than doubled compared to the previous comparable year. By 2004, Swiss tourists were the third largest group of visitors to Masoala by nationality (after Malagasy and Americans). While numbers are still low in absolute terms (see Figure 14.4), it is clear that the advance publicity for Masoala in Switzerland has inspired more Swiss tourists to visit the park in 2003 than in previous years. Numbers in 2004 and 2005 were considerably higher (although those for 2005

include visits to the park by 457 passengers of cruise ships whose contributions to the local economy are negligible). If one attributes 200 additional visitors in 2004 to the publicity provided by the Zoo Zürich exhibit, they would represent a direct contribution of about $172 000 to the national economy of Madagascar. There is considerable potential for these numbers to rise in future.

Several conclusions can be drawn from the above information. First, experience since the exhibit opened in 2003 indicates that visitors are inspired to make significant voluntary contributions to support *in situ* conservation at Masoala in the form of cash, pledges, and other forms of giving. One of the goals of the continuing alliance between Zoo Zürich and Masoala National Park is to maximize the use of the exhibit and of associated mechanisms such as the Friends of Masoala group and the electronic news bulletin to raise money in Europe in support of park management. Zoo Zürich has committed to contribute $100 000 per year to Masoala, $75 000 of which will be spent on park running costs. The remaining $25 000, and as much as possible of any additional funds raised, will be used to establish a trust fund aimed at covering park management costs in the long term. Given an annual operating budget of $300 000–400 000 and an annual return on investment of 5%, that fund will need to be endowed at the level of $7–8 million.

Less easily quantifiable are a second group of indirect benefits that contribute to *in situ* conservation in Madagascar through a range of mechanisms including generating support among local communities through the sale of handicrafts, individual projects such as the Masoala book, increased involvement in the Masoala region by other Swiss institutions, and the value of the exhibit as an outpost of Madagascar in Europe and a source of inspiration for zoo visitors. Although putting a dollar value on this second category of linkages is very difficult, several of them are likely to feed into other, more easily quantifiable ones, such as direct financial contributions, increased tourism, or even improved conservation effectiveness on the ground at Masoala. For the future, it is vital that the evolution of these various indirect linkages is monitored in such a way that conservation benefits can be properly identified and demonstrated as the basis for future decisions and for evaluating the full range of the exhibit's contributions to *in situ* conservation from year to year.

Finally, we have classified tourism-related benefits as a separate group in their own right. It is clear that each foreign visitor is worth many times more to the economy of Madagascar than their contribution to conservation through park entry fees, guides, and related expenditures. While a large proportion of that income may not contribute directly to conservation or protected area management, causing some people to question the benefits of such earnings for conservation, most Malagasy are likely to conclude that it is all the more effective for exactly that reason. The fact that Masoala National Park is making a growing contribution to the local and national economy by attracting new foreign tourists is a major argument in its favor. As a result of the publicity that the Zürich exhibit has brought, more and more Swiss travel agencies now offer Masoala as a destination. As long as its benefits extend to local communities, increased tourism can be a major factor in assuring the park's survival into the future. While it is true that the potential negative impacts of tourism need to be carefully managed at both national and local levels, we anticipate that it will be possible to increase tourist numbers slowly so that local infrastructure can develop gradually without becoming overwhelmed. As the infrastructure improves, there is no reason why numbers of tourists at Masoala each year should not reach into the thousands, each visitor making corresponding contributions to the national economy as well as to the conservation of the National Park. According to the INSTAT figures cited above, every 1160 additional tourists per year inspired to visit Madagascar will generate an extra $1 m per year for the Malagasy economy.

CONCLUSION

We suggest that there are several general conclusions to be drawn from this Masoala case study.

First, the alliance between a European zoo and a protected area in a developing country is in the interests of both parties because it addresses the need of the zoo to increase the impact of its activities on *in situ* wildlife conservation and that of the protected area to address its long-term funding needs. Our experience at Masoala has been that the multiplicity of potential linkages and the long-term nature of the needs mean that the relationship between zoo and protected area functions best if it is regarded as a true partnership, not simply as a contractual arrangement between donor and grantee. This case study may be a useful model both for the development of future zoo exhibits and as a potential funding mechanism for protected areas in developing countries.

Second, although the various linkages developed between Zoo Zürich and Masoala National Park during the creation of the exhibit contributed significantly to the concept and design of the Hall, what are most important when it comes to the impacts of the exhibit on *in situ* wildlife conservation are the on-going linkages between the zoo and the National Park on the ground. As implied in the previous paragraph, these have led in particular at Masoala to a sense of shared responsibility for the financial sustainability of the park, which also increases its overall viability in the long term.

Third, increases in tourism inspired by Masoala Hall contribute to both conservation and economic development in Madagascar. While some may question the value of non-conservation-related revenues when it comes to their conservation impact, most Malagasy will come to exactly the opposite conclusion. The fact that Masoala National Park and the Zoo Zürich exhibit are contributing to the local and national economy is a major argument in their favor, and, as long as the benefits reach the communities around the park, one of the main factors that will assure its survival into the future. The main challenge now is to monitor and quantify those contributions and then market them clearly as the sorts of benefits for conservation that can be gained from a close partnership between a "First World" zoo and a protected area in a country such as Madagascar.

Fourth, while it is important to note that money spent on building the Masoala Hall would not otherwise have been available for conservation in Madagascar, the exhibit contributes in so many ways to *in situ* conservation that the two should not in any case be considered as mutually exclusive options. An investment in a zoo exhibit at the heart of Europe can be a very effective way of contributing to conservation in the long term through the range of direct and indirect linkages described here. Considered in this light, one of the principal achievements of the Zoo Zürich–Masoala partnership has been to derive significant benefits for conservation in Madagascar from funding whose primary goal was to enhance the zoo and the city of Zürich with a major new tourist attraction.

Finally, none of the above mechanisms and linkages would be possible without commitments by all partner institutions to support the management of Masoala National Park in the long term. While Malagasy national institutions such as ANGAP have little choice in the matter, international organizations frequently are unwilling to make commitments to individual sites beyond the grant periods established by their donors. In this case, the Wildlife Conservation Society has identified the Masoala landscape as one of many long-term sites around the world which it has made an open-ended commitment to protect. Given the scale and nature of the investment that the new exhibit in Switzerland required, such a commitment on the part of its main partner was a prerequisite for Zoo Zürich's choice of Masoala as the focus of the undertaking.

References

Chape, S., Blyth, S., Fish, L., Fox, P., & Spalding, M. (2003). *2003 United Nations List of Protected Areas.* Gland: IUCN, Cambridge: UNEP-WCMC, ix + 44 pp.

Hatchwell, M. (1998). Plan de Gestion du Parc National de Masoala. Unpublished report. Antananarivo.

Institut National de la Statistique (INSTAT) (2002). *Journée Africaine de la Statistique 2001: Les Points Saillants.* Antananarivo, Madagascar: INSTAT.

IUDZG/CBSG (IUCN/SSC) (1993). *The World Zoo Conservation Strategy: The Role of the Zoos and Aquaria of the World in Global Conservation.* Brookfield, IL: Chicago Zoological Society.

James, A., Gaston, K., & Balmford, A. (1999). Balancing the Earth's accounts. *Nature,* **401**, 323–324.

Keck, A. (2001). The national conservation finance planning process in Madagascar. In *Conservation Finance Guide.* Conservation Finance Alliance, 2002.

Kremen, C., Razafimahatratra, V., Guillery, P., Rakotomalala, J., Weiss, A., & Ratsisompatrariv, J.-S. (1999). Designing the Masoala National Park in Madagascar based on biological and socio-economic data. *Conservation Biology,* **13(5)**, 1055–1068.

MacKinnon, J., MacKinnon, K., Child, G., & Thorsell, J. (1986). *Managing Protected Areas in the Tropics.* Gland: IUCN.

Miller, B., Conway, W., Reading, R. P., Wemmer, C., Wildt, D., Kleiman, D., Monfort, S., Rabinowitz, A., Armstrong, B., & Hutchins, M. (2004). Evaluating the conservation mission of zoos, aquariums, botanical gardens, and natural history museums. *Conservation Biology,* **18(1)**, 86–93.

Myers, N., Mittermeier, R. A., Mittermeier, C. G., da Fonseca, G. A. B., & Kent, J. (2000). Biodiversity hotspots for conservation priorities. *Nature,* **403**, 853–858.

Rübel, A., Hatchwell, M., & MacKinnon, J. (2003). *Masoala: The Eye of the Forest. A New Strategy for Rainforest Conservation in Madagascar.* Stäfa: Th. Gut Verlag.

WAZA (2005). *The World Zoo and Aquarium Conservation Strategy: Building a Future for Wildlife.* Liebefeld-Bern: WAZA.

Wheater, R. W. (1995). World Zoo Conservation Strategy: a blueprint for zoo development. *Biodiversity and Conservation,* **4**, 544–553.

Wilkie, D. S., Carpenter, J. F., & Zhang, Q. (2001). The under-financing of protected areas in the Congo Basin: so many parks and so little willingness to pay. *Biodiversity and Conservation,* **10**, 691–709.

15

In situ and *ex situ* conservation: blurring the boundaries between zoos and the wild

LESLEY A. DICKIE, JEFFREY P. BONNER, AND CHRIS WEST

INTRODUCTION

For much of their history zoos have worked in isolation, building collections of animals, often based on the interests and preferences of zoo managers. These collections were put together primarily for the purpose of academic study (Wemmer and Thompson 1995) and recreation, and often driven more by civic pride rather than any conservation objectives. However, zoos now operate in a different conservation context, with species declining at an unprecedented rate (Balmford *et al.* 1998). As previous authors have suggested (Conway 2003, Anon 2005), zoos of the 21st century have a moral obligation to re-focus their efforts to helping conserve wildlife. While this conservation focus has consisted largely of breeding insurance populations of threatened species in captivity (Soulé *et al.* 1986), recently there has been a re-assessment of the areas in which zoos can additionally contribute to conservation. Many now actively participate in or manage their own field projects, which thus become in a sense *in situ* outposts of the zoos themselves, frequently blurring the boundaries where captive and wildlife management meet. But zoos are also re-appraising exactly what they mean by "zoo conservation." What do we do, as a community, that contributes to conservation?

Zoos must be aware of and balance these activities, *in* and *ex situ*, with their impact in the wild, asking themselves fundamental questions such as, "*How do we effectively pursue conservation goals?*" and "*Do we know how our activities impact on wild populations, both positively and negatively?*". The tension between presenting a world-class zoo, a financially costly undertaking, and conservation initiatives can be difficult to resolve. There are ever-increasing demands on the revenues generated by individual zoos. The linkages between the wild and captivity are complicated, and consensus as to what zoo-based conservation actually consists of, which species should be held in captivity, how they are sourced and the conservation role they play has been difficult to reach. Reaching consensus is a challenge within, and especially between, regional zoo associations. Yet zoos are uniquely placed to be integrated centers for conservation, providing a portal for the public to understand the meaning of conservation in a wider sense. Incorporating their own behaviors into the framework of conservation will lead the public to make better informed decisions, but only if zoos, as institutions, are clear about what they really contribute, in total, to conservation, and in doing so foster a relationship of trust with those visitors.

WHAT IS "ZOO-BASED CONSERVATION"?

Other authors have discussed the increasing activity of zoos in *in situ* conservation, either through partnerships with other organizations or through their own endeavors (see Chapter 20) and this will not be elaborated further here. Instead the conservation ability of zoos in the realm of their *ex situ* activities will be discussed. A fundamental philosophical question challenges all zoos that wish to be active in regard to conservation and to fulfil conservation objectives. Field activities can be easily and handily labeled "conservation," regardless of whether any attempt has been made to measure the real impact, and cost-effectiveness, of those actions (see Chapter 21). But what do we do within our zoos that we can, with honesty, label conservation?

The definition of conservation used in the *World Zoo and Aquarium Conservation Strategy* (WAZA 2005) is as follows: *Conservation is the securing of long-term populations of species in natural ecosystems and habitats wherever possible.* There is a plethora of definitions for conservation and over-examination of these is unproductive, however it is suggested that a subset of the WZACS definition as applied to activities *within* zoos could read as: *Zoo-based conservation is the use of living collections of animals in captivity to secure the long-term survival of species in natural ecosystems and habitats, through* ex situ *management, research, fundraising, education, awareness and advocacy.*

Examining how we describe our work in zoos is reflective of entrenched views. The terminology in conservation and the responsible zoo community is value-laden and zoos may well be de-valuing their overall contribution to conservation as a result.[1] By using terminology such as "direct" to describe field activities and "indirect" to label activities in the zoo we are assigning a hierarchy of worth. Yet, there are actions carried out in the captive setting that can have profound impacts on the future survivorship of species. Examining common synonyms for direct, words such as "absolute," "categorical," and "explicit" abound. If we do the same with "indirect," the synonyms are far from positive: "ambiguous," "erratic," and "incidental." Yet, the reality of the situation may be much more uncertain. Can we see through the haze to at least describe our work with more clarity?

By changing terminology it may be possible to measure our activities with greater objectivity, and this may also have benefits to non-zoo conservation organizations. In this chapter we suggest that the word "intent" be used as a descriptor for all conservation activities, whether in the zoo or field. Is there conservation "intent" to a specified action? Here the meaning of intent conforms to that of purpose.

However, this is an idea fraught with baggage that may be wilfully misunderstood by the unscrupulous. It is untenable and retrograde to follow the route of claiming that all activities carried out by a zoo, or the simple existence of the zoo itself, are conservation. What then would distinguish the conservation-oriented zoo from the profit-oriented one that carries out no conservation activities? Clear intent must underlie all activities and be identified. The commercial activities of a zoo such as the entry fee, the zoo shop or the catering facilities may have the specific intent of providing income with which to maintain a collection of animals in a captive setting, regardless of their conservation impact. Many zoos have little or no government support, local or national, and therefore commercial imperatives are an important factor. However, it is *how* the collection of animals is utilized after that point that is important.

Intent also gives us a terminology that can be used to describe our activities separately from the need to evaluate their impact. While the evaluation of conservation programs is a burgeoning and long-overdue field of endeavor (Kleiman *et al.* 2000, Saterson 2004, Struhsaker *et al.* 2005), the inability to effectively measure conservation impact within short time frames (a difficulty inherent in much conservation work) can lead to

[1] This refers to progressive zoos seeking to improve their conservation credentials *and* impact.

spurious criticism. By declaring what the intent of the activity is, regardless of the short-term difficulties of measuring success, gives a sense of purpose to those activities, whether *ex* or *in situ*.[2] This discussion touches on the "USP" or *unique selling point* of zoos as conservation non-Governmental organizations (NGOs). Unlike other international conservation NGOs, zoos have a physical site that can be visited. People can interact in a much more personal way, meeting the people who implement conservation, both in the zoo and the wild (Bonner 2006) and therefore the zoo and the activities in the zoo are the living embodiment of the activities across the organization, regardless of where those activities are undertaken. The public can be told about field activities, see pictures. But they don't see it for themselves: they don't see the field crickets in the field; they see them, for example, in London Zoo *before* they are released into the wild as part of a breed-and-release program. This is the strength of a zoo visit at its best.

It would be beneficial, though maybe inherently subjective, to ensure that as a community we understand what we mean by zoo-based conservation. This consensus does not exist currently and appears to be an area of discussion that remains uncomfortable for many people. This discussion will grow and needs to examine all areas of the zoo, but fundamental to this question will be the species that are held in captivity. What animals do we maintain and why?

WHEN IS *EX SITU* MANAGEMENT CONSERVATION?

For the majority of zoos, it is likely that their conservation activities arise from, or are stimulated by, their own living collections. When addressing the composition of zoo populations, the first question that must be asked is "*what is the purpose of the collection?*" It could be argued that we choose animals that the public like to see and exhibit large, charismatic vertebrates in preference to others (Balmford *et al.* 1996), and that exhibit value overrides conservation need. It is clear that financial imperatives do play a part in the function of a zoo, and emblematic species are believed to be a greater attraction for visitors, though this has been disputed (see Balmford 2000).

Alternatively, do we choose good storybook animals, with readily accessible education messages that link to the demands of national school

[2] While evaluation of conservation programs may be problematic in the short term, it remains an essential component for any organization engaged in conservation and should be incorporated into conservation planning wherever possible.

curricula, or animals that are abundant in captivity and therefore readily sourced, or animals with low maintenance costs? Do we fit species around the conservation messages that we want to convey? It would appear that few zoos articulate clear intent relating to biodiversity conservation when they devise collection plans – indeed it is not known how many zoos have collection plans in the first place (Hutchins *et al.* 2005). It should also be remembered that zoos have multiple functions, and tension between the different functions needs to be tempered.

Many zoos make claim that maintaining certain threatened species in captivity is, in isolation, conservation. But is it (Wiese *et al.* 1994)? If the threatened species serves no function other than existing in captivity, does it really benefit population security in the wild? This ultimately depends on the species and there is no "one size fits all" answer. At a basic level, maintaining assurance populations of a threatened species in captivity ensures that in the worst-case scenario – the complete extirpation of the species from the wild – a fragmented population (whether managed or not) existing in zoos could be used to reintroduce that species back to the wild, once the threats that had caused the original decline had been extinguished or minimized to manageable levels. This methodology has previously been implemented for a few species such as the Arabian oryx (*Oryx leucoryx*) (Stanley Price 1989), and, in the case of near-extinction in the wild, the golden lion tamarin (*Leontopithecus rosalia*) (Dietz *et al.* 1994). These examples of the "ark" function of zoos are relatively rare (Beck *et al.* 1994), and previous over-reliance on this argument by zoos has been unhelpful, and, it could be argued, detrimental to the holistic role of zoos (see Chapter 11). Yet, where there is a clear need for captive management, for example as part of an IUCN-recommended action, then zoos have finely honed expertise and the Convention on Biological Diversity 1992 (Glowka *et al.* 1994) and the IUCN Technical Guidelines on the Management of *ex situ* Populations for Conservation (2002) enshrine the role of zoos in this respect (see Chapter 3).

The frogs' last croak?

Perhaps the most striking example of conservation through *ex situ* management is one that is yet to be achieved. Plans are currently ongoing to mobilize the world's zoos behind a single idea, that of managing a large number of threatened species of amphibians in captivity in response to the daunting findings of the Global Amphibian Assessment (Stuart *et al.* 2004). While it had been noted since the late 1980s that amphibian populations appeared to be declining (Blaustein and Wake 1990, Wyman 1990), often in pristine habitats with no obvious cause (so-called enigmatic declines) (Lips 1998), the true scale of the problem had remained obscure. Stuart *et al.* (2004) found that 43% of amphibian species are declining in population size, with 32% (1856 species) threatened with extinction. This far exceeds the 25% of mammal or 12% of bird species facing extinction, frightening though those statistics remain (IUCN 2006). Causes have now been identified to explain most, if not all, amphibian declines and include habitat loss, climate change, hunting and the spread of chytrid fungus (Berger *et al.* 1998, Andreone *et al.* 2005, Daszak *et al.* 1999, Cunningham *et al.* 2005).

A traditional conservation response to widespread decline in a species may have followed a number of routes: designation of protected areas; development of community initiatives to change hunting practices; legal enforcement to halt trade. These more traditional responses are, however, largely futile in relation to the current patterns of decline for many species of amphibian. The chytrid fungus, with its apparently unstoppable progressive inter-continental spread, requires a re-evaluation of conservation response. Maintenance of assurance populations in captivity may be the only route to survival for hundreds of species of amphibian, until a future point where chytrid is, if ever, eradicated from, or controlled in, the environment. In comparison to many other taxa, amphibians are excellent candidates for breed-and-release programs: they have high fecundity (once the individual species code to breeding them in captivity has been cracked), they are inexpensive to maintain, and have few behavioral problems (Bloxam and Tonge 1995).

However, the complexity of such an operation should not be underestimated. Coordinating a global captive action plan for amphibians (being undertaken jointly by the Conservation Breeding Specialist Group and Amphibian Specialist Group of IUCN and WAZA) requires accurate information as to which species to target, that these target species can be sourced from the wild or already exist in captive management, that the required funds are available, that husbandry expertise exists or can be developed, that facilities exist or can be developed, and so on (Zippel 2006). As with the IUCN *ex situ* management guidelines (IUCN 2002), the proposals so far recommend that regional zoo associations work with threatened species in their own geographic area, and thereafter lend assistance to captive management of "exotic" species from regions with less developed infrastructure and funding. In WAZA over 1200 zoos are represented. If each could maintain and breed five threatened species of amphibians in their individual collections it is possible that zoos could be instrumental in saving an entire group of vertebrates. But, if zoos fail to meet this challenge it will signal that it may not be only amphibians that face an uncertain future. Consequently the appropriate resources, and more importantly willingness to act, have to be found by many zoos, not by just a few leading the way.

There are clearly important contributions to conservation as a result of *ex situ* management, however, the species for which this is most appropriate should be identified and, where possible, the conservation integrity of any zoo improved by working with these target species in the captive setting. Both institutional and regional collection plans need to be developed, with clear intentions as to their purpose. Zoos can accommodate a multiple focus within their collections, and this could be viewed as one of their great strengths. However, if zoos are to improve and increase their conservation output, they need to focus more of their resources on conservation-dependent species.

Open doors

It is also apparent that, in range countries, the interactions between zoos and the wild have the additional dimension of the increasing role of zoos as recipients of individuals or shipments of threatened species, either handed into the zoo by the public or as a result of confiscations by local authorities (Cuaron 2005). Where the species is exceptionally threatened every individual may be critical and the difficulties faced by under-resourced developing-world zoos in providing appropriate management can be enormous. Additionally the cultural dimension of the leadership of zoos is a topic seldom discussed, but one that can have significant ramifications for the conservation output of zoos. Developing-world zoos, or some city zoos in the developed world, may have government-imposed civil servants as managers, who may or may not have the interest or skills required to successfully lead a zoo, or to respond to the particular challenges that a limited budget brings. Such personnel may also be subject to regular re-deployment within wildlife authorities or other branches of the government, resulting in disjointed zoos with little direction. How can conservation objectives be achieved under such circumstances?

The *World Zoo and Aquarium Conservation Strategy* (WAZA 2005) suggests that better resourced zoos "twin" with their less able counterparts, providing expertise, technical support, and funds where appropriate. As zoos in the "frontline" these institutions have a key role, not only in the housing of threatened taxa, but also in communicating with local peoples whose impact on locally rare or endemic species may be profound. There is an existing world "family" of zoos in the form of WAZA and, where possible and welcomed, helping out other members, or zoos that could be productive members, should be seen as a responsibility, and a key role for improving the conservation infrastructure in the developing world.

SOURCING ANIMALS FROM THE WILD – A CONSERVATION ACTION?

When collection plans are in place on paper there follows the task of making that plan a reality. Well-established breeding programs, regional and international in scope [(e.g., European Endangered Species Program (EEPs) and Species Survival Plans (SSPs), etc.)] are most commonly the route of acquisition and (equally important) disposition of animals. This is generally not contentious and zoos have done well to explain to the general public about the efficacy of captive management programs. What zoos have failed to do, in this attempt to explain their role, is to convey coherently the much more complex argument surrounding imports of animals from the wild. There is now an almost immediate revulsion about sourcing animals from the wild, yet the amphibian discussion above demonstrates that acquiring wild animals for conservation purposes is sometimes not only the lesser of two evils, but the best and most realistic way to ensure that species survive in the short term (depending on the generation time of the species).

Zoos need to develop a more open discourse, amongst themselves and the public, in relation to wild acquisitions. Clearly delineating the argument into its constituent parts would assist substantially in this process and includes: the conservation need of the focal species (including recommendations of noted authorities such as the IUCN), the status of the focal species in the wild, the ability of zoos to manage the species in captivity, and the method of sourcing the individual animals. It should be apparent that sourcing animals from commercial dealers without due caution is untenable for the modern conservation zoo. If there is a clear conservation need to work with a species, zoos, in partnerships with other conservation bodies, should work to bring the species into captivity within the framework of government and regional protocols and agreements. Relying on a commercial middleman is fraught with difficulties, and perhaps regional zoo associations should seek to develop their own professional sourcing teams that can be called upon to undertake such delicate operations, with transparent procedures that adhere to the highest ethical and welfare standards. There are existing examples of ethical sourcing of wild animals for commercial interest that could be useful in developing further protocols for zoos.

The Marine Stewardship Council (MSC) has developed a certification system for commercial fisheries to assist in the development of sustainable fisheries for the future (MSC 2005). The MSC provides a certification system that can be independently assessed and a chain of custody program that enables tracking of fish from source to customer. Could a similar system be put in place for the sourcing of wild animals for conservation purposes in zoos? The total numbers of animals sourced for zoos would be substantially less than that sanctioned for commercial interests (and even when fish stocks are dangerously low commercial imperatives dictated by governments often override conservation needs). Such a system may be able to make good use of existing and developing global zoo database systems and would provide a transparent mechanism with which to inspire public trust in the system.

Another concept of sourcing from the wild for conservation purposes is that of extractive reserves, first described by Conway (1998) and Redford (1998). Designating a natural habitat as an extractive reserve, and thereby managing it for animal production, has multiple benefits. The intense activity this would require ensures a high level of activity inherent in conserving a suitable environment. Surveys of the habitat, the species present, ecological interactions, and sustainable off-take of target species would generate valuable research data. In addition, the hiring of staff (as employees of an individual zoo or zoo association) dedicated to the whole function of the reserve would negate the problems and pitfalls of going through middlemen. Meta-population management, with movement of animals in both directions, from the reserve to zoo populations, would improve the genetic diversity of the individual populations. Perhaps most significantly, developing the extractive reserve in partnership with local peoples, managing it for their sustainable extractive needs in addition to that of the zoos, could promote closer harmony between human development needs and those of conservation imperatives. This concept has not been tested and it is likely that individual zoos are not best placed to do so. Rather, zoo consortiums, regional associations of zoos, or WAZA would be preferred to oversee such a concept and develop it in practice. The reluctance to do so is likely more linked to worries of public perception and condemnation than the utility of the idea.

CAPTIVE MANAGEMENT . . . AND THEN WHAT?

While *ex situ* management can be used to great effect in the conservation of species (through management solely in captivity or through meta-population management with transfer of individuals from the wild) this remains only one area where zoo collections can actively assist in conservation and link to *in situ* activities. More commonly, in tandem with the idea of providing an assurance population, animals housed in zoos can function to provide an inspirational, educational, and flagship role.

Flagships for conservation

Although the idea of flagship species has received criticism in the literature (Leader-Williams and Dublin 2000) it does present an easy identifier for the general public and often stimulates an emotional reaction. People don't tend to weep for the process of desertification but they are often inspired to act by the plight of charismatic species, and particularly individuals. This common emotional response can be used to good effect for conservation purposes by zoos, engaging zoo visitors in a connection with threatened species, through exhibition in zoos. In addition, the aspect of presenting "expected" species (Ward *et al.* 1998) should be borne in mind if zoos are not to alienate their core funding base, the public. However, this is not without some controversy and discussion. Some totem species remain contentious in zoos with elephants the most obvious example. Does breeding elephants in captivity really contribute to their safety and viability in the wild? Quite clearly the answer, at the present time, is no.[3] In isolation, simply maintaining elephants in captivity does not impact on their conservation status in the wild. However, it presents a conservation *opportunity* for zoos.

Elephants remain a storybook species and one that can be used to focus attention on ecosystems under threat, acting as "translators" or focal animals for complex conservation problems (in and out of range countries). Elephants in zoos also generate the opportunity to assist field conservation through scientific research, fundraising, technology development, and training and technology transfer (Hutchins and Keele 2006). Much of what we understand about reproductive physiology of African elephants has been discovered in zoos, and can now be applied to management problems in range countries, where elephant "space" in the wild is becoming increasingly rare (see Chapter 12 for an extended discussion of research in zoos).

While the Asian elephant population in North American and European zoos is still some way from being entirely self-sustaining (Belterman 2004,

[3] The lessons of the extinction of the passenger pigeon (*Ectopistes migratorius*) and the near extinction of the American bison (*Bison bison*), both originally profusely abundant species, should serve as reminders that situations change and unlikely species may become vulnerable due to unforeseen events. A continuing expansion of the human population across Asia and increasing demand for resources by an increasingly urbanized planet will, if unchecked, lead to a continued decline in suitable habitat for the Asian elephant. What then the importance of a captive population?

Faust *et al.* 2006), great strides have been made in elephant care in the recent past and births in European zoos are now relatively commonplace, which should in time diminish the need for wild imports, an issue that often prompts heated and unproductive debate.

Generating funds for the wild

The potential conservation fundraising abilities of zoos are substantial. The monies spent on field conservation worldwide are relatively small in comparison to the billions spent by consumers on disposable and luxury lifestyle items each year, and the total sums requested by conservation initiatives remain modest. Zoos have an important role to play in diverting disposable incomes into conservation projects where possible. Fundraising through the medium of living captive collections allows zoos to target their monies increasingly to *in situ* conservation (see Chapter 20). This fundraising can be targeted at the general visiting public and/or more focused at wealthy donors, and zoos are becoming far more professional in relation to both aspects. In addition a few *ex situ* management program coordinators have taken the opportunity to raise monies for field conservation through their programs. One such example is that of the fossa (*Cryptoprocta ferox*).

Madagascar is primarily associated with lemurs, the endemic primate fauna, yet there are numerous other unique and charismatic species found on the island. The relatively small captive population of the fossa, the largest of the endemic *Carnivora*, has stimulated interest and been instrumental in raising funds for *in situ* conservation, and both captive and wild research, through the auspices of the Fossa Fund administered by Zoo Duisburg. This has been achieved solely through exhibiting this unusual, threatened species in captivity. Where zoos wish to receive fossa from the EEP they pay a "conservation surcharge," wherein the sum of approximately £1000 is payable on receipt of the animals. This is not payment for the animals *per se* and does not go to the institution relinquishing the animals. Rather the funds are transferred to the Fossa Fund, where they are administered by the program coordinator. This ensures that the animals are not requested lightly and that zoos make a conscious decision to donate funds. This is an innovative idea that has now been copied in other programs.

The above is an example of fundraising for a specific purpose *between* zoos. More generally, living collections can be used to raise monies from visitors for specific *in situ* conservation activities (see Chapters 17 and 19). Informing the more financially able developed world, inspiring zoo visitors to care and to provide financial support to conservation efforts, is an important role of any developed world zoo collection. The example of the Congo

Gorilla Forest at the Bronx Zoo exemplifies the possibility of linking a zoo exhibit of charismatic animals to *in situ* conservation. Perhaps most importantly for the long-term aims of inspiring caring in zoo visitors, the Congo Gorilla Forest asks its visitor to make active choices about which conservation programs their dollars will fund, described by Conway (1999) as a zoo experiment in "participatory conservation." Visitors travel through a Congo experience, meeting amazing animals along the trail, they learn of the challenges to the survival of the animals and their habitats, and they meet WCS scientists (on film) who work in the Congo to alleviate these problems. By this stage of the exhibit the visitor is engaged and involved. Only at the end, and therefore at a point when they are more aware of the challenges facing the species they have just viewed and to which they are hopefully more emotionally connected, are they asked to make a choice about how they want their entrance fee spent. Overwhelmingly visitors take the time to make an active choice, becoming involved in *in situ* conservation through the portal of the zoo in a far more meaningful way than previously would have been available (see also Chapter 14).

Awareness followed by action

Zoos in the "frontline," particularly in the developing world, can use their voice to highlight the problems of unsustainable use to the people who utilize the resources. In addition these zoos should strive to find local solutions to local problems. Ivoloina, the zoo and forest station managed by the Madagascar Fauna Group (MFG) (see Chapter 18), runs classes at which more sustainable agricultural practices are taught. MFG also uses the zoo to convey conservation messages to local peoples, who directly impact on threatened species. This approach appears to be making some progress with government agencies, both local and national, requesting further training specific to conservation and sustainable-use practices. Ivoloina also receives confiscated and donated rare species. MFG zoos should therefore be able to feel that Ivoloina is a partner zoo in Madagascar and an outpost of their own zoo activities (see Chapter 18).

It is not only in the developing world that zoos can use their animal collections to convey conservation messages to their local constituents. Reading (Chapter 6), Balmford (Chapter 9), and Sterling (Chapter 4) make abundantly clear that zoos have not yet fully committed to understanding the effectiveness of the interpretive and education messages they convey. However, zoos should remain optimistic. The opportunity to have live animal "advocates" for species, protected areas, and even ecosystems is one not conventionally available to other conservation groups. Tapping into the emotional response that people have for individual animals in zoos and transferring that caring to wider conservation issues is one of the key roles of the 21st century zoo. Blurring that line to ensure that zoos are linked to specific field sites, either the zoo's own or a partner organization, further positions zoos in a unique niche to stimulate positive action for conservation.

Providing a link to field conservation activities, and *in situ* initiatives, through innovative and inspiring zoo graphics (see Chapter 5) is an exceptionally important tool for zoos. Yet too often this aspect of zoos is underfunded, under-examined, and treated as "window dressing" instead of as an exciting opportunity to connect with visitors for the benefit of conservation. If this is the most common way that zoos communicate with their visitors, then zoos need to invest, not only in the production of good-quality graphics, but in testing those graphics (Chapter 9).

In addition the public communication of science, much discussed and puzzled over by the academic community, could find a powerful outlet through the non-threatening atmosphere of the zoo, particularly if research carried out by or in zoos has an applied conservation focus. Can zoos become a focal point in their local area for communicating science? Zoos are egalitarian places, with an attendance that reflects a broader and more even social spectrum than that of museums and similar attractions (Museums, Libraries and Archive Council 2004). Issues of global sustainability, and consequent impact on biodiversity, need to be conveyed, understood, and assimilated into the lives of all, not just a more literate and educated elite. Can important scientific discoveries that remain largely hidden in journals, and read solely by fellow professionals, find a vast, new, and excited audience if the means to "translate" this world were developed in zoo graphics? The global population is becoming more urban than rural and in tandem with communicating science, zoos as green spaces may also have a hither to unknown function in the psyche of an increasingly urban human population, that ultimately benefits conservation. Recent research suggests that access to a "green" view could enhance recovery of patients in hospitals (Ulrich 2002) and the benefits of exercise may well be enhanced by exercising in a rural as opposed to an urban environment (Pretty *et al.* 2005). Are there similar benefits for conservation education?

WHERE DO ZOOS GO NEXT IN FURTHER MERGING *EX* AND *IN SITU* ROLES?

One possible future challenge for zoos as they move toward being holistic conservation NGOs is in losing the very essence of what they can achieve by segregating conservation from the animal collection. By separating research, conservation and animal departments from each other, this may institute a level of division that could hamper efficient and smooth integration. All zoo staff (regardless of the department they work in) need to be stakeholders in a common conservation vision that stretches from the urban zoo to field sites around the world. When a zoo has a committed staff, motivated by a common goal and fulfilling their part of the conservation jigsaw, bringing with them individual expertise strengths in cooperation with colleagues, and building a strong support base via their visitors and partners, then that zoo has truly become an integrated force for conservation.

Perhaps the greatest challenge for zoos is not to become "jacks of all trades and masters of none." Zoos can be highly effective at conservation in the field and in the zoo; they can be powerful advocates for behavioral change in their staff and visitors, leading to a more sustainable world; and they can inspire people to value wildlife for its intrinsic worth as well as its bountiful benefits for humanity. But what zoos should be seeking is to be masters of holistic conservation, the effective combination of all these activities, a goal that is simply unattainable for any other type of conservation NGO.

References

Andreone, F., Cadle, J. E., Cox, N., Glaw, F., Nussbaum, R. A., Raxworthy, C. J., Stuart, S. N., Vallan, D., & Vences, M. (2005). Species review of amphibian extinction risk in Madagascar: conclusions from the global amphibian assessments. *Conservation Biology*, **19**(6), 1790–1802.

Anon (2005). *Introduction. Building a Future for Wildlife: The World Zoo and Aquarium Conservation Strategy*. Bern: World Association of Zoos and Aquariums.

Balmford, A. (2000). Separating fact from artifact in analyses of zoo visitor preferences. *Conservation Biology*, **14**(4), 1193–1195.

Balmford, A., Mace, G. M., & Leader-Williams, N. (1996). Designing the ark: setting priorities for captive breeding. *Conservation Biology*, **10**(3), 719–727.

Balmford, A., Mace, G. M., & Ginsberg, J. R. (1998). The challenges to conservation in a changing world: putting processes on the map. In *Conservation in a Changing World*, eds. G. M. Mace, A. Balmford, & J. R. Ginsberg. Cambridge: Cambridge University Press, pp. 1–28.

Beck, B. B., Rapaport, L. G., Stanley Price, M. R., & Wilson, A. C. (1994). Re-introduction of captive born animals. In *Creative Conservation: Interactive Management of Wild and Captive Animals*, eds. P. J. S. Olney, G. M. Mace, & A. T. C. Feistner. London: Chapman and Hall, pp. 265–286.

Belterman, R. (2004). Status of the Asian elephant EEP. *EAZA News*, **47**, 11–13.

Berger, L., Speare, R., Daszak, P. *et al.* (1998). Chytridiomycosis causes amphibian mortality associated with population declines in the rain forests of Australia and Central America. *Proceedings of the National Academy of Sciences of the United States of America*, **95**, 9031–9036.

Blaustein, A. R., & Wake, D. B. (1990). Declining amphibian populations: a global phenomenon. *Trends in Ecology and Evolution*, **5**, 203–204.

Bloxam, Q. M. C. & Tonge, S. J. (1995). Amphibians: suitable candidates for breeding-release programmes. *Biodiversity and Conservation*, **4**, 636–644.

Bonner, J. (2006). Captive to wild: how do zoos make that leap and where should they focus their resources. In *Proceedings of the World Association of Zoos and Aquaria Annual Meeting*, 2–6 October, New York, ed. P. Dollinger. Washington, D.C.: WAZA.

Conway, W. (1998). Zoo reserves: a proposal. In *Proceedings of the AZA Annual Conference*, Tulsa Zoo & Living Museum, Tulsa, Oklahoma. Chicago, IL: AZA, pp. 54–58.

Conway, W. (1999). Congo: a zoo experiment in participatory conservation. In *Proceedings of the AZA Annual Conference*, Minnesota Zoo, Minneapolis, Minnesota. Chicago, IL: AZA, pp. 101–104.

Conway, W. (2003). The role of zoos in the 21st century. *International Zoo Yearbook*, **38**, 7–13.

Cuaron, A. D. (2005). Further role of zoos in conservation: monitoring wildlife use and dilemma of receiving donated and confiscated animals. *Zoo Biology*, **24**(2), 115–124.

Cunningham, A. A., Garner, T. W. J., Aguilar-Sanchez, V. *et al.* (2005). Emergence of amphibian chytridiomycosis in Britain. *Veterinary Record*, **157**, 386–387.

Daszak, P., Berger, L., Cunningham, A. A., Hyatt, A. D., Green, D. E., & Speare, R. (1999). Emerging infectious diseases and amphibian population declines. *Emerging Infectious Diseases*, **5**, 735–748.

Dietz, J. M., Diet, L. A., & Nagagata, E. Y. (1994). The effective use of flagship species for conservation of biodiversity: the example of lion tamarins in Brazil. In *Creative Conservation: Interactive Management of Wild and Captive Animals*, eds. P. J. S. Olney, G. M. Mace, & A. T. C. Feistner. London: Chapman and Hall, pp. 32–49.

Faust, L. J., Thompson, S. D., & Earnhardt, J. M. (2006). Is reversing the decline of Asian elephants in North American zoos possible? An individual-based modelling approach. *Zoo Biology*, **25**, 201–218.

Glowka, L., Burhenne-Guilman, F., Synge, H., McNeely, J. A., & Gündling, L. (1994). A guide to the Convention on Biological Diversity. *Environmental Policy and the Law*, Paper No. 30. Gland: IUCN.

Hutchins, M. & Keele, M. (2006). Elephant importation from range countries: ethical and practical considerations for accredited zoos. *Zoo Biology*, **25**, 219–233.

Hutchins, M., Willis, K., & Wiese, R. J. (2005). Strategic collection planning: theory and practice. *Zoo Biology*, **14**(1), 5–25.

IUCN (2002). Technical guidelines on the management of *ex situ* populations for conservation. Prepared by IUCN/SSC Conservation Breeding Specialist Group. Gland: IUCN.

IUCN (2006). *2006 IUCN Red List of Threatened Species*. Gland: IUCN The World Conservation Union, www.redlist.org.

Kleiman, D. G., Reading, R. R., Miller, B. J. *et al.* (2000). Improving the evaluation of conservation programs. *Conservation Biology*, **14**(2), 356–365.

Leader-Williams, N. & Dublin, H. T. (2000). Charismatic mega-fauna as "flagship species." In *Priorities for the Conservation of Mammalian Diversity: Has the Panda had its Day?*, eds. A. Entwistle & N. Dunstone. Cambridge: Cambridge University Press, pp. 52–81.

Lips, K. (1998). Decline of a tropical Montane amphibian fauna. *Conservation Biology*, **12**(1), 106–117.

MSC (2005). Marine Stewardship Council Guidance to potential or actual clients: The MSC Fishery Assessment & Certification Process. http://www.msc.org/html/content_465.htm.

Museums, Libraries and Archive Council (2004). Unpublished report.

Pretty, J., Peacock, J., Sellens, M., & Griffin, M. (2005). The mental and physical outcomes of green exercise. *International Journal of Environmental Health Research*, **15**(5), 319–337.

Redford, K. (1998). The role of extractive reserves in wildlife conservation. In *Proceedings of the AZA Annual Conference* 1998, Tulsa Zoo & Living Museum, Tulsa, Oklahoma. Chicago, IL: AZA, pp. 103–105.

Saterson, K. A., Christensen, N. L., Jackson, R. B. (2004). Disconnects in evaluating the relative effectiveness of conservation strategies. *Conservation Biology*, **18**(3), 597–599.

Soulé, M., Gilpin, M., Conway, W., & Foose, T. (1986). The millennium ark: how long a voyage, how many staterooms, how many passengers? *Zoo Biology*, **5**, 101–113.

Stanley Price, M. R. (1989). *Animal Reintroductions: The Arabian Oryx in Oman.* Cambridge: Cambridge University Press.

Struhsaker, T. T., Struhsaker, P. J., & Siex, K. S. (2005). Conserving Africa's rain forests: problems in protected areas and possible solutions. *Biological Diversity*, **123**, 45–54.

Stuart, S. N., Chanson, J. S., Cox, N. A., Young, B. E., Rodrigues, A. S. L., Fischman, D. L., & Waller, R. W. (2004). Status and trends of amphibian declines and extinctions worldwide. *Science*, **306**, 1783–1786.

Ulrich, R. S. (2002). Health benefits of gardens in hospitals. In *Proceedings of the Plants for People International Symposium.* Floriad, The Netherlands.

Ward, P. I., Mosberger, N., Kistier, C., & Fischer, O. (1998). The relationship between popularity and body size in zoo animals. *Conservation Biology*, **12**(6), 1408–1411.

WAZA (2005). *The World Zoo and Aquarium Conservation Strategy: Building a Future for Wildlife.* Liebefeld-Bern: WAZA.

Wemmer, C. M. & Thompson, S. (1995). A Short History of Scientific Research in *The Ark Evolving; Zoos and Aquariums in Transition*, ed. C. M. Wemmer. Front Royal, VA: Smithsonian Institution, Conservation and Research Center, pp. 70–94.

Wiese, R. J., Willis, K., & Hutchins, M. (1994). Is genetic and demographic management conservation? *Zoo Biology*, **13**, 297–299.

Wyman, R. L. (1990). What's happening to the amphibians? *Conservation Biology*, **4**(4), 350.

Zippel, K. (2006). Amphibian conservation primer for zoo directors. Unpublished report. The Association of Zoos and Aquaria/CBSG.

16

Beyond the ark: conservation biologists' views of the achievements of zoos in conservation

NIGEL LEADER-WILLIAMS, ANDREW BALMFORD, MATTHEW LINKIE, GEORGINA M. MACE, ROBERT J. SMITH, MIRANDA STEVENSON, OLIVIA WALTER, CHRIS WEST, AND ALEXANDRA ZIMMERMANN

INTRODUCTION

Conservation comprises actions that directly enhance the chances of habitats and species persisting in the wild. Globally, the collective challenge of conserving biodiversity is underpinned by the 1992 Convention on Biological Diversity, Article 1 of which recognizes its three main objectives as:

- maintaining biological diversity;
- allowing sustainable use of its components; and
- promoting equitable sharing of its benefits.

In turn, most conservation biologists recognize that zoological collections have a key and active role in this collective challenge. The importance of zoological collections in complementing *in situ* measures through *ex situ* conservation is underpinned through Article 9 of the Convention on Biological Diversity (Glowka 1993), and through internationally endorsed policies such as the IUCN Technical Guidelines (2002).

Most zoo professionals also recognize their role in this collective challenge. Indeed, the original *World Zoo Conservation Strategy* (IUDZG/CBSG 1993) stated the core objectives for all zoos that wish to make a substantial contribution to conservation as:

- actively supporting conservation of endangered species through coordinated programs;
- offering support and facilities to increase scientific knowledge that benefits conservation; and,
- promoting an increase of public and political awareness of the need for conservation.

All responsible zoos were obliged to play a role in achieving the goals of the *World Zoo Conservation Strategy* (IUDZG/CBSG 1993), and to use this strategy as a basis upon which to formulate their mission statements and conservation objectives. In some instances, licensing further obliges zoos to become fully engaged in achieving conservation objectives. For example, the European Community Zoos Directive of 1999 requires all licensed collections to make some contributions to conservation, education and research (EC 1999).

Nevertheless, despite these encouraging sentiments and increasing requirements to contribute to conservation, it may not always be clear which direction zoological collections should follow. For example, the scope for species conservation through captive breeding is limited by the availability of space and resources; and problems of husbandry, reintroduction, cost, domestication, and disease (Conway 1986, 2003). Instead, many zoo professionals advocate focusing on flagship species that raise public awareness and support for *in situ* conservation (Hutchins *et al.* 1995, Conway 2003, and others).

So is it possible to evaluate the different approaches that zoos may take? Several qualitative studies assess the achievements of zoos and aquariums (e.g., Tribe and Booth 2003, Conway 2003, Miller *et al.* 2004, chapters in this volume), and identify many of the questions that need to be asked. However, there are few quantitative assessments of these different approaches. In turn, this offers conservation biologists a clear role to work with zoo professionals to provide balanced and constructive analysis. In response to this opportunity, an independent research group, the "Zoo Measures Group" was formed, which contained both conservation biologists and zoo professionals, to begin the process of quantifying the impact of conservation measures taken by zoos.

This chapter, which is one of three offered by the group, seeks to answer the following questions:

- do coordinated programs focus on a wide range of threatened species, and follow rational criteria for selecting species?

- are the benefits of zoos, which rely on species from biodiversity-rich countries, equitably distributed between countries? and
- do zoos implement the *World Zoo Conservation Strategy* (IUDZG/CBSG 1993) through efforts to promote education and public awareness, research and *in situ* conservation?

Two other chapters propose a method for measuring the overall conservation impact of zoos (Chapter 21), and their educational impact in particular (Chapter 9).

SETTING PRIORITIES FOR CAPTIVE BREEDING

Given tight constraints on zoo space and resources, we previously identified the need for rational criteria to identify threatened taxa to include in captive breeding programs (Balmford *et al.* 1996). Our analysis suggested that species in coordinated captive breeding programs should mainly comprise small-bodied species for reasons of breeding and cost; and species for which habitat remains to ensure realistic chances of reintroduction. Nevertheless, at the time most zoos were doing the opposite, and on balance tending to keep a disproportionate number of mammals, mammals from larger-bodied orders in particular, and a disproportionate number of species with very limited prospects of reintroduction. As it is now 10 years since the data for that analysis were gathered, this section evaluates whether there have been any changes in the species included in coordinated breeding programs over the past decade.

First, we compare the numbers of threatened species in the wild with those in coordinated captive programs in 1992/93 (Balmford *et al.* 1996) and 2003 (Figure 16.1). We gathered data compiled by the European Association of Zoos and Aquaria Association of Zoos and Aquariums (EAZA) and the (AZA), and from the IUCN Red List. However, there have been some changes in the way data were recorded between 1992/93 and 2003. Both EAZA and AZA have categorized species held in coordinated captive breeding programs under new designations since 1993. EAZA now lists species under both Europäisches Erhaltungszucht–Programm or European Endangered Species Program (EEPs) and European Studbooks (ESBs), while AZA lists species under both Species Survival Plans (SSPs) and Population Management Plans (PMPs). The data for 2003 include species held under all four designations. Additionally, there was a new Red List system in 2003, and we needed to make the threatened categories as comparable as possible for 1992/93 and 2003. Therefore, we re-worked the 1993 data to

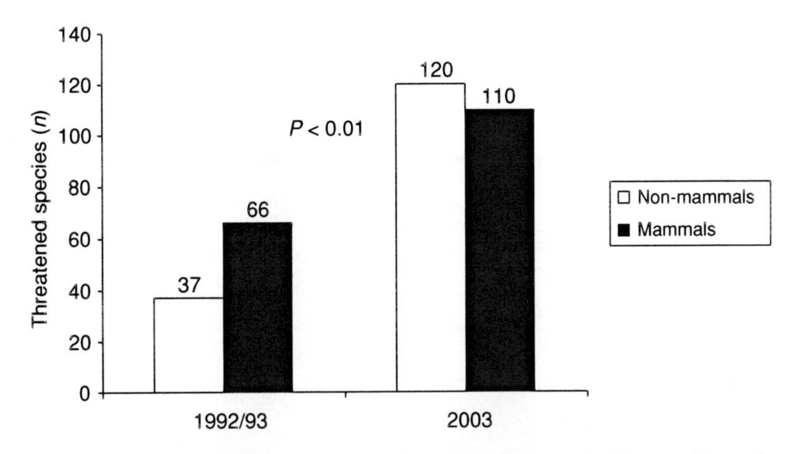

Figure 16.1 The increase in numbers of threatened species held in coordinated captive breeding programs between 1992/93 and 2003, based on data from Balmford *et al.* (1996), Hilton-Taylor (2003), EAZA (2003) and AZA (2003). The sample sizes are shown above each column, and the change in the proportions of non-mammals to mammals was significant ($\chi^2 = 7.54$, df $= 1$, $P = 0.006$)

include species that were listed as Endangered, Vulnerable, Rare and Indeterminate, but to exclude species listed as Insufficiently Known (Groombridge 1993). The 2003 data included all species that were listed as Extinct in the Wild, Critically Endangered, Endangered and Vulnerable (Hilton-Taylor 2003).

There was a marked increase in the numbers of threatened species held in coordinated captive breeding programs between 1992/93 and 2003 (Figure 16.1). The number of mammals held in coordinated programs had close to doubled, while the number of non-mammals had more than trebled. When the proportions of threatened species in the wild were compared with those held in coordinated breeding programs, the skew evident in 1992/93 toward threatened mammals (Figure 16.2a) was less marked but still very significant in 2003 (Figure 16.2b). Thus, nearly 80% of species that were listed as threatened in the wild are non-mammals, but there were similar numbers of threatened mammal and non-mammal species in coordinated captive breeding programs (Figure 16.2b).

Looking across different mammalian Orders, threatened species in 2003 were still more likely to be held in coordinated breeding programs if they belonged to larger-bodied Orders (Figure 16.3). Threatened species from 11 mammalian Orders were held in coordinated captive programs, an increase of two small-bodied Orders, Dasyuromorphia and Xenarthra, since 1992/93 (Balmford *et al.* 1996). Nevertheless, very few threatened

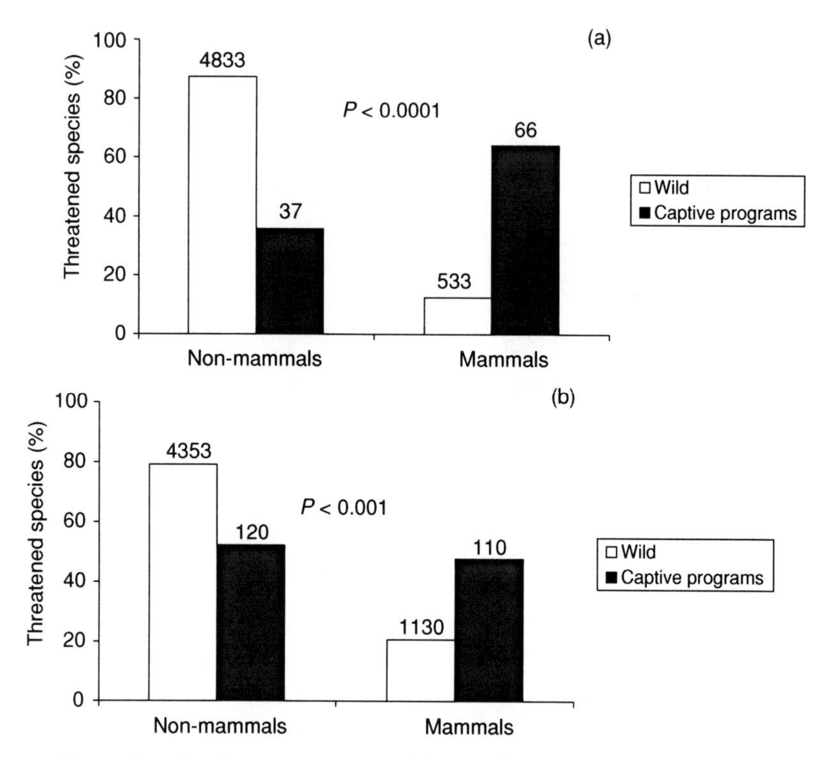

Figure 16.2a, b The representation of threatened species in the wild and in coordinated captive breeding programs (a) in 1992/93 (modified from Balmford *et al.* 1996), and (b) in 2003, based on data from Hilton-Taylor (2003), EAZA (2003) and AZA (2003). The sample sizes are shown above each column. The clear columns and the shaded columns each add up to 100% in both (a) and (b). The skewed representation of mammals as compared with non-mammals still remained highly significant in 2003 ($\chi^2 = 96.22$, df $= 1$, $P < 0.001$)

species among small-bodied Orders were held in coordinated programs, while some 20% of larger-bodied Orders such as Primates, Carnivora and Artiodactyla, together with all threatened Perissodactyla, were held in coordinated breeding programs.

Among threatened species of non-mammals, there was a disproportionate representation of birds and reptiles held in coordinated breeding programs in 2003, compared with the numbers of threatened taxa listed for each class in the wild (Figure 16.4). Furthermore, it should be noted that the true skew was much greater than represented here as, of these groups, only birds had been subject to a comprehensive threat assessment at the time of this analysis. Of the non-mammals listed as threatened in 2003 (Hilton-Taylor 2003), these mostly comprised birds, molluscs, fishes,

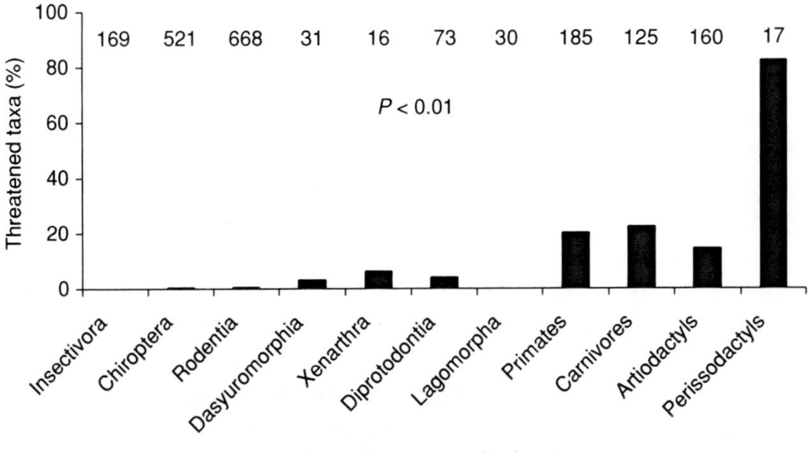

Figure 16.3 The representation of different mammalian orders in coordinated captive breeding programs in 2003, arranged by increasing body mass, based on data from Balmford *et al.* (1996), Hilton-Taylor (2003), EAZA (2003) and AZA (2003). The number of threatened taxa for each order in the wild is shown above the column, while each point shows the proportion of each Order that was represented in coordinated captive breeding programs. The increasing representation among large-bodied species was significant ($r_s = 2.55$, $P < 0.01$)

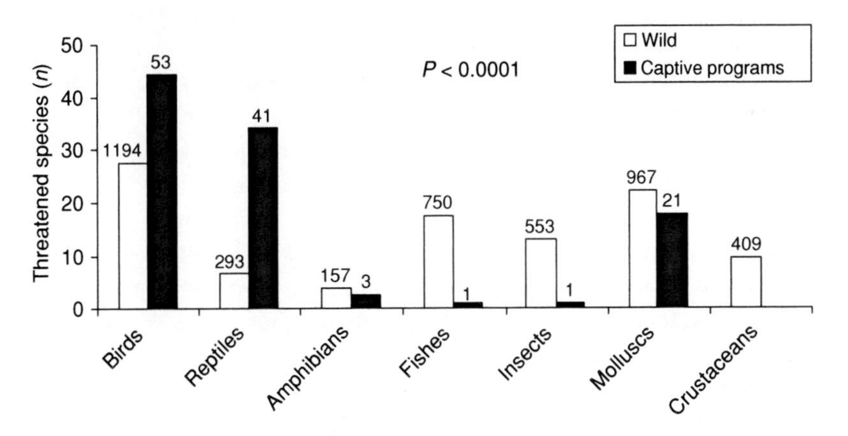

Figure 16.4 The representation of threatened species in the wild and in coordinated captive breeding programs for different classes of non-mammals in 2003, based on data from Hilton-Taylor (2003), EAZA (2003) and AZA (2003). The sample sizes are shown above each column. The clear columns and the shaded columns respectively each add up to 100% across all classes. The representation among classes in the wild and in coordinated captive breeding programs differed very significantly ($\chi^2 = 173.15$, df = 6, $P < 0.0001$)

insects, and crustaceans. However, among the non-mammal species held in coordinated captive breeding programs, there was an over-representation among birds and reptiles, and an extreme under-representation among amphibians, fishes, insects, and crustaceans. The skew evident in these data will have now worsened, given that the number of threatened amphibians has since been fully assessed (IUCN *et al.* 2006). Nevertheless, there are very encouraging signs of appropriate representation among threatened molluscs, primarily due to the different coordinated programs for breeding species of *Partula* snail in captivity (Pearce-Kelly 1994).

In terms of "reintroducability," if zoos are to focus on threatened species that stand a realistic chance of reintroduction, coordinated programs should avoid focusing on species facing large-scale and irreversible habitat loss. Instead, they should concentrate on species facing more reversible threats, such as over-hunting. There was no evidence of this in 1993 for carnivores, primates, and ungulates (Balmford *et al.* 1996). For example, there was no difference in the proportions of threatened species of carnivore affected by habitat loss that were held within, and not within, coordinated captive breeding programs in 1992/93 (Figure 16.5a). Likewise, there was no difference in 2003: zoos still appear to be selecting threatened species for coordinated breeding programs at random with respect to the threats they face in the wild. This has become a more serious issue since 1992/93 because the proportion of threatened species affected by habitat loss has increased relative to those affected by more reversible threats (Figure 16.5b).

In summary, this section shows some very positive signs of movement in the 10 years since we published our recommendations on selecting species for coordinated captive breeding and reintroduction programs (Balmford *et al.* 1996). There are now more coordinated captive breeding programs (Figure 16.1), and these programs now show a better balance toward non-mammals (Figure 16.2). For example, there is now an impressive showing of classes such as molluscs (Figure 16.4). Nevertheless, there still remained a continued skew to large mammals (Figure 16.3), and to birds and reptiles (Figure 16.4). Furthermore, there remains a continued lack of emphasis on seeking to include species that are more reintroducable into coordinated captive breeding programes (Figure 16.5). Taking these different data in combination, the successful reintroduction of different species of captive-bred *Partula* snail into the wild offers a good model of selecting appropriate species for cost-effective captive breeding and greater returns per species reintroduced, than opting for those larger-bodied species that are harder both to breed and to reintroduce to the wild because they faced irreversible habitat loss (Balmford *et al.* 1996).

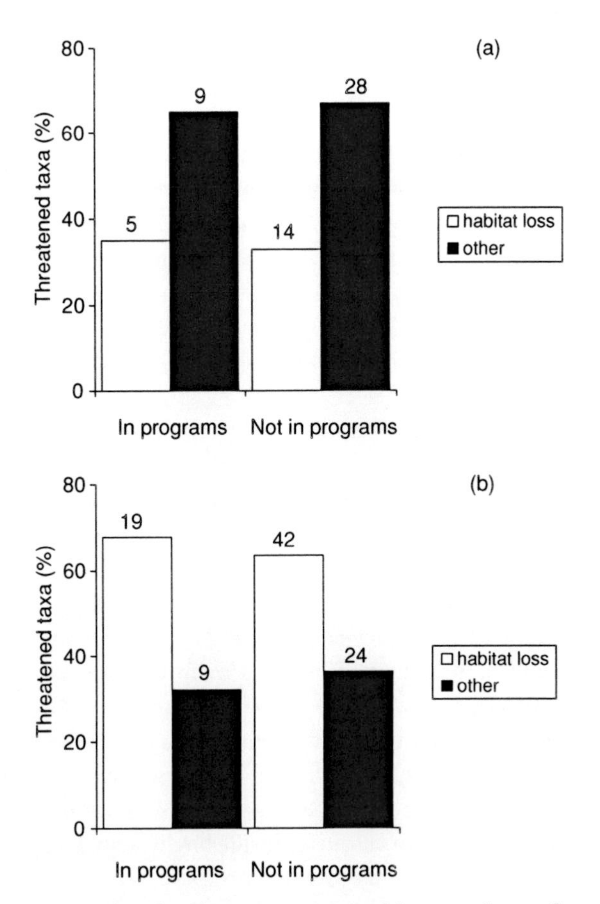

Figure 16.5a, b The representation of threatened taxa of carnivores held and not held in captive programs, subdivided by whether or not they faced more irreversible threats from habitat loss, as compared with more reversible threats say from over-hunting. Data are shown (a) for 1992/93 (based on Balmford *et al.* 1996), and (b) in 2003, based on data from Hilton-Taylor (2003), EAZA (2003) and AZA (2003). There was no difference (both $P > 0.05$) in representation in either 1992/93 or 2003, while a comparison between 1992/93 and 2003 shows that more carnivores faced more irreversible threats arising from habitat loss in 2003

THE POWER OF VISITOR NUMBERS AND EQUITY IN BENEFITS FROM ZOOS

The *World Zoo Conservation Strategy* recognizes the opportunity of zoos to influence large numbers of visitors (IUDZG/CBSG 1993). However, are there any patterns to zoo visitation that should determine additional responsibilities for certain zoological collections? For example, are there issues relating to the equitable sharing of benefits arising from biodiversity, as required in Article 1 of the Convention on Biological Diversity (1992), particularly for zoological collections in developed countries? To explore such questions, this section examines data on the distribution and visitation rates of 875 zoos which report to the *International Zoo Yearbook* (Olney and Fisken 2003), and compares these data with measures of biodiversity, level of threat to biodiversity, and wealth. We recognize that these data may be biased by spatial variation in zoo reporting rates, and note the possible different implications of the analysis, depending on whether or not this proves to be the case.

In terms of location of zoos, the numbers of zoos reporting in different countries were corrected by country population size, to give a national measure of the number of zoos per million people. There were many fewer zoos per million people in countries which overlap with Conservation International's biodiversity hotspots (areas of globally high endemism and threat: Myers *et al.* 2000), as compared with countries outside hotspots (Figure 16.6a). Furthermore, there was a negative correlation between the number of zoos per million people and combined bird and mammal species richness in different countries ($r_s = -0.398$, $P < 0.001$, $N = 101$).

In terms of visitation to zoos, the total number of visits reported to zoos in each country was summed and corrected by population size, and measured as $\log_{10}$ number of visits per million people. There were many fewer zoo visits per million people in countries inside biodiversity hotspots as compared with countries outside hotspots (Figure 16.6b). Furthermore, there was a negative correlation between the number of zoo visits per million people and combined bird and mammal species richness in different countries ($r_s = -0.311$, $P < 0.05$, $N = 66$).

In terms of combining aspects of location, visitation, and national wealth, there was a positive correlation ($r_s = 0.674$, $P < 0.001$, $N = 109$) between the number of zoos per million people and $\log_{10}$ per capita gross domestic product (GDP). Likewise, there was a positive correlation between $\log_{10}$ number of visits per million people and $\log_{10}$ per capita GDP (Figure 16.7).

In summary, this section suggests that most zoos are located, and most zoo visits occur, in richer countries (Figure 16.7), while there are fewer zoos and zoo visits in countries rich in biodiversity, and in countries with threatened biodiversity (Figure 16.6), at least based on the data reported to the *International Zoo Yearbook* (Olney and Fisken 2003). If these data are representative, the analysis suggests that benefits that are associated with zoos in developed countries need to be shared more equitably with poorer and

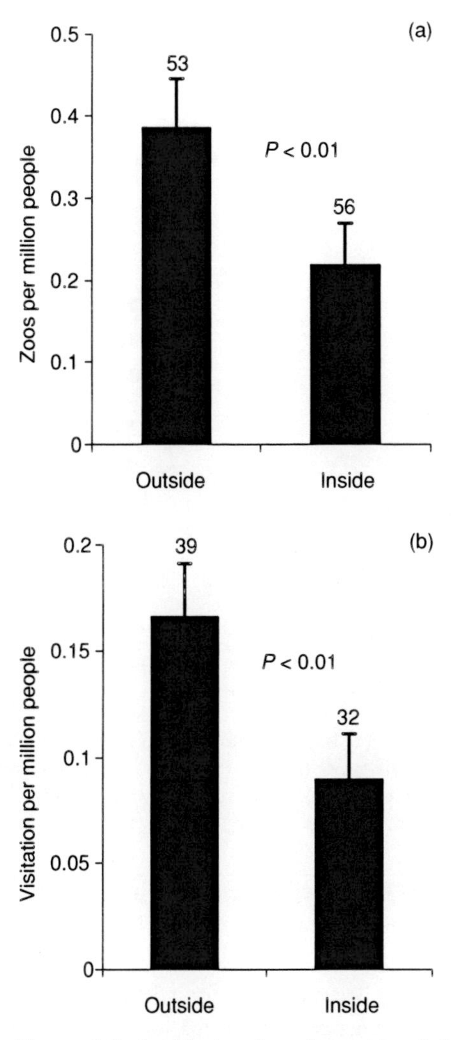

Figure 16.6a, b The location of zoos (a) and the extent of visitation to zoos (b), both corrected for national population size, in countries outside and inside hotspots, based on data from *International Zoo Yearbook* (Olney and Fisken 2003) and Myers *et al.* (2000). The number of zoos located inside and outside hotspot countries differed significantly ($z = 2.56$, $P < 0.01$) as did the extent of visitation ($z = 2.62$, $P < 0.01$)

biodiversity-rich countries, that in turn have provided many of the animals on which zoos in developed countries depend. Furthermore, if zoos are to play a role in public education among countries with rich and threatened biodiversity, there appears a clear need to encourage more zoo visits in poorer developing countries.

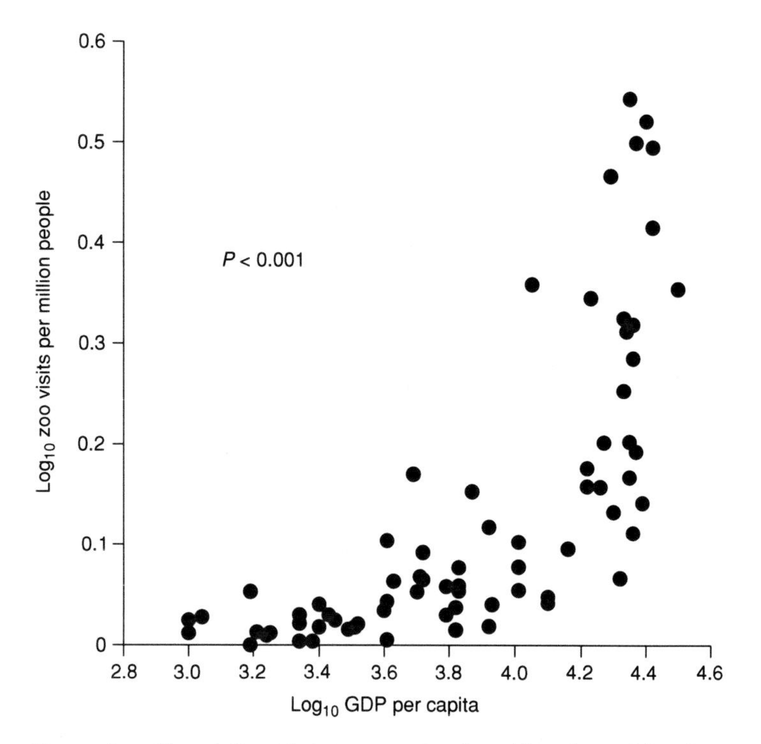

Figure 16.7 The relationship between national wealth and zoo visits, shown as $\log_{10}$ GDP per capita and $\log_{10}$ zoo visits per million people, based on data from the *International Zoo Yearbook* (Olney and Fisken 2003). The correlation was highly significant ($r_s = 0.721$, $P < 0.001$, $N = 71$)

At the same time, the analysis presented here may be affected by disproportionate under-reporting to the *International Zoo Yearbook* (Olney and Fisken 2003) among zoos in poorer but biodiversity-rich countries. If this proves to be the case, the analysis is still useful in flagging the need to ensure that levels of reporting are equalized across the international zoo community (IUDZG/CBSG 1993). In turn, this may require a considerable focus by zoos in developed countries on building the capacity of zoo professionals in biodiversity-rich, developing countries. Without standardized levels of central reporting, the world's zoo community will never be able to measure its impact upon the collective challenge of conserving biodiversity globally.

DO MOST ZOOS IMPLEMENT THE WORLD ZOO CONSERVATION STRATEGY?

Zoological collections worldwide adopted a well-established and widely recognized *World Zoo Conservation Strategy* (IUDZG/CBSG 1993), but is it

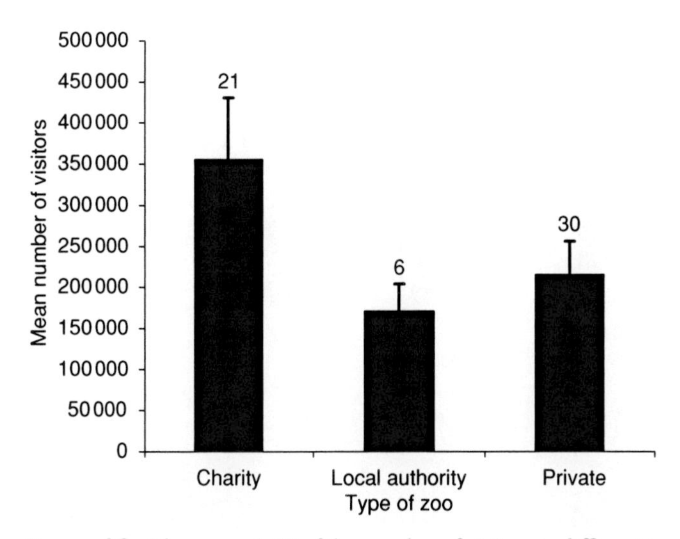

Figure 16.8 The mean ± SE of the number of visitors to different types of zoo in Britain and Ireland, subdivided according to whether they were run by a charity, by a local authority or under private management, based on data from the *International Zoo Yearbook* (Olney and Fisken 2003). There was no significant difference in levels of zoo visitation across different types of zoo ($P > 0.05$)

being widely implemented? While there are numerous good examples of implementation, some qualitative assessments suggest that the strategy might not have been fully implemented (e.g., Miller *et al.* 2004). However, there are few data available globally to make such an assessment, so this section undertakes a more local analysis, using the data compiled by the Federation of Zoological Gardens for Great Britain and Ireland [now known as the British and Irish Association of Zoos and Aquariums (BIAZA)] (Lees 1994, Stevenson 2001), supplemented with data on British and Irish zoos from the *International Zoo Yearbook* (Olney and Fisken 2003). Specifically, this section examines the relationship between visitation, which we assume serves as a reasonable surrogate for gate receipts and income, and the extent to which different types of zoological collection fulfil their obligations both to the *World Zoo Conservation Strategy* (IUDZG/CBSG 1993), and to the European Community Zoos Directive of 1999 (EC 1999). Both the Strategy and the Directive require zoos, whether run by a charity, a local authority or under private management, to make some contributions to education, research, and conservation.

In terms of visitation to British and Irish zoos, there were wide variations in, but no differences between, the mean annual number of visitors to different types of zoological collection, whether run by a charity, by a local authority or under private management (Figure 16.8). This suggests

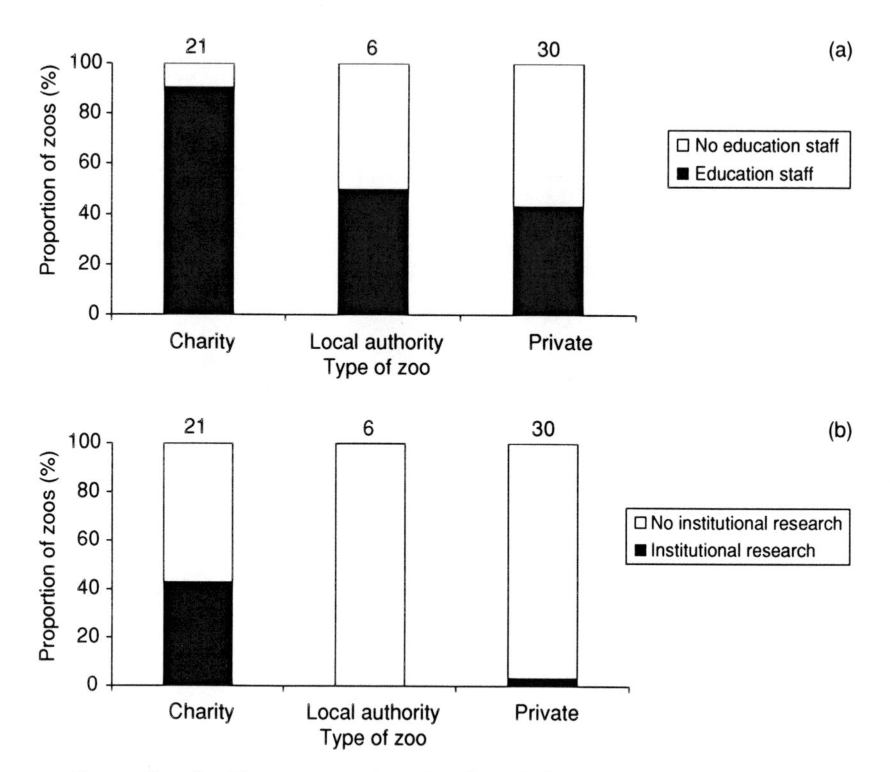

Figure 16.9a, b The representation of (a) formal education programs, and (b) formal research programs, supported by different types of zoo in Britain and Ireland, subdivided according to whether they were run by a charity, by a local authority or under private management, based on data from the *International Zoo Yearbook* (Olney and Fisken 2003). Because of small sample sizes, zoos run by a local authority and under private management were combined, and the differences between types of zoo were significant for both formal education programs ($\chi^2 = 10.66$, df = 1, $P = 0.0011$), and for formal research programs ($\chi^2 = 14.73$, df = 1, $P < 0.001$)

that gate receipts, and zoo income, are roughly comparable across different types of zoological collections.

In terms of their education and research missions, all Federation members claim to provide informal educational opportunities. However, there was a difference between types of zoo in terms of whether they support a dedicated and formalized education program, staffed by a minimum of at least one part-time education officer (Figure 16.9a). Most zoos run by a charity support such a program, while this was the case for many fewer zoos run by a local authority or under private management. Many fewer Federation members claim to undertake or to support research. However, there

Table 16.1
Growth in in situ *conservation projects supported by members of the Federation of Zoological Gardens for Great Britain and Ireland (based on data from Lees 1994, Stevenson 2001)*

Years	Members	Projects	Countries
1989–1994	30	69	36
1997–2001	43	177	62

was a difference between types of zoo in terms of whether they support a dedicated research program, staffed by a minimum of at least one part-time research officer (Figure 16.9b). Some 50% of zoos run by a charity support a dedicated program, while virtually no zoos run by a local authority or under private management do so.

There was encouraging growth in the number of *in situ* conservation projects supported by Federation members between two surveys in 1989–1994 and 1997–2001 (Lees 1994, Stevenson 2001). While the number of Federation members had increased by some 50%, the reported number of *in situ* conservation projects in which members were involved had more than doubled, while the number of countries in which Federation members support *in situ* projects had almost doubled (Table 16.1). This suggested real growth in the number of projects in which individual members were involved, over and above a straight increase in numbers of Federation members reporting the projects in which they were involved.

The surveys suggested that Federation members spent £10 million on *in situ* conservation projects, and that £5 million was contributed directly by Federation members, from 1997 to 2001 (Stevenson 2001). However, there was no relationship between $\log_{10}$ annual visitation and $\log_{10}$ funds spent by individual Federation member zoos (Figure 16.10a). Put another way, the numbers of visitors, and the likely gate receipts of different zoos did not determine the sums each invested in *in situ* conservation projects. Instead, there was a marked difference in their involvement in conservation projects between different types of zoological collections (Figure 16.10b). Thus most zoos run by a charity were involved in *in situ* conservation projects, while this was the case for many fewer zoos run by a local authority or under private management.

In terms of support provided to different classes of animal by Federation members involved in *in situ* conservation projects, there was a marked difference between the types of project supported within Britain and Europe, and outside Europe, measured in terms of numbers of projects

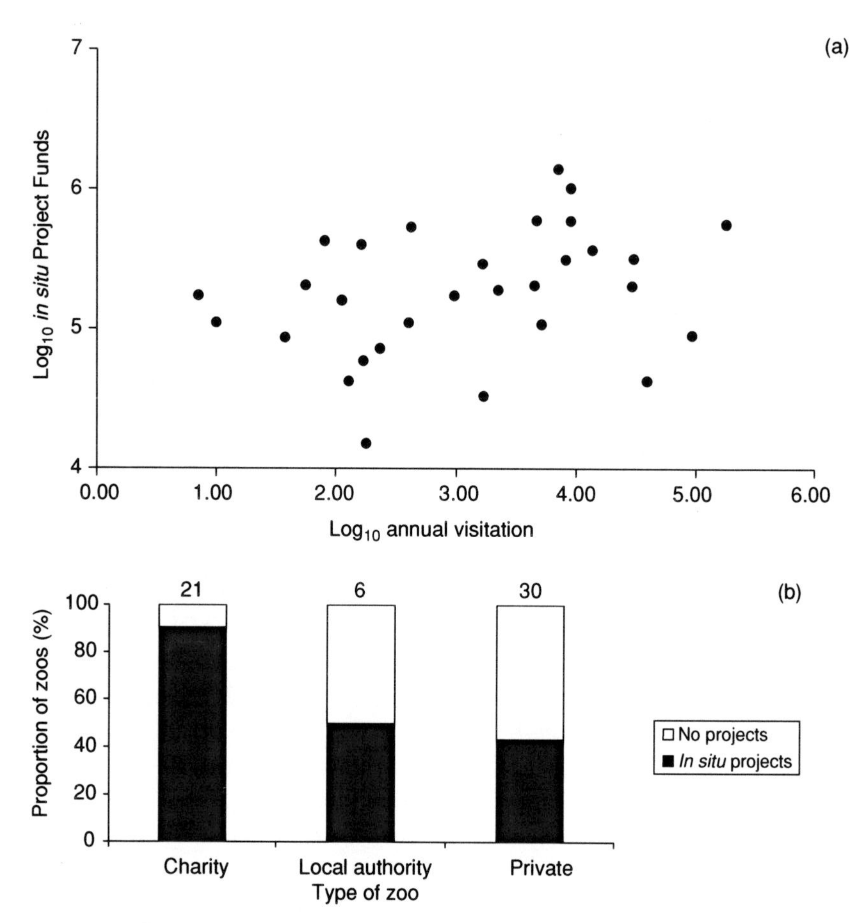

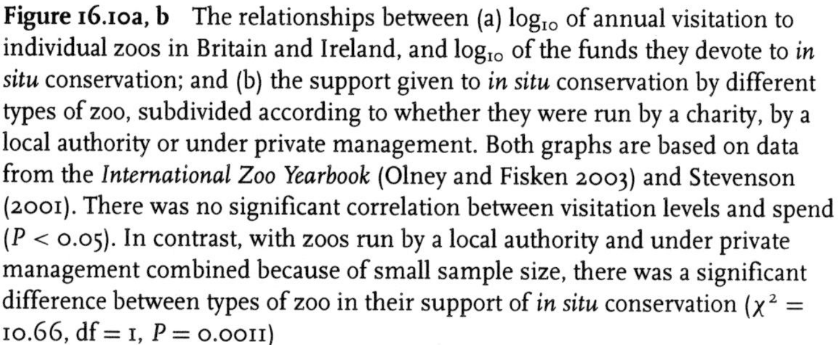

Figure 16.10a, b The relationships between (a) $\log_{10}$ of annual visitation to individual zoos in Britain and Ireland, and $\log_{10}$ of the funds they devote to *in situ* conservation; and (b) the support given to *in situ* conservation by different types of zoo, subdivided according to whether they were run by a charity, by a local authority or under private management. Both graphs are based on data from the *International Zoo Yearbook* (Olney and Fisken 2003) and Stevenson (2001). There was no significant correlation between visitation levels and spend ($P < 0.05$). In contrast, with zoos run by a local authority and under private management combined because of small sample size, there was a significant difference between types of zoo in their support of *in situ* conservation ($\chi^2 = 10.66$, df $= 1$, $P = 0.0011$)

(Figure 16.11). Among species-based *in situ* conservation projects supported in Britain and Europe, there was considerable involvement with projects on insects, crustaceans, amphibians, and small mammals. Put another way, close to home, Federation members were involved with species that can be bred cost-effectively in captive breeding programs (Balmford *et al.*

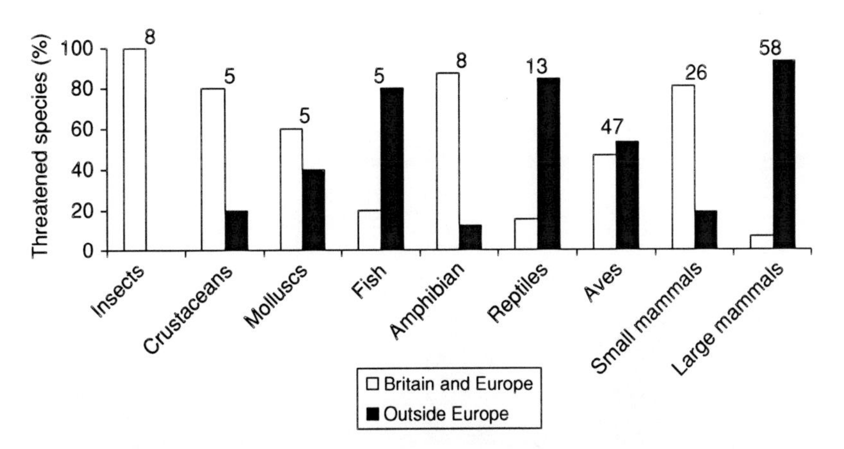

Figure 16.11 The support given by Federation members from 1997 to 2001, for *in situ* conservation projects in Britain and Europe, and outside Europe, involving different classes of species, based on data from Stevenson (2001). The total sample sizes of *in situ* projects are shown for each class, and the clear and the shaded columns add up to 100% separately for each class. With insects, crustaceans and amphibians, and fish and reptiles, combined into two classes because of their small sample sizes, the differences in representation between different classes in Britain and Europe, and outside Europe, was highly significant ($\chi^2 = 66.14$, df $= 5$, $P < 0.001$)

1996). Furthermore, several species of insect such as the wartbiter cricket have been successfully reintroduced into sites in Britain (Pearce-Kelly *et al.* 1998). In contrast, most *in situ* conservation projects away from home focused on large mammals, birds and reptiles (Figure 16.11), those same classes that were over-represented in captive breeding programs for threatened species (Figures 16.3, 16.4). However, in terms of their location (Figure 16.12), the numbers of *in situ* conservation projects supported by Federation members outside Britain were more likely to be inside a hotspot country.

In summary, this section shows that many zoos within the Federation of Zoological Gardens for Great Britain and Ireland, now BIAZA, were not fulfilling the mission of the *World Zoo Conservation Strategy* (IUDZG/CBSG 1993), in terms of their efforts to promote education and public awareness, research and *in situ* conservation. Indeed, there were striking differences between different types of zoo, whether run by a charity, a local authority or under private management (Figures 16.9, 16.10b). Nevertheless, there were signs that Federation members were providing increasing levels of support to *in situ* projects (Table 16.1). In particular, Federation members were better supporting cost-effective projects close to home (Figure 16.11), and better supporting projects in biodiversity-rich countries overseas (Figure 16.12).

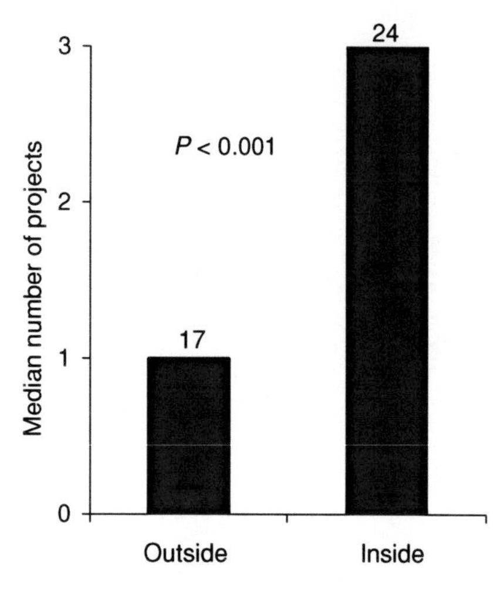

Figure 16.12 The representation of countries outside Britain in *in situ* conservation projects supported by Federation members from 1997 to 2001, based on data from Stevenson (2001) and Myers *et al.* (1998). The sample size of countries where projects were based is shown above each column, which show, respectively, the median number of projects supported in different countries outside and inside hotspots. The representation of *in situ* conservation projects inside and outside hotspot countries differed significantly ($z = 3.258$, $P < 0.001$)

However, more support was still given to mammal projects outside Britain and Europe (Figure 16.11), which in turn mirrors the focus placed by other conservation organizations on charismatic flagship species away from home (Leader-Williams and Dublin 2000).

CONCLUSIONS AND NEXT STEPS

This analysis has provided a more quantified inventory of coordinated programs than the qualitative analyses available hitherto (Conway 2003, Tribe and Booth 2003, Miller *et al.* 2004). As such, it provides a useful step forward in identifying gaps in the implementation of *ex situ* strategies and integrated conservation programs of zoos. Nevertheless, this analysis does not measure the effectiveness or impact of actions taken by zoos in achieving conservation success. This is a challenge not only for zoos but for conservation organizations in general, and considerable intellectual effort is now being spent on measuring success in conservation (e.g., Margoulis and Salafsky 1998, Salafsky *et al.* 2002). Therefore, the Zoo Measures Group sought to take forward ways of measuring the impact of the actions of zoos.

Consequently, two other chapters in this book propose a method for measuring the overall conservation impact of zoos (Chapter 21), and their educational impact in particular (Chapter 9).

References

AZA (2003). www.aza.org

Balmford, A., Mace, G. M., & Leader-Williams, N. (1996). Designing the ark: setting priorities for captive breeding. *Conservation Biology*, **10**, 719–727.

Conway, W. G. (1986). The practical difficulties and financial implications of endangered species breeding programmes. *International Zoo Yearbook*, **24/25**, 210–219.

Conway, W. G. (2003). The role of zoos in the 21st century. *International Zoo Yearbook*, **38**, 7–13.

EAZA (2003). www.eaza.net

EC (1999). *Council Directive 1999/22/EC of March 29, 1999 Relating to the Keeping of Wild Animals in Zoos.* Official Journal of the European Communities. 09/04/1999.

Glowka, L. (1993). *A Guide to the Convention on Biological Diversity.* Gland: IUCN.

Groombridge, B. (1993). *The 1994 Red List of Threatened Animals.* Gland: IUCN.

Hilton-Taylor, C. (2003). *The 2003 Red List of Threatened Species.* Gland: IUCN.

Hutchins, M., Willis, K., & Wiese, R. (1995). Strategic collection planning: theory and practice. *Zoo Biology*, **14**, 2–22.

IUCN (2002). *Technical Guidelines on the Management of* Ex Situ *Populations for Conservation.* Gland: IUCN. http://www.iucn.org/themes/ssc/pubs/policy/exsituen.htm.

IUCN, Conservation International, and NatureServe (2006). *Global amphibian assessment.* www.globalamphibians.org.

IUDZG/CBSG (1993). *The World Zoo Conservation Strategy: the Role of Zoos and Aquaria of the World in Global Conservation.* Brookfield, IL: Chicago Zoological Society.

Leader-Williams, N. & Dublin, H. T. (2000). Charismatic megafauna as "flagship species." In *Has the Panda had its Day? Future Priorities for the Conservation of Mammalian Biodiversity*, eds. A. Entwistle & N. Dunstone. Cambridge: Cambridge University Press.

Lees, C. (1994). Conservation projects 1989–1994. Unpublished report of the Federation of Zoological Gardens Great Britain and Ireland, London.

Margoluis, R. & Salafsky, N. (1998). *Designing, Monitoring, and Managing Conservation and Development Projects.* Washington, D.C.: Island Press, 382 pp.

Miller, B. *et al.* (2004). Evaluating the conservation mission of zoos, aquariums, botanical gardens, and natural history museums. *Conservation Biology*, **18**, 86–93.

Myers, N., Mittermeier, R. A., Mittermeier, C. G., Fonseca, G. A. B., & Kent, J. (2000). Biodiversity hotspots for conservation priorities. *Nature*, **403**, 853–858.

Olney, P. J. S. & Fisken, F. A. (2003). Zoos and aquariums of the world. *International Zoo Yearbook*, **38**, 247–356.

Pearce-Kelly, P. (1994). Invertebrate propagation and re-establishment programmes: the conservation and education potential for zoos and related institutions. In *Creative Conservation: Interactive Management of Wild and Captive Animals*, eds. P. J. S. Olney, G. M. Mace, & A. T. C. Feistner. London: Chapman and Hall, pp. 329–337.

Pearce-Kelly, P., Jones, R., Clarke, D., Walker, C., Atkin, P., & Cunningham, A. A. (1998). The captive rearing of threatened Orthoptera: a comparison of the conservation potential and practical considerations of two species' breeding programmes at the Zoological Society of London. *Journal of Insect Conservation*, **2**, 201–210.

Salafsky, N., Margoluis, R., Redford, K. H., & Robinson, J. G. (2002). Improving the practice of conservation: a conceptual framework and research agenda for conservation science. *Conservation Biology*, **16(6)**, 1469.

Stevenson, M. (2001). Report on review of *in situ* projects supported by members of the Federation of Zoos. Unpublished report of the Federation of Zoological Gardens for Great Britain and Ireland, London.

Tribe, A. & Booth, R. (2003). Assessing the role of zoos in wildlife conservation. *Human Dimensions of Wildlife*, **8**, 65–74.

PART IV

Direct involvement of zoos in *in situ* conservation

17

Zoo-based fundraising for *in situ* wildlife conservation

SARAH CHRISTIE

INTRODUCTION

Public perception of zoo support for conservation in the latter part of the 20th century has largely been limited to their educational role and their potential to supply animals for reintroduction. Zoo conservation support can, however, be categorized into a total of four broad types, adding fundraising for the field and the generation of data and skills useful for field conservation to awareness-raising and the genetic lifeboat. Fundraising has been particularly neglected, both inside and outside the zoo community, until recently; during the 1980s and 1990s, statements from the World Association of Zoos and Aquariums (WAZA – at that time known as the International Union of Directors of Zoological Gardens or IUDZG) and the IUCN on the conservation roles of zoos and aquariums did not include it. Collection and analysis of data on this aspect of zoos' conservation work have also lagged behind documentation of the more traditional areas.

This omission has perhaps been due to a certain distaste for simply supplying funds, which have been felt to be "just money" and not linked to the unique skills of zoos. Such an attitude is perhaps unwise; there are few things that are more significant to the world at large than money, and there is great PR potential in reporting large contributions. And in fact, the zoo world does have a unique advantage for fundraising, not possessed by any other group of conservation NGOs – live animals that can be seen, heard, smelt and, occasionally, touched. Almost any animal can inspire interest and wonder if presented right, but there are clear advantages in selecting creatures of power, grace and beauty as a fundraising focus, especially where one is able to bring the visitors face to face with the animals. The tiger is an exceptionally good choice for focusing a zoo fundraising appeal, and considerable information is available on the results of such efforts to date. Hence, the tiger is used as the primary example here, but the general principles do of course also apply to other taxa.

A review of relevant documents reveals how the global conservation community has viewed the relative importance of the various types of conservation support generated over time. The first edition of the *World Zoo Conservation Strategy* (IUDZG/CBSG 1993) lists three clear conservation objectives for zoos: managed breeding programs to conserve the gene pool; increasing scientific knowledge; and increasing awareness of conservation issues. These cover three of the four support categories mentioned, but there is no reference to fundraising at all. Ten years on, in the *World Zoo and Aquarium Conservation Strategy* under review at the time of writing (WAZA 2005), there is no corresponding statement of conservation objectives, but there *are* two clear references to fundraising: first, in the population management chapter, and second in the chapter on conservation of wild populations. At the same time, the view taken by the wider conservation community has also developed. The IUCN's 1987 *Policy Statement on Captive Breeding* (IUCN 1987) was rightly viewed at the time as a breakthrough in linking zoos and field conservation bodies, but its exclusive focus on the genetic lifeboat seems dated today. By the time the 2002 *Technical Guidelines on the Management* of In Situ *Populations* were issued (IUCN 2002), a much broader view was in evidence with nine points under Goals including fundraising.

Generation of funds is now acknowledged as a valid contribution to conservation. The zoo community has begun to encourage and support this activity, and to gather data on it, in the same way as participation in breeding programs, research and education is encouraged, supported, and documented. Potential roles for the regional zoo associations here include help with generation of funds, compilation of data on supply of funds, provision of information on available projects, and quality control of such projects. All of these are linked, but only the first two will be examined here as they are the most relevant to fundraising.

REGIONAL INITIATIVES TO ENCOURAGE AND SUPPORT GENERATION OF FUNDS

Such cooperative initiatives are not of course confined to regional zoo associations; there are many coalitions of zoos pooling their resources to meet common conservation goals, e.g., the Madagascar Fauna Group (see Chapter 18). However, it is the official regional association initiatives that most accurately reflect the policies and aims of the world zoo community. There are three regional mechanisms so far in use – trust funds, regional funding pools, and regional conservation campaigns.

TRUST FUNDS

The obvious advantage of a trust fund over a central funding pool is that once established the fund is self-sustaining, continuing to yield an income year after year without further top-up. However, it is necessary to invest large sums in order to generate income visible to the naked eye and at present the AZA is the only zoo association that has been able to do so. Their Conservation Endowment Fund (CEF) was established in 1984 and reached $8 million in 2000, mostly through members' contributions and fundraising events, with 20% coming from corporate donors. After compensating for inflation and covering administration costs, the CEF generates $150 000–$200 000 annually. The funds are available exclusively for supporting AZA members' own projects, both *in situ* and *ex situ*, in the areas of education, research, training, breeding, and field conservation. The records for the fund do distinguish between *in situ* and *ex situ* projects – 90 *in situ* projects have been supported as of 2003 – but it is not possible to separate actual funds provided from the values assigned to in-kind support such as staff time.

Figure 17.1 shows how centrally administered AZA conservation funds have been spent, including contributions from Ralston Purina (1991 to 1996) and Disney (1995 to the time of writing) as well as from the CEF. Approximately one-third of this central funding since 1991 – about $1.1 million – has gone to *in situ* projects. Within this, mammals get the biggest share as one would expect, closely followed by reptiles. However, if one lumps reptiles with amphibians as herpetofauna or "herps" their share exceeds the mammal slice. Remembering that these funds are only available for the work of AZA members and their partners, this unusual bias may reflect the fact that it takes fewer resources to run the average herp project than the average mammal project. Figure 17.2 shows this central funding distribution over time. The 2-year gap in the CEF contributions represents a moratorium to allow the fund to be put on a firm financial footing. Over this whole period, about half of the total funds came from the CEF itself and a quarter each from the other two sources.

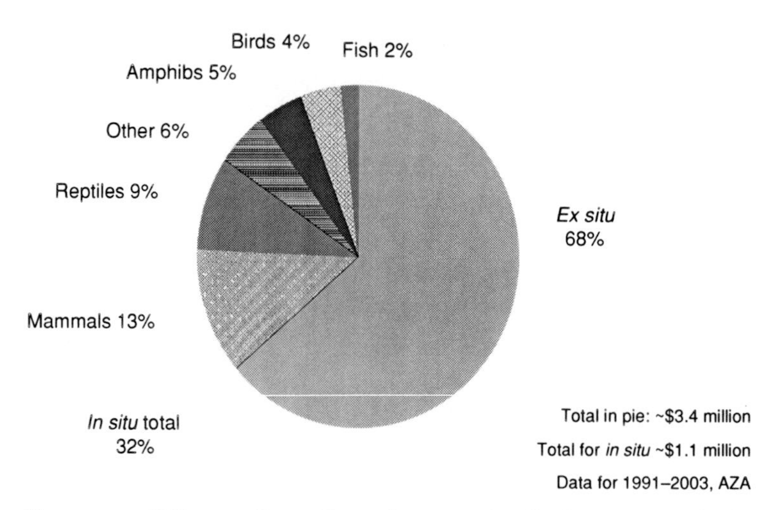

Figure 17.1 AZA expenditure of central conservation funds 1991–2003, by *in/ex situ* and by taxonomic group within *in situ*

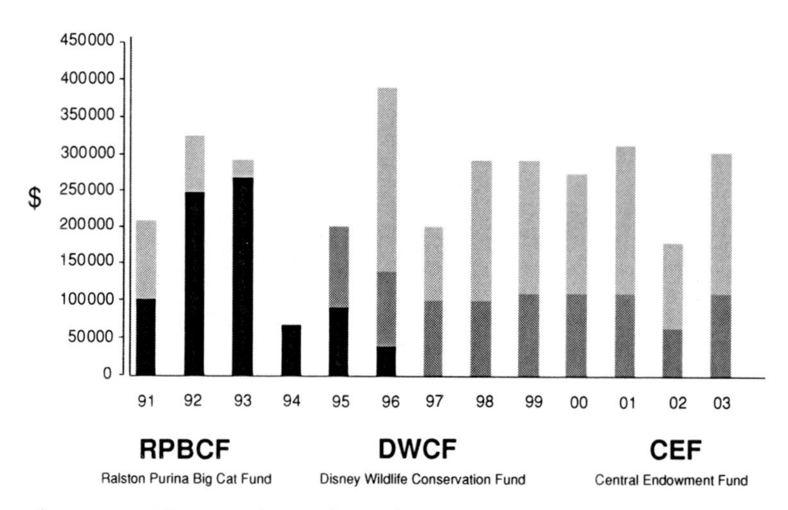

Figure 17.2 AZA expenditure of central conservation funds 1991–2003, by year. Darkest shading indicates data for RPBCF; medium shading for DWCF; lightest shading for CEF

Though EAZA does not have an endowment fund of its own along these lines, it has together with IBAMA (the Brazilian wildlife protection agency) set up a small endowment fund for the Atlantic Rainforest with part of the money generated through the second EAZA Conservation Campaign. Sponsorship for additional funds is being sought.

REGIONAL FUNDING POOLS

The Australasian Regional Association of Zoological Parks and Aquariums (ARAZPA) has established the ARAZPA Wildlife Conservation Fund to provide an avenue for members (particularly those not currently involved in conservation) to contribute to *in situ* conservation projects. Since 2001 over AU$100 000 has been raised for *in situ* conservation projects operating in Australasia and Southeast Asia. EAZA has plans for the establishment of an EAZA *in situ* Conservation Fund, which is included in its Strategy for the 21st Century Action Plan 2002–2004. At the national level, the Dutch Zoo Federation has a similar setup: its conservation foundation, "Zoos Help," has spent €700 000 in 7 years on some 30 *in situ* projects. Eighty percent of this has come from the 12 members, and the rest from sponsors. In 2004, the Zoos Help budget was €146 000.

COORDINATED REGIONAL CAMPAIGNS

The 1990s saw the first two nationally organized zoo conservation campaigns – both of them for tigers – with the Dutch Zoo Federation raising €25 000 in 1992 and the British and Irish Association of Zoos and Aquariums (BIAZA – at that time known as the Federation of Zoological Gardens for Great Britain and Ireland) raising €130 000 in 1996. The European Association of Zoos and Aquariums (EAZA) began regional campaigns in 2000/2001 and the fourth – on the conservation of tortoises and freshwater turtles – was launched in 2004. The first EAZA campaign, targeting the illegal bushmeat trade in West Africa, focused on political lobbying through a petition rather than on fundraising but nevertheless generated €80 000 during the campaign year, while the second, for Atlantic Rainforests in 2001/2002, had produced over €287 000 by 2004.

The EAZA Tiger Campaign was launched in September 2002 and was adopted by the Euro-Asian Regional Association of Zoos and Aquariums (EARAZA) and ARAZPA in 2003/2004, as well as extended for a second year by EAZA and taken up at a national level by several national zoo associations within the EAZA region. By September 2004 the 21st Century Tiger account had received over €515 000 from the two regions, with a further €170 000 reported held at participating zoos in Europe and Australasia pending submission and fundraising activities still continuing in some zoos. By the time all funds raised under the campaign had been received in 2005, the total was over €750 000.

While the AZA has not mounted any official regional campaigns, it does support cooperative initiatives such as the Butterfly Conservation Initiative (BFCI). The BFCI is dedicated to the conservation of North American butterflies and the habitats that sustain them, with a focus on recovery, research, and education. The initiative was created to meet AZA members' stated desire for "packaged" conservation programs highlighting local species of interest. BFCI is funded by membership pledges from 50 AZA facilities (as of September 2004), who participate in butterfly conservation activities on many levels, including recovery plan implementation, habitat restoration, population monitoring and other research, education and outreach through butterfly gardens, festivals, and other programming. Since its inception in 2001, this project has generated over $125 000 from pledges alone, and has begun serving as a prime example of the power of zoo and aquarium collaborative efforts for local community conservation.

WHAT MAKES A SUCCESSFUL CAMPAIGN?

A species focus is of course an essential ingredient for a zoo-based campaign as all activities are based around the living animals; even where a campaign targeted a particular habitat as the Atlantic Rainforest one did, a charismatic taxon – lion tamarins – was chosen as a species focus. Another factor common to recent regional campaigns was a "campaign pack" of information, advice, and pictures, including material aimed at marketing and education departments. Experience to date indicates a number of additional key factors contributing to campaign participation and popularity among zoos:

- A widely held taxon present in many zoos
- A charismatic taxon popular with public, media, and staff
- A taxon that inhabits a range of habitat types in a range of different countries
- A "landscape" taxon whose conservation will have knock-on effects for other taxa
- A range of target projects in different countries and with different goals.

All these factors affect the number of zoos that participate in each campaign. Using the EAZA Tiger Campaign as an example, most sizeable zoos do hold tigers, but some that do not were happy to participate. Apenheul, which primarily focuses on primates, chose a campaign project from India that happened to be in key lion-tailed macaque habitat, while Harewood

Bird Gardens and others were happy to set up a tiger display and collect donations even in the absence of a clear link between their own exhibits and the campaign projects, because of the popularity of the tiger theme with visitors. Selection of projects is also important in ensuring that campaign funds are well spent and that there is something available for all tastes; in the case of the Tiger Campaign, zoos could choose between anti-poaching, wildlife trade investigations, local awareness, ecological research, fire-fighting, forest guard training, conflict resolution, and habitat restoration projects, and between four countries in different parts of Asia.

Once the campaign is running and zoos have signed up, a second set of factors comes into play, including:

- A good flow of news, reports, and images from the target projects
- The personal touch – connections with the projects
- Saleable items, particularly for children
- Public acknowledgement of zoo success.

For zoos, probably more than for any other group of donors, a steady flow of news, reports and above all images is necessary to maintain impetus. Zoos need to be able to stimulate donations and to promote their provision of support through onsite graphics, articles in their in-house magazines and websites and, where possible, items in their local and national media. Zoo staff who take on the extra work associated with campaigns also appreciate long-term and personal links to specific projects, which can be fostered through support for an individual working in the field and through visits by project staff to the zoo. The EAZA Atlantic Rainforest Campaign's adopt-a-field-worker option provided monthly personal reports from each field worker to each zoo wishing to adopt one; in return, the field projects received a firm amount of $5000 for each such arrangement. Seven zoos took up these adoptions, and all of these have become long-term commitments scheduled to run until at least 2007. The EAZA Tiger Campaign also encouraged projects to provide personal letters and reports but with much less success, probably because there was no adopt-a-field-worker scheme and so provision of funds was not directly linked to provision of personal reports. Visits to zoos were also very difficult to organize, partly because of the long lead-in time needed to set up any public event; project staff were unable to commit to travel sufficiently far in advance, and the language barrier was also a problem, with most project staff speaking only English among European languages, and that often as a second language.

Saleable campaign products (key-rings, badges and t-shirts) were very popular with participating zoos, but involved a heavy administrative burden in coordinating orders, payments, and shipments. This is significantly reduced if the items are donated or sponsored and so can be distributed without charge with the money collected later, after sale; the EAZA Tiger Campaign was able to use 100 000 tiger key-rings donated by Esso UK plc (now Exxon-Mobil) in this way. The planning group for the 2004/5 EAZA Shellshock Campaign (for freshwater turtles and tortoises) went a step further and arranged for direct liaison between participating zoos and suppliers of specially designed merchandise, with the added bonus of free delivery anywhere in the EAZA region, thus removing themselves entirely from the administrative loop on this aspect.

As well as information and images, individual zoos appreciate public acknowledgement of success. Credit can be given in regional zoo association publications and meetings, and the tiger campaign added to this a system of certificates – bronze, silver, gold, and platinum – for fundraising success. These can be formally presented at conferences and displayed to the public outside the relevant enclosures, and have been very well received. The 2004/5 EAZA Shellshock Campaign followed suit with its own certificate system.

All of the activities above require administration. Standard following up of grant funds – setting up and monitoring money transfers and ensuring the projects provide reports and accounts on time – is quite time-consuming in itself, and when one adds production of the initial campaign information packs and the necessity to track and credit hundreds of small bank transfers from individual zoos and manage the constant flow of information, the commitment of time and energy becomes very significant. The evolution of administration of the EAZA Conservation Campaigns nicely illustrates the learning curve on this. The 2001 Bushmeat Campaign was administered almost entirely by the EAZA Executive Office, with assistance from a Campaign Planning Group made up of appropriate individuals from EAZA institutions. This was a heavy burden on the office, which even in 2004 continued to put time into relevant ongoing issues. The Atlantic Rainforest Campaign was administered jointly by the EAZA Office, the Campaign Planning Group and Copenhagen Zoo, with the very significant change that funds were managed at Copenhagen who also dealt with project liaison through their existing connection to IBAMA. The EAZA Tiger Campaign was administered primarily by ZSL through the vehicle of 21st Century Tiger, again of course working in collaboration with the EAZA Office and a Campaign Planning Group. 21st Century Tiger is a fundraising coalition between ZSL and Global Tiger Patrol, based at ZSL and run by a half-time administrator; it has supplied well over £1 m to tiger field projects since 1997. This existing infrastructure meant a much lower workload for the EAZA Executive Office. A significant number of zoos are interested in continuing to use 21st Century Tiger as a channel for conservation support after the end of the formal campaign, and so representatives from EAZA and ARAZPA have been added to its Project Review Panel, to ensure zoo association input into not only the spending of the existing campaign funds but also selection of projects for zoo support in future.

Key recommendations for a successful campaign therefore include ensuring it has wide appeal to both zoos and the public, provision of plentiful information and images to participants throughout, linking with an existing infrastructure capable of carrying most of the administrative load, providing a supply of campaign merchandise, and rewarding individual zoos for successful efforts.

RECORDING AND REPORTING ZOO FUNDING FOR CONSERVATION

Compilation of data on zoos' efforts is the second major task regional zoo associations need to undertake in this area, and there are very strong imperatives for so doing. All zoos of course stand to gain enormously from the positive effects that would result from presentation of the full dataset, and legal or zoo association membership requirements for conservation support are also set to become a major motivator for data supply. In the US, UK and Australasia, initiatives to make reporting of conservation support a condition of regional zoo association accreditation are under way, and there is legislation in place in the US which refuses CITES permits for Appendix One species to any zoo unable to demonstrate a commitment to conservation support for the taxon. Despite these pressures, all those who have been involved in collection of such data so far agree that getting blood out of stones is child's play in comparison.

PROBLEMS AND PROGRESS

The first problem is just to persuade institutions to provide the data. A large and varied range of reporting requirements is placed on all zoos nowadays, and it is tedious and time-consuming to check financial records for conservation support and enter the information in a database. But even this is only the beginning; experience to date shows that no two zoos will define a project in the same way or give it the same name. For example, one often finds the name of an institution, an individual or a taxon entered in the project name field instead of any possible project name; the data compiler must resolve these anomalies, make decisions about what does or does not constitute a discrete project, and assign and edit the records accordingly.

Then there is double reporting; funds often move from source zoos to a central collecting point in another zoo – e.g., Heidelberg Zoo's West African Primate Conservation Action (WAPCA) or ZSL's 21st Century Tiger – and are sent on from there to the target projects. Hence they will be reported twice, once by Heidelberg/ZSL and once by the source zoo, and such instances must be identified and corrected. This then leads to the next problem, of attribution, as each tranche of funds can only be credited to one source. Exchange rates and differing financial years also muddy the waters. In order to achieve any accuracy, therefore, there must be knowledgeable and painstaking personal oversight at multiple levels. Data entry cannot be left entirely to the participating zoos, and data checking cannot be left entirely to automated routines in the database software.

Despite all these problems, progress has been made in several regions and Table 17.1 compares the work done to date. The AZA are furthest along with data collection and have chosen a simple structure with limited fields for their Annual Report on Conservation and Science (ARCS) database. Their simple field structure, stringent reporting requirements and online data entry interface have given them the highest response rate amongst the regional zoo associations compiling these data. The downside to this format is that the more limited reporting options lead to a lower level of detail in the dataset than found elsewhere, e.g., the database does not distinguish between funds contributed and value of in-kind support supplied.

EAZA have designed a more complex Access database. Although many of the fields are not compulsory at present, the potential is there for the future. The decision to use a CD ROM rather than online data entry has added to the costs in terms of database design time, as routines to combine the individual datasets had to be written in addition to the database itself. Teething problems with the data entry module, plus the lack of any legal or accreditation pressures, have meant that their response rate is the lowest among the associations collecting these data. However, the data compilation routines do contain automated processes for identifying and resolving anomalies, a major time-saver for the human compiler. Figures shown in Table 17.1 below for EAZA are drawn only from the electronic data returns (32 zoos).

Table 17.1
Progress with data compilation

	AZA (ARCS)	EAZA	ARAZPA	BIAZA
Date of survey	2002 ongoing	2002 ongoing	2001 one-off	2000 one-off
Data collection method	Paper/electronic initially, now online	Data collection CD	Paper/electronic returns	Paper/electronic returns
Response rate	70%	29%	37%	49%
Automated data checking mechanisms	No	Yes	No	No
Level of accuracy	Low without considerable admin. effort	Potential exists for high accuracy when work completed	Low	High
Funding recorded separately from value of in-kind?	No	Yes	Yes	Yes
Period covered	Figures supplied for 2002 only	1991–2002 and through paper surveys in 2003	Not analyzed	1997–2001
No. of projects recorded	1240	~500	~120	177
Amount of funds	US$ 6.4 million (includes in-kind value)	€ 9 million	AU $5 million	£10 million
Overlap with other regional data	No	Yes (BIAZA)	No	Yes (EAZA)

ARAZPA carried out a paper survey in 2001 but have not compiled the results. BIAZA carried out a survey in 2000, using a field list intermediate between the AZA and EAZA. They had a smaller set of zoos to target and their work is included here alongside the three continental regions as the dataset is in fact the best yet available; it has been thoroughly checked for double reporting, correct identity of project, and correct attribution of funds, and it separates in-kind support from funds supplied. No other zoo associations have attempted to gather these data as yet.

It is clear that there are already considerable difficulties in combining the available data, despite being in the early stages:

- The two AZA datasets (CEF records and ARCS database) overlap to an unknown extent (we do not know what proportion of the projects funded by CEF are recorded in the ARCS database).
- The EAZA and BIAZA datasets overlap to an extent that will remain unknown until the EAZA data are analyzed (probably considerable as the most significant UK players are likely to have reported to both).
- The ARCS database does not allow separation of straightforward funding supply from values assigned to in-kind support, while the others do.

And, of course, no region has yet obtained anything close to a complete dataset, though BIAZA probably has the majority of the necessary information for the period its survey covered. In short, the data so far available are so incomplete, incompatible, and overlapping that it is impossible to produce any kind of meaningful global summary at this stage. The problem of incompatibility in particular should be addressed before the work gets very much further, or the world zoo community will remain unable to make clear summary statements about its conservation contributions even after regional data have been compiled.

ZOO CONSERVATION SUPPORT DATA FROM ELSEWHERE

The lack of good information within the zoo world is extremely frustrating if one wants to make a case for the significance of conservation funding supplied by zoos. However, there are very good data on at least two taxa available from other sources.

TIGERS

Over the last five years ZSL has compiled a comprehensive and accurate database on tiger conservation projects around the world. This database was

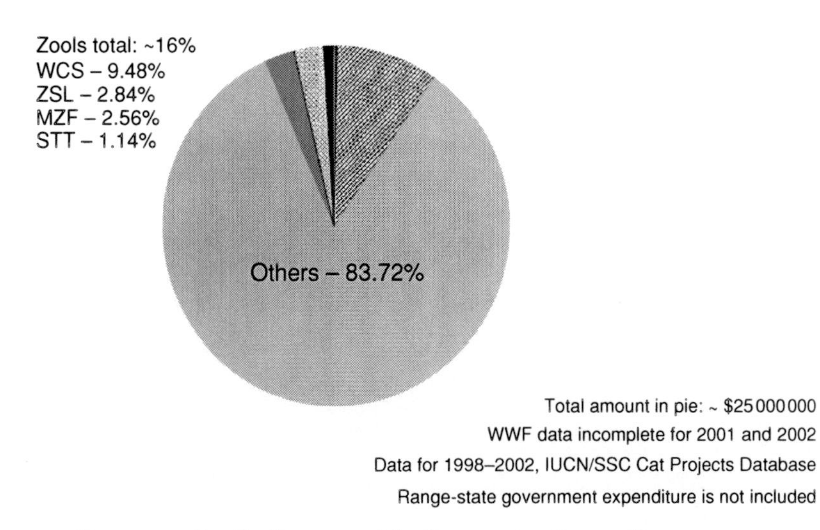

Figure 17.3 Zoo funding support for tiger conservation 1998–2002

designed by Double R Consulting (who also produced the EAZA Conservation Database) and is in the process of being adapted to include other felid taxa for the IUCN/SSC Cat Specialist Group. It is available online and the URL will be made public when measures to exclude incomplete financial records from analysis have been completed. It has online data entry with a password-protected system enabling the database manager to assign editing rights for individual projects to appropriate people, and a reporting interface which allows analysis of funding supply as well as searches for individual projects of interest.

Figure 17.3 shows the proportion of NGO tiger funding supplied by zoos from 1998 to 2002 inclusive, with the single rider that data entry for WWF in 2001 and 2002 was incomplete at the time of analysis. It can be seen that zoos have supplied or channeled about 16% of the total NGO budget for tiger conservation in that period. This includes grants obtained by zoos from other sources, e.g., the Save the Tiger Fund, and all of the relevant funding from ZSL's 21st Century Tiger. WCS is here included as a zoo-based organization, despite the relative independence of their field program from their zoos in comparison to the other zoos involved, but their contribution is specified so that the figure can be quoted with or without them. They are the most significant contributor globally, and the other major players are ZSL with 21st Century Tiger, Minnesota Zoo with Ron Tilson's long-standing Sumatran Tiger Conservation Programme (STCP), and South Lakes Wild Animal Park's Sumatran Tiger Trust, which also funds the STCP.

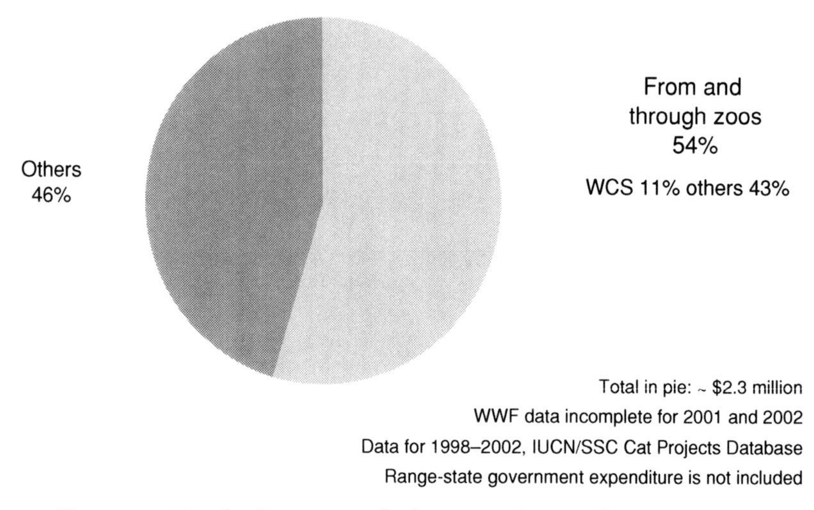

Figure 17.4 Zoo funding support for Sumatran tigers 1998–2002

As is evident above, zoos are particularly significant for Sumatran tigers. Figure 17.4 shows that they supplied or channeled a massive 54% of the global NGO budget for this taxon between 1998 and 2002. The same four contributors remain the most significant, but Minnesota Zoo rather than WCS is the single largest channeler of funds. It is highly likely that the zoo share of this pie will increase with the inclusion of 2003 and 2004 data, as these will include all the EAZA Tiger Campaign funds that went to Sumatran tiger projects.

AMUR LEOPARDS

Zoo support for Amur leopards is equally significant, as can be seen in Figure 17.5. Data here come from the Tigris Foundation, which has been very active in both stimulating and channeling zoo funding for this little-known taxon, and from WCS Russia and Phoenix, the other major implementers of the work. Again, accurate data for WWF are not available, and the figure used for them represents a well-educated guess by conservationists on the ground.

The slice of this pie currently provided by zoos is 58%, with EAZA zoos contributing 29%, WCS 25%, and other AZA zoos the remainder. Again, an increase is feasible over the next few years with the launch in 2004 of a push to raise more money from both EAZA and AZA zoos through the two regional managed programs. This effort will use the same mechanisms as

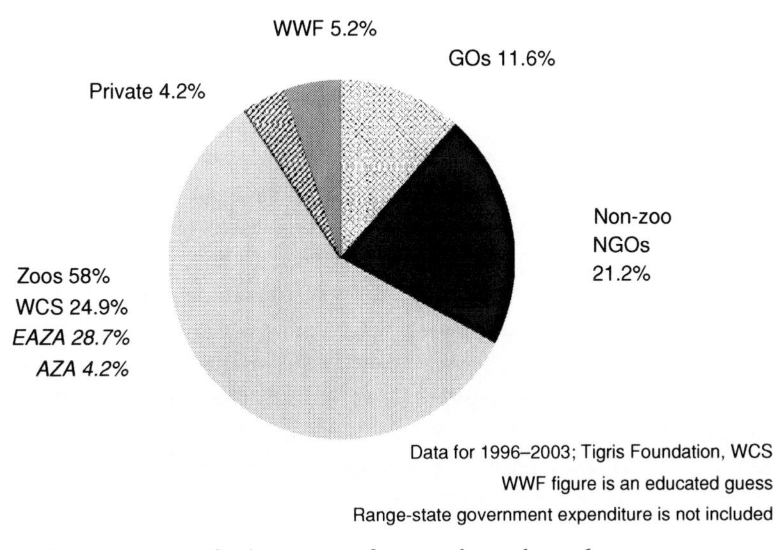

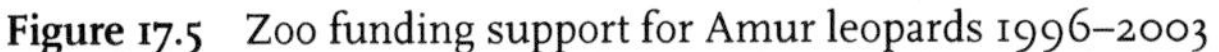
Figure 17.5 Zoo funding support for Amur leopards 1996–2003

the EAZA Tiger Campaign, e.g., ongoing provision of reports and images, and award certificates for success in fundraising. In addition, the role of zoos in Amur leopard conservation may perhaps reach well beyond the provision of funds, as the Amur leopard is probably the only large cat for which there is real potential for reintroduction from zoo stocks in the foreseeable future. Russian and international conservationists are agreed that reintroduction is a necessary conservation action and that zoo leopards can be used for this purpose. Discussion over suitable sites and methodologies is ongoing, as are efforts to secure the necessary political support and long-term funding. There is a long way to go yet; but if zoos can provide the majority of the funds for field projects and also supply cats for release, they will have a cast-iron case for being essential to the Amur leopard's survival in the wild.

NEXT STEPS

It is obvious from the work done to date that zoos are highly significant in supplying funding for conservation work. There is simply no arguing with the significance of the zoo funding figures for Sumatran tigers and Amur leopards given here, and it would doubtless be possible to say the same for other taxa too – *if only the data were available.* However, it is not going to be easy to get to this point. It is of course up to each regional zoo association to decide its own course on this, but a few basic suggestions are made here to support the argument that we simply must begin implementing plans to quantify the significant conservation contributions made by zoos around the world.

- Considerably increase the availability of regional resources for data collection and compilation.
- Work towards an online database in Europe and Australasia.
- Coordinate globally in order to ensure that regional databases are compatible in at least the major fields. WAZA and the Committee on Interregional Conservation Cooperation (CIRCC) might both be suitable vehicles for this. EAZA has offered to make its database available to other zoo regions.
- Encourage the development of centralized initiatives; both zoo coalitions such as the French CEPA (Conservation of Species and Populations of Animals) and the Butterfly Conservation Initiative in the US, and regional funds such as those at the AZA, the Dutch Zoo Federation and ARAZPA. Such initiatives make data collection infinitely easier, though it is acknowledged that they do not suit all purposes.

CONCLUSION

Fundraising is now recognized as a valid and important part of zoos' contributions to conservation. There is enormous potential around the world to increase both the funding contributions zoos make to field projects and public awareness of such support. These two factors are inter-related: the more public credit zoos get for fundraising, the keener individual zoos, and zoo associations, will be to participate. In order to fulfil this potential, improvements in coordination and communication are needed at all levels, and existing administrative resources must be expanded to absorb the additional strain.

Coordinated regional or national conservation campaigns can greatly increase both the flow of funds and the ease of data compilation. Key planning points here include ensuring wide public appeal and linking with other groups to spread the administrative load involved in handling funds and providing saleable campaign products. The EAZA Tiger Campaign took the first steps in both these arenas. The EAZA Shellshock and Rhino campaigns developed the commercial links and funding administration, respectively, to the point where the workload for zoo personnel became more manageable.

Such increased centralization will assist in data collection, but will never come close to capturing the full funding dataset. To do this, it is essential that zoos be encouraged to compile and supply their data, and that coordination be improved at global as well as regional levels. Regional zoo associations can assist with the first by making not only conservation contributions but also data submission on them a condition of membership. But unless steps are also taken globally, now, to ensure that data collected are compatible across the regions, public statements such as "last year, the world's zoos supplied seven million euros to field projects around the world" will remain out of reach. It is in everyone's interests to ensure that, in this arena, zoos work together at the international level on collection and compilation of data. If they remain unable or unwilling to do so, they will do a considerable disservice both to themselves and to the conservation causes they espouse.

ACKNOWLEDGEMENTS

Grateful thanks are due to the various representatives of the regional zoo associations who provided the necessary information on progress in their regions as well as helpful comments on this chapter: Katrine Garn, Corinne Bos, and Koen Brouwer of EAZA; Caroline Lees and Claire Ford of ARAZPA; Ruth Allard, Michael Hutchins, and Michael Souza of the AZA; Olivia Walter and Miranda Stevenson of BIAZA; and Koen Brouwer and Wim Verberkmoes of the Dutch Federation/Zoos' Help. Michiel Hötte of ZSL and the Tigris Foundation provided additional data on Amur leopard conservation funding, while Russell Rose of Double R Consulting designed and built both the IUCN/SSC Cat Projects Database and the EAZA Conservation Database, and was helpful in producing the zoo-based analysis of the tiger projects data used in this chapter. The IUCN/SSC Cat Specialist Group Cat Projects Database is maintained by ZSL with funding support from the UK's Department of the Environment, Farming and Rural Affairs, the Bosack and Kruger Foundation, and WWF-US. Kristin Nowell of the Cat Action Treasury was instrumental in obtaining the Bosack and Kruger grant.

References

IUCN/CBSG (1987). *Policy Statement on Captive Breeding*. Gland: IUCN.
IUCN (2002). *Technical Guidelines on the Management of* Ex Situ *Populations for Conservation*. Gland: IUCN. http://www.iucn.org/themes/ssc/pubs/policy/exsituen.htm.

IUCN/SSC Cat Projects Database (unpub.). ZSL, London, UK.
IUDZG/CBSG (1993). *The World Zoo Conservation Strategy: the Role of Zoos and Aquaria of the World in Global Conservation.* Brookfield, IL: Chicago Zoological Society.
WAZA (2005). *World Zoo and Aquarium Conservation Strategy: Building a Future for Wildlife.* Liebefeld-Bern: WAZA.

18

The Madagascar Fauna Group: what zoo cooperation can do for conservation

LEE DURRELL, DAVID E. ANDERSON, ANDREA S. KATZ, DEAN GIBSON, CHARLES R. WELCH, EVA L. SARGENT, AND INGRID PORTON

The Madagascar Fauna Group (MFG) is an independent, international consortium of about three dozen western zoos that collectively apply their resources and expertise to overcoming the biodiversity crisis in Madagascar. This chapter outlines briefly the history of the MFG, how it works, and highlights its strategy, activities, and lessons learned. The MFG is a good model for zoos that want to undertake effective conservation action.

Madagascar is truly a biological wonderland, with species endemism running at 100% in many groups (Jenkins 1987). Lemurs are the ambassadors, but the more humble denizens of this naturalist's paradise are no less remarkable. Sadly, the Great Red Island has been cut, burned, and eroded. Pristine habitats have been invaded by exotics. Many native species are declining, and the human inhabitants struggle to survive in their daily lives.

Madagascar has drawn missionaries, conservationists, and aid agencies like a magnet. Zoos, however, had little real input into the country until the 1980s when, after years of separate and often frustrating initiatives by zoos, the Malagasy government asked the international zoo community as a whole to assist with endangered species propagation, habitat protection, research, and training. The series of meetings that followed, often deftly facilitated by the late Ulysses Seal of the IUCN Conservation Breeding Specialist Group (CBSG), culminated in the establishment of the

MFG in 1988 with about a dozen zoos and a few universities from the US and Europe as founding members (Sargent and Anderson 2003).

The founders devised a set of by-laws to manage the group, significantly including a basic annual subscription cost for members, and developed a "protocol of collaboration" with the Malagasy government to define its in-country role and operating rules. There are now approximately two dozen American members, half as many from Europe and one each from Australia and Africa. Nearly all the members are zoos; the few others are the universities who joined at the beginning.

The MFG is currently governed by a steering committee of 11 people chosen to represent their institutions, not themselves, and chaired by the zoo administering communications, finances, and other paperwork. The committee meets annually, together with its scientific advisors and Madagascar-based technical advisors, to evaluate the work and plan for the future.

In the early days the MFG's strategy to meet the requests of the Malagasy government was fairly simple and based on what zoos at the time did best – providing technical and financial support to help develop the two in-country zoological collections, and funding several short-term field studies on threatened species of interest to members to help them match conservation efforts to their collections. It is worth noting here that what partly motivated some of the first members was the expectation that helping the zoos *in* Madagascar would facilitate the export of stock to zoos *outside* Madagascar, to enter into managed breeding programs, of course, but also to enhance the gate and the reputation of the receiving zoos themselves. It is also worth noting that other members joined to make sure that zoos were behaving honorably and not trying to cut individual deals with the government just to try to get animals out. In fact, only eight specimens have ever left Madagascar under the MFG's protocol, and all members' attention soon became focused on what could be done for conservation in-country.

Shortly after the MFG was established, it began to address the other requests of the government by planning a comprehensive program to protect an important rainforest reserve, the centerpiece of which was the release of zoo-bred lemurs to reinforce a wild population. Associated with this program and the work at the two zoos were activities to build conservation capacity and ethics through professional training, primary education, and partnerships with government agencies and local non-Governmental organizations (NGOs). A recent, very ambitious MFG project was to organize a conservation planning workshop with Malagasy and foreign experts to identify country-wide conservation, management, and research priorities and to recommend appropriate actions.

In the 15 years since MFG began, it has experienced successes and setbacks and learned a lot about working in a developing country with a unique culture and complex politics. A lot has also been learned about working together in an international group.

ACTIVITIES AND OUTCOMES IN MADAGASCAR

Efforts to develop the two distinct in-country zoos have had widely contrasting results.

The Parc Zoologique d'Ivoloina was created on a small site within a 400-ha forestry station just north of the major port on the east coast, under the ægis of one of the more under-resourced ministries. It was virtually destroyed by a cyclone in 1986. The MFG was asked by the government of Madagascar to work with the relevant ministerial department and the Duke University Primate Center to rehabilitate the site as a threatened species breeding station and a conservation education facility for the region. Two expatriate technical advisors were provided with substantial management authority. They worked with government colleagues on a master plan, including the restoration and development of the grounds, animal enclosures and facilities, the training of staff in animal husbandry, zoo management and environmental education, and feasibility studies relating to reintroduction and re-stocking options for captive-bred and confiscated lemurs.

Parc Ivoloina re-opened to the public in 1990, and has since made enormous progress (Katz and Welch 2003). It holds genetically important stock as a reservoir for several global breeding programs, with a current collection of lemurs, tortoises, and a few other reptiles and amphibians. It is a nationally recognized facility for teacher training in environmental education and a regional base for public awareness about conservation issues; it prepares local primary school children to advance to secondary school; it demonstrates agroforestry techniques to local farmers and has implemented an active reforestation program on site. The Parc enjoys some financial self-sufficiency from the modest entrance fees and the sale of various zoo merchandise. With tremendous support from the local community and government, its visitors average more than 10 000 annually, of which 70% are Malagasy nationals. Recently, the MFG was awarded a 25-year contract to manage the zoo and the forestry station at Ivoloina, a clear sign of the esteem in which the MFG is held by the authorities.

The other zoo is sometimes referred to as the national zoo – Parc Botanique et Zoologique de Tsimbazaza, a 12-ha oasis of trees, lakes, and lawns in the middle of the crowded capital city, most of whose inhabitants will only there ever see a lemur or a chameleon or a baobab. Tsimbazaza maintains a large botanical and animal collection, which includes mammals, birds, reptiles and amphibians, as well as an entomology division. Its visitors number about 150 000 annually, of which most are Malagasy nationals, and the Education Department provides ongoing environmental education programs for the urban community as well as for the surrounding villages. Established in colonial times, but now under the ægis of another under-resourced ministry, Tsimbazaza has found it difficult to make the transition from menagerie to modern, conservation-minded zoo. The government asked the MFG to assist Tsimbazaza as it had Ivoloina. However, senior management personnel were already firmly in place, thus, unlike at Ivoloina, significant management authority by the MFG was not possible (Gibson *et al.* 2001). As a result, one expatriate technical advisor was employed in 1991 to work with staff on husbandry training, facility improvement, staff development, and public education. The MFG provided funds or in-kind aid for these and, indeed, for basic operations, as needed. The advisor also worked to transfer skills in zoo management to the senior staff, including bringing in measures to develop a modest income stream.

There were some dramatic improvements in animal care and breeding, the quality of exhibits and signage and the degree of staff interaction with colleagues elsewhere. However, the fundamental, long-lasting changes at management level required for Tsimbazaza to reach its full potential have been slow and inconsistent. In attempting to match the technical advisor's skills to what the MFG and the senior staff felt was needed, the advisor was replaced on three occasions, providing a total of only 6 years of assistance with unavoidable gaps in time between each advisor, a situation in sharp contrast to that at Ivoloina. Combined with several changes at high level in Tsimbazaza personnel, the lack of continuity was clearly not conducive to success, and in 1998 the MFG reluctantly decided to suspend support to Tsimbazaza pending solutions to the long-standing problems. To date these have not been found.

The Betampona Natural Reserve is one of the few remaining tracts of lowland rainforest in northeast Madagascar and is the site of MFG's highest-profile project, the release of captive-born black and white ruffed lemurs to reinforce a remnant of a wild population (Britt *et al.* 2003b). Seven years after the first release in 1997, 3 of the 13 released ruffed lemurs are still surviving and monitored regularly. Four offspring are also surviving, of which three are the result of outbreeding between a releasee and a wild resident. The ability of the captive-bred lemurs to survive for years without provisioning, to successfully integrate and to reproduce with the resident population have all been demonstrated (Britt *et al.* 2003a and pers. comm.)

The release project is the centerpiece of a comprehensive program to protect the Reserve itself. The components include surveys and field research on a variety of taxa from mammals to plants, support for Malagasy graduate students, and education and training in natural resources management for local residents, as well as active protection from illegal incursion into the Reserve. A tree nursery and a sustainable forestry demonstration plot have been established at the base camp at the edge of the Reserve, which are intended for the eventual reforestation of degraded areas of the Reserve and for local community reforestation projects. Also opportunities for employment, sustainable micro-development projects, special environmental awareness events, assistance with school construction, and teacher training are made available to the inhabitants of villages surrounding the Reserve, as are education and agricultural outreach schemes, based on those developed at Ivoloina.

In 2000 planning began in earnest for a CAMP (Conservation Action Management Plan) workshop to be held in Madagascar in 2001, in partnership with the National Association for the Management of Protected Areas and the University of Madagascar and facilitated by the CBSG. It was a rare opportunity for Malagasy and foreign experts to sit down with other stakeholders and examine available data on mammals, reptiles, amphibians, and fish. Together they updated the IUCN Red List categories for known and recently discovered taxa and planned a long-term program of research and conservation priorities. This collaborative process was impressive, and the government reaffirmed its commitment to conservation, putting a number of species back on the conservation agenda. The results have now been delivered to all appropriate authorities, libraries, and conservation organizations in Madagascar, and the MFG intends to fund some of the recommended activities over the next few years (Conservation Breeding Specialist Group 2002).

ACTIVITIES AND OUTCOMES OUTSIDE MADAGASCAR

What does the MFG do outside of Madagascar to gather the resources needed for its in-country programs? As mentioned earlier, it is a fundamental tenet of the MFG that each member pays a basic annual subscription. Some founding members began with a subscription of $9000, as did a few later members, but it became clear that this sum was not within the means of all potential participants, and even loyal members had to reduce their donations from time to time depending on their financial circumstances. The MFG is sensitive to members' financial situations, and few members have ever abandoned the Group. Much valuable support is received from donations in kind, such as computer and veterinary supplies, zoo and field agent uniforms, and school materials. Staff from member zoos also have the opportunity to work in Madagascar, which is of real help to a program as well as an edifying experience for the person concerned. Examples include providing vet training at Tsimbazaza, improving exhibits at Ivoloina, and collecting data on the released lemurs at Betampona.

What have the in-country programs cost? In 1996 the MFG's budget was $108 000; in 2003 it was $250 000, an increase of 130%. However, membership numbers rose by only 30% during that time, and dues revenues, always variable, rose by only 20%. The additional funding has come from extra substantial donations by a few members, private foundations and corporations, and a few generous individuals. But in spite of variation in the membership revenues and the worrisome reliance on soft money, the budgeted expenditure has always been covered, usually carrying forward funds into the next fiscal year.

MFG members are kept informed in several ways. After each annual steering committee meeting, minutes, reports, management accounts, and budgets are sent to the members. An informal meeting is held in the US in conjunction with the Association of Zoos and Aquariums (AZA) annual meeting. A periodic, but irregular bulletin has been replaced by an email listserve and the MFG website (www.savethelemur.org), and reports from the Ivoloina and Betampona projects are circulated by email. Members are provided with press releases and information packets and helped with graphics about Madagascar. Speakers and materials are offered for special events.

LESSONS LEARNED

Most of the objectives of MFG's numerous activities in Madagascar over the last 15 years have been achieved, and all of the ongoing activities are showing good rates of progress.

The factors leading to success are as follows:

- Many institutions working together toward a common goal, which lends credibility to activities and gives collective strength.
- Assembling a large group of scientific advisors who have direct conservation and research experience in Madagascar and actively participate in

the annual working meetings. Steering Committee members, who are primarily zoo directors with limited in-country experience, are guided by these experts who insert reality into the finances, politics, and logistics required to achieve the objectives.

- Maintaining the same full-time staff in-country over time, which develops the cultural and political sensitivities we must have as foreigners, promotes stable working relationships with in-country partners and permits projects to be built up gradually, ensuring their solid foundation before expansion. The exception was the work at Tsimbazaza – staff discontinuity was one reason the effort had so little success.
- Delivering funding reliably and sufficiently, which allows continuity and longevity of staff, as well as meeting the less glamorous needs of conservation quickly and efficiently.
- Maintaining member loyalty, but needing to improve "buy-in."

The disproportion between the increase in budget and the increase in membership revenues over the years indicates this lack of "buy-in." There is a loyal membership which has been important to the MFG's long-term financial security, which, in turn, has built the confidence of the granting agencies and other special funding sources. Because sharing operating costs is demonstrably more effective than paying for them individually, the MFG will continue to emphasize this to its members. However, to address the budgetary and marketing challenges now faced by modern zoos a well-structured program can also offer discrete, attractive components with which donors can be identified. The truth is that zoo directors are in a better position to allocate a portion of their zoo's budget and/or raise conservation monies if their zoo takes credit for a *specific* project, i.e., assumes ownership of that project.

A STRENGTHENED STRATEGY FOR THE FUTURE

The July 2003 Steering Committee meeting in Zürich was a milestone for the MFG. After 15 years of leadership by San Francisco Zoo, the mantle was handed to Saint Louis Zoo. For a number of people long involved in the MFG, the meeting was their last, whereas for others it was their first. The participants updated and clarified the vision, mission, and strategy of the MFG, producing a document which can serve as a robust blueprint for all zoos interested in working together to make a positive impact on in-country conservation.

The vision statement is simple, yet reflects the membership's heartfelt and ultimate goal:

The zoos and aquariums of the world will unite with Madagascar to halt extinction of Malagasy wildlife.

The mission statement emphasizes the importance of collaborative action:

The Madagascar Fauna Group mobilizes zoos, aquariums and related conservation organizations worldwide to conserve the wildlife of Madagascar.

The nine-point strategy guides our actions both outside Madagascar and within the country:

- connecting zoos, aquariums, and their vast constituencies to the flora and fauna of Madagascar
- encouraging development of exciting and informative exhibits and programs
- stimulating the flow of funds from zoos and their constituencies to conservation in Madagascar
- setting priorities for research and conservation action
- building partnerships with the Government of Madagascar and other organizations
- acting to conserve the most endangered species
- protecting sites to maintain the diversity of wildlife
- building the capacity for conservation within Madagascar
- fostering a conservation ethic among the people of Madagascar.

The MFG's operational approach as a dues-paying membership organization with a permanent in-country staff was reaffirmed as a successful one. However, it was agreed that the long-held notions of governance and membership required a fresh approach. To that end, these were redefined and matched to the conservation program. All members still receive credit for the group's broad conservation achievements, but also assume ownership of a specific component of the conservation work. This is a new blueprint for the management of other zoo groups wanting to undertake regional conservation. There are three categories of MFG membership:

- *Managing Members* pay an annual subscription of $10 000, sit on the Steering Committee, and select one from a list of major conservation programs, including the research, monitoring and survey work at Betampona and environmental education at Ivoloina, each with a full-time Malagasy member of staff heading up each team. Also on the list are the new "Conservation Contract Programmes," a creative and experimental initiative that makes it possible for zoos to partner directly with villages in Madagascar to protect their forests. Managing Members have their own webpage on the MFG website and regularly receive interpretative materials for their own educational and marketing purposes.

- *Sponsoring Members* pay $5000 per year, vote on MFG matters, and select a project from another list, including the biodiversity surveys recommended by the CAMP in 2001 and captive breeding, education or forestry projects at Ivoloina, each with full- or part-time staff to support. Supporting Members are well recognized on the website and regularly receive interpretative materials.
- *Contributing Members* pay $2500 per year and, like all members, receive credit for the MFG's broad conservation achievements.
- *Friends of the MFG* are valued groups, businesses, and individuals who donate less than $2500 and who provide supplies, in-kind services, and/or funding for subcomponents of a conservation program. They can download information and updates from the website from a restricted Friends page.

How do other zoo groups differ from the MFG? There are some excellent examples of regional conservation programs or networks which are sponsored financially and in-kind by zoos from outside the countries concerned. They include those overseen by CBSG South Asia for seven countries in South Asia (Walker 2003 and pers. comm.), Fauna and Flora International (FFI) for the Philippines (Oliver 2003 and pers. comm.), the Zoological Society for the Conservation of Species and Populations (ZGAP) in cooperation with Stiftung Artenschutz for Vietnam (Wirth 2002, pers. comm. and www.zgap.de), West African Primate Conservation Action (WAPCA), which is coordinated by Heidelberg Zoo for two countries in West Africa (www.wapca.org), and the International Committee for the Conservation and Management of the Lion Tamarins (ICCM) for Brazil (Mallinson 2001). Another initiative for Brazil, slightly different from the above examples, is the endowment fund created from the income from the EAZA Atlantic Rainforest Campaign, supported by 97 European zoos in 2001–02 (see Chapter 19). Finally, AZA has promoted the concept of Conservation Action Partnerships (CAP), which aim to coordinate AZA zoos' in-country conservation initiatives (www.aza.org).

In all of these regions the work ranges from modest projects to which single zoos are sometimes matched, to comprehensive programs supported by several zoos involving multiple conservation actions, such as species recovery, habitat protection and restoration, research, planning workshops, environmental education, professional training, and advocacy. In all cases the appropriate government agencies are viewed as partners, and local involvement is keenly sought. However, the "foreign" zoos themselves seem little involved in strategic direction and decision-making, and funds are usually raised on a project-by-project basis and/or in sums the individual zoos can afford at the time, creating problems with financial security.

The success of the work in all these regions seems to be underpinned by the energies of a few charismatic individuals. This is not to say that MFG does not have charismatic individuals nor that those individuals in the other regions will not try to ensure that the work endures long after they have gone. The difference is that the MFG is a body of zoos bonded together by rules of governance, a tradition of regular financial input, and a steering committee formed from the member zoos themselves which determines strategic direction and budgets. According to the Oxford English Dictionary, the word "consortium" was first used in the 1800s to describe a union of Italian banks which agreed to provide equal amounts of money to the national exchequer. It may be unflattering to compare the MFG to a bunch of old Italian banks, but the true meaning of "consortium" does not seem to apply to any zoo group but the MFG.

The MFG was founded on the basic principle that uniting zoos and aquariums under one umbrella would significantly increase the contribution any one institution could make on its own. Zoos are a powerful vehicle for initiating and sustaining a conservation action loop: enabling in-country conservation which, when communicated to visitors, tells them their support is important. Unlike any other conservation organization, zoos and aquariums can hook visitors by showing them wonderful creatures in stunning natural settings and then give them very real opportunities to contribute to the animals' survival in the wild. A relevant example is the extraordinary new exhibit at Zürich Zoo – the re-creation of an eastern Madagascar rainforest – and already the number of Swiss visitors to Madagascar's east coast has increased! More and more zoos are striving to link their collections to in-country conservation efforts but have stumbled when faced with identifying and administering effective programs in countries physically, culturally, and politically distant from their own home ground. The MFG has been successful in this regard because it supports an in-country staff dedicated to the conservation projects shared by the members. But in a fast-paced world of soundbites and intensive competition for the recreational dollar, the MFG's membership strategy has been re-formulated so a member can easily highlight and market its individual contribution while, in reality, remaining a collaborative partner.

ACKNOWLEDGEMENTS

After 15 years of operation, the MFG has a long list of diverse supporters, from its member zoos to private foundations, from universities to businesses, from zookeepers (and their children!) to movie stars. It would be impossible to name them individually here, but to all of them we are most grateful for funds, in-kind aid, technical expertise, logistical support, and collaborative spirit. Special warm thanks are due to our in-country partners, both within government and from non-Governmental organizations, and to our own dedicated and talented Malagasy staff. Finally, we would like to recognize the helpful input of Joanna Durbin and John Hartley to drafts of this paper, as well as the ready willingness to share information about other zoo cooperative programs by William Oliver, Sally Walker, and Roland Wirth.

References

Britt, A., Welch, C., & Katz, A. (2003a). Project Betampona update. *Lemur News*, **8**, 6.

Britt, A., Iambana, B. R., Welch, C. R., & Katz, A. S. (2003b). Re-stocking of *Varecia variegata variegata* into the Réserve Naturelle Intégrale de Betampona. In *The Natural History of Madagascar*, eds. S. Goodman & J. Benstead. Chicago, IL: University of Chicago Press, pp. 1145–1151.

Conservation Breeding Specialist Group (SSC/IUCN) (2002). *Evaluation et Plans de Gestion pour la Conservation (CAMP) de la Faune de Madagascar: Lémuriens, Autres Mammifères, Reptiles et Amphibiens, Poissons d'eau douce et Evaluation de la Viabilité des Populations et des Habitats de* Hypogeomys antimena (*Vositse*). Apple Valley, MN: CBSG.

Gibson, D., Katz, A., & Welch, C. (2001). Madagascar Fauna Group: important sociocultural issues. In *The AZA Field Conservation Resource Guide*, eds. W. G. Conway, M. Hutchins, M. Souza, Y. Kapetanakos, & E. Paul. Atlanta, GA: Zoo Atlanta, pp. 190–197.

Jenkins, M. D., ed. (1987). *Madagascar: An Environmental Profile.* Gland: IUCN.

Katz, A. & Welch, C. (2003). Parc Ivoloina. In *The Natural History of Madagascar*, eds. S. Goodman & J. Benstead. Chicago, IL: University of Chicago Press, pp. 1555–1559.

Mallinson, J. (2001). Saving Brazil's Atlantic rainforest: using the golden-headed lion tamarin *Leontopithecus chrysomelas* as a flagship for a biodiversity hotspot. *Dodo*, **37**, 9–20.

Oliver, W. (2003). Philippines Biodiversity Conservation Programme. Unpublished report. Cambridge: Fauna and Flora International.

Sargent, E. L. & Anderson, D. (2003). The Madagascar Fauna Group. In *The Natural History of Madagascar*, eds. S. Goodman & J. Benstead. Chicago, IL: University of Chicago Press, pp. 1543–1545.

Walker, S. (2003). CBSG South Asia Activity Report September 2002 – December 2003. Unpublished report. IUCN: Conservation Breeding Specialist Group.

Wirth, R. (2002). Species recovery programmes in the Philippines and Vietnam. In *Proceedings of the 1st Forum about in-situ Conservation Projects supported by European 2005*, 12–14 June 2004, Centre de Congrès, Anmgers, France. Organized by Zoo de Doué-la-Fontaine.

19

Zoo coalitions for conservation

DAVID A. FIELD AND LESLEY A. DICKIE

INTRODUCTION

There are myriad ways in which zoos have supported or participated in conservation projects and initiatives. Zoos have made single initiatives, joined forces with non-Governmental organizations (NGOs), formed alliances with other zoos in small groups, or been part of wider regional or international coalitions. Any coalition possesses a number of process-related features that must be present for a coalition or partnership to function in reality. Focus, communication, ownership, and individual strengths are some of the features that require clear definition. But are there particular benefits that can be gained when working with a number of partners rather than pursuing single initiatives? And do zoos fill a specific niche in coalitions and partnerships?

More zoos are now actively engaged in *in situ* conservation, and any zoo that is attempting to offer such conservation provision should be encouraged and supported. However, there are always lessons to be learned from the methods previously employed, and, by reviewing past exercises, future conservation initiatives should be better equipped to deliver their stated aims. In this chapter we will not restrict the term "coalition," but use it in the sense of any institutions (two or more) working together.

WITH WHOM SHOULD ZOOS WORK?

Zoos are not homogeneous organizations. They differ markedly both within and between countries. A number have explicit conservation focuses, some educational and some research-based. The majority have a combination of the two and a number may not yet have clearly defined their focus. Some zoos are charities; others, trusts or private concerns. In addition many are maintained by local authorities, and these basic organizational structures may have ramifications for the ways in which they address conservation activities (see Chapter 20). For zoos to build effective coalitions, partner identification is fundamental. Whilst zoos can be idiosyncratic organizations, coalitions can also be highly variable. The combination of partners can give a particular emphasis to the focus of the project, or even change the focus entirely. Coalitions can be large and global, small and local, and developed at regional, national or international levels. Different individual organizations may have widely differing agendas while ostensibly appearing to have the same conservation aims. For example, can proponents of sustainable-use strategies work coherently with advocates of a protectionist, "fortress conservation" outlook (Berkes 2004)? Can these divergent viewpoints be accommodated within a single project?

There must be a clear and step-wise decision process by which each partner at the table can articulate their objectives and find the common ground on which the partnership should proceed, including the responsibilities and commitments of each partner (Mallinson 1991). Each partner may also bring a unique strength which enhances the chances of achieving stated objectives. However, underlying individual goals should not be allowed to obscure the wider objective. "Goal substitutions" can derail projects, which may ultimately impact on species survivorship (Clark and Brunner 1996).

Zoos partner with zoos, with NGOs, with statutory bodies, research organizations, governments, private individuals, or with a combination of agencies (Walker *et al.* 2005). Such coalitions can be informal or subject to memorandums of understanding (MOU), which more clearly define the aims and objectives for the benefit of all parties. There may be organizational differences between such disparate parties, differences of scale and cultural viewpoints, which need to be accommodated. The last point is of particular importance in relation to the dominance of large, developed-world conservation agencies and their work in the less developed, but mega-diversity countries. Chapin (2004) suggested that some conservation agencies, with large budgets and global reach, may not be fully integrating the needs, and skills, of local agencies and communities. In addition, staff on the ground may find themselves operating within very different frameworks, and employing alternative strategies in the face of immediate difficulties, which may not tally with policy set at higher organizational levels. The Chapin paper generated much discussion, with some of the global NGOs referred to in the paper vigorously refuting the claims made.

Whether or not this conservation "cultural imperialism" is real in terms of action or perceived to be due to poor communication, it highlights some of the actual difficulties in achieving functional partnerships where disparities exist (see also Romero and Andrade 2004). In addition Metrick and Weitzman (1998) suggest that, as of yet, clear definitions for biodiversity preservation have not been established, a requirement if local needs are to be integrated and met in conservation planning. Such objectives must be clear to allow measurement of performance, to ascertain what results were generated and whether they contributed to the fulfilment of initial conservation aims (see Chapter 21).

By far the most common zoo coalition is in the form of national and regional zoo organizations [e.g., BIAZA, EAZA, AZA, African Association of Zoos and Aquaria (PAAZAB), etc.]. Zoo associations are not primarily concerned with conservation; rather they act as representatives of their members on national or regional matters, provide membership services, and act as a focal point for issues surrounding zoos within their area of representation. However, many now have specific conservation committees, designed to help their members reach conservation objectives, providing a forum for advice and stimulating activity.

CORE STRENGTHS OF ZOOS FOR CONSERVATION COLLABORATION

Zoo participants in larger coalitions of predominantly non-zoo organizations bring many specific core areas of expertise, namely the care of animals in captivity and all that that entails. Where some zoos may lack first-hand experience of *in situ* conservation, all responsible zoos will be participants in collaborative *ex situ* breeding programs, wherein recommendations made by appointed program coordinators are carried out by individual zoos. This can be at regional [e.g., EAZA, South East Asian Zoos Association (SEAZA), etc.] or international level, with International Studbooks (ISBs) coordinated by WAZA. In conjunction with breeding programs, zoos, through individuals, collaborate to produce husbandry guidelines for varied taxa, with husbandry lessons learned at one institution passed on to others. This unique position of zoos, visitor attractions which may be in competition with neighboring zoos whilst still working in partnership to meet *ex situ* breeding demands, means that they are well equipped to pursue collaborative projects, as it is an intrinsic function of any responsible zoo.

Zoos can also provide a public face for conservation activities which may not be available to other conservation NGOs. Having a base, the zoo itself, which the public can visit, can promote projects which may not reach the wider community in any other way, particularly when less charismatic species, not deemed as newsworthy, are targeted. Science, a discipline area that is often exclusionary, can be communicated to a wider audience through zoo interpretation. Linking zoos with researchers carrying out conservation *in situ*, or by zoo-employed researchers, can help reach this "non-traditional" audience (Nadkarni 2004). This can be particularly powerful in spreading a single coherent message over a large geographic area (e.g., EAZA campaigns, see later in this chapter).

This is important in two main contexts. In the developed world, commonly zoo visitors are targeted by requests for conservation funding and can be asked, as consumers, to re-evaluate their personal impact on the environment. In the developing world, in contrast, visitors may directly impact on species survival in their local area. The provision of funds from the more economically able nations of the world is vital to species survival. However, communities in mega-diversity countries must be engaged in the conservation of their local biodiversity, regardless of whether they are able to provide funds. Collaborative zoo organizations, such as the International Zoo Educators (IZE), can assist developing-country zoos to produce educational material that may have very real conservation benefits for threatened species. During the 2004 WAZA annual meeting in Taipei it was also suggested that larger, financially able zoos "twin" with zoos in developing countries, providing technical and practical support. It was also recommended that where developed-country zoos initiate or participate in *in situ* conservation they actively involve, where possible, the local zoo, again ensuring that in-country zoos can gain greater involvement in the preservation of their native fauna (WAZA 2005). This type of zoo–zoo coalition can be relatively informal but may benefit the developing–country zoo if the ties are formally ratified, allowing their role to be recognized locally, nationally, and internationally.

THE NEED FOR LOCAL COMMUNITY COLLABORATION IN CONSERVATION COALITIONS

Local community buy-in is of particular relevance where conservation activities are contentious. The reintroduction or maintenance of carnivores is an area of conservation activity that requires consummate social advocacy whether in the developed or developing world. Clark *et al.* (2001) reviewed the difficulties associated with the Grizzly Bear (*Ursus arctos horribilis*) Recovery Plan in the US, an initiative led by a coalition of varied organizations. It highlighted the difficulties when scientific evidence blends into, or is influenced by, strong individual stances. Whilst this was not a zoo-associated coalition it does provide valuable lessons for all such conservation activities, reminding us that conservation actions are subject to human value judgements, and that these become exceptionally strained where public safety issues are relevant (Hart and O'Connell 2000, Sekerak *et al.* 2002, Thirgood *et al.* 2005).

Conservation action may also impinge on the livelihoods of local people. Changing social attitudes may be necessary for conservation action to be successful, whether this is in relation to the use of traditional medicines or animals prized as vital sources of protein (Rose 2001). In addition conservation alliances may find themselves in conflict with human development agendas (Davies 2002, Berkes 2004), with different groups having differing objectives which may hinder action by both. Davies (2002) suggests, in relation to the bushmeat trade in Africa, that conservation and development groups should work in conjunction with one another to more readily reach objectives. Bushmeat, an increasing conservation focus, is extremely important to the poorest peoples who are the focus of development agency goals. Areas of overlap that bring benefits to both groups should be investigated, in an attempt to both conserve populations of wild animals and address underlying problems of food security. This form of alliance, across general subject areas, is likely to be of growing importance and utility, in the future, as people and animals come into increasing conflict (Jackson and Wangchuk 2000, Balmford *et al.* 2001).

Where good relationships have been established with local peoples for the benefit of conservation these need to be communicated and the process incorporated where possible into other projects. Steinmetz *et al.* (2006) illustrate how a workshop process in their focus area in Thailand led to the identification of shared problems, resulting in a more engaged local population. Schwartzman and Zimmerman (2005) highlight, in relation to the Brazilian Amazon, that indigenous peoples' "*egalitarian common-property resource management regimes*" may be beneficial to conservation aims, particularly where some level of resource use is a part of the planned objectives. This may have further application in other locations and highlights the role of the anthropological sciences in conservation.

In-country political structures can also affect conservation aims. Navigating different cultural values and the interplay this has with political process is an area where most zoos are inexperienced. Partnering with local zoos and conservation NGOs gives some insight, and therefore improved ability, in successfully assisting with conservation aims. Lewis (2005) gives an extensive overview of how some of the above issues have affected tiger conservation in India in the 1970s (Lewis 2005).

The Conservation Breeding Specialist Group (CBSG) of the IUCN has considerable knowledge and experience of bringing together local peoples, conservation NGOs and other interest groups in Population, Habitat and Viability Analysis workshops (Byers and Seal 2003) and other processes. Many zoos are members of the CBSG and may have participated in their workshop processes. Applying these experiences to partnerships could lead to improved goal identification and communication.

The case studies below demonstrate three zoo coalitions, two of which involved EAZA campaigns (but see also Chapter 18): the Lion Tamarins of Brazil, the EAZA Bushmeat Campaign and the West African Primate Conservation Action (WAPCA) alliance.

LION TAMARINS OF BRAZIL: A CASE STUDY

This is the longest running example of a conservation coalition that will be discussed directly in this chapter. There are four species of lion tamarin which are endemic to the Mata Atlântica, the Atlantic Rainforest of Brazil. These are the golden lion tamarin (*Leontopithecus rosalia*), the golden-headed lion tamarin (*Leontopithecus chrysomelas*), black lion tamarin (*Leontopithecus chrysopygus*), and black-faced lion tamarin (*Leontopithecus caissara*). Today as little as 2% of their original forest remains with the result that all the lion tamarins are considered to be under various levels of threat (Kleiman and Mallinson 1998).

In the history of lion tamarin conservation there are examples of zoos working as individual institutions, as loose networks of finance providers, and as a European-wide focused coalition. A captive management and research program for *L. rosalia* was established by the Smithsonian National Zoological Park (SNZP) in the early 1970s, while in 1983 a reintroduction program for captive-born *L. rosalia* was begun. Also in 1983 the Golden Lion Tamarin Conservation Program was established and was fully administered by the SNZP until its activities came under the remit of the AMLD – Golden Lion Tamarin Association, a Brazilian NGO created in 1992.

In these early years, lion tamarin conservation owes an enormous debt to two institutions, the Durrell Wildlife Conservation Trust (DWCT, formerly the Jersey Wildlife Preservation Trust) and the SNZP. Both these institutions contributed not only considerable financial resources but also substantial technical and administrative support. Durrell hosted and trained a number of key researchers in lion tamarin conservation. Equally the initiatives of the SNZP have had a major impact, in particular their considerable allocation of resources to establishing the research and management agenda for the captive program. The SNZP also took the lead in establishing those partnerships in and outside Brazil that resulted in the Golden Lion Tamarin Conservation Program and additionally provided the support for the development of the infrastructure for the AMLD. In 1981 an International Recovery and Management Committee (IRMC) was established to coordinate and oversee the conservation efforts for the golden lion tamarin (Kleiman and Mallinson 1998). A single IRMC was established for all the *Leontopithecus* species and international zoo representation at this level was through SNZP and the Durrell Wildlife Conservation Trust. This International Committee acts in an advisory capacity to IBAMA (the Brazilian Institute for the Environment and Natural Renewable Resources). Zoos are therefore integrally involved in the decision-making process for lion tamarin conservation, which has important value when seeking funds and commitment from zoos. The Lion Tamarins of Brazil Fund (LTBF) was established for *Leontopithecus* species in 1991 and this fundraising initiative targeted zoos by requesting donations to assist in conservation projects for the four lion tamarin species *in situ*.

Due to the success of this initial appeal, and recognizing the increasing importance of interactive management between captive and wild populations, the International Committee(s) decided to continue the appeal on an annual basis. Therefore, zoos that were already participating in the development of the scientifically managed captive populations were also able to contribute funds to aid the conservation of the remnant wild populations of lion tamarins. The LTBF initiative represented the first time that an international fund had been established to which all holders of individuals of an endangered species maintained outside the country to which it is endemic were requested to contribute annually to aid conservation in the wild. By March 2001 the LTBF had, with three matching grants from the Margot Marsh Biodiversity Foundation's Primate Action Fund, raised over US$ 250 000.

The year 2001 saw the launch of the EAZA Rainforest Campaign. Following the success of the Bushmeat Campaign the year before (see following case study), this was seen as an opportunity to target a large number of zoos and get them to work together on a focused conservation theme. This campaign had the full support of IBAMA and was extremely successful in its application, with 97 zoos across Europe from 20 different countries participating. The level of participation ranged from putting up posters to transforming parts of the zoo into a South American landscape, while others had major one-off events. Over €250 000 was raised by this campaign, 2.5 times the original goal, representing almost as much as had been raised by the LTBF since 1991. The money raised was sufficient to establish an endowment fund for long-term financial support of lion tamarin conservation activities. Therefore, this coalition of 97 European zoos had not only covered the costs of ongoing field projects but allowed a long-term commitment to be made, an essential foundation for successful conservation planning.

Zoo involvement in lion tamarin conservation has therefore been at a number of different levels:

- As independent institutions such as DWPT and SNZP initially
- In conjunction with NGOs such as the ICCM
- As finance providers through the LTBF
- As part of a wider coalition through the Rainforest Campaign.

THE EAZA BUSHMEAT CAMPAIGN 2000/2001

Wild game has long been an important source of protein to some of the poorest peoples in the developing world (Fa *et al.* 2003). However, due to a number of factors (increasing human populations, access to firearms, habitat loss, and habitat fragmentation) (Thibault and Blaney 2003) bushmeat hunting is becoming unsustainable in a number of regions, and is now believed to be a major threat to wildlife (Cowlishaw *et al.* 2005, Refisch and Koné 2005). The focus of the first EAZA campaign in 2000/2001 was bushmeat. In contrast to subsequent campaigns, political advocacy was the main tool used by this coalition. This stemmed from the setting up of an EAZA Bushmeat Working Group in 1999. The campaign was run in conjunction with the International Fund for Animal Welfare (IFAW). While the EAZA campaign focused on the problems of illegal hunting and food security in Africa, it acknowledged that bushmeat hunting was becoming unsustainable in a number of countries in both Asia and South America.

The three main objectives were to bring awareness of the bushmeat crisis to an international audience, to raise funds for bushmeat-related projects, and to collect signatures for an international petition against the illegal bushmeat trade. This petition requested that the EU, a major development donor to Africa, ask African leaders to ensure that wildlife protection laws are enforced. It also asked that sustainable means of development be implemented, ensuring that wildlife resources could continue to be used, highlighting that this was not only an issue of wildlife conservation, but of people and their security. A total of 162 zoos participated in this campaign, with over 1.9 million signatures collected, a feat called "*a major demonstration of participative democracy*" by the Irish MEP Proinsias de Rossa, who sponsored and helped draft the report. A bushmeat hearing entitled "Planet of the Apes" was held at the EU Parliament in Brussels in November 2001, with presentations from the steering group of the campaign, and the handing over of the petition to the parliament. This was followed by the presentation of the petition to the EU Development Ministers a few days later and to the Committee on Petitions in January 2002. A reply from the Petitions Committee was received in January 2003 and finally in January 2004 there was a majority vote in favor by the EU Parliament to adopt the campaign demands.

However, the campaign steering committee continues the work initiated during the single year of the larger campaign. Whilst adopting the EAZA Bushmeat Report the Parliament failed to give specific measures as to how it would proceed. In December 2004 Bryan Carroll, Chair of the EAZA Bushmeat Working Group, commented on the Commission's response to the European Parliament Resolution adopting the report (A5-0355/2003), asking for specific actions. This is likely to be ongoing for a considerable period of time and demonstrates the need for continuity in conservation action. It also demonstrates that, while 162 zoos participated in the campaign, there must be individual "champions" who continue to coordinate and drive such coalition action.

WEST AFRICAN PRIMATE CONSERVATION ACTION (WAPCA)

WAPCA is an example of a smaller, single-focus conservation coalition between 11 European zoos, the Zoological Society for the Conservation of Species and Populations (ZGAP) and Conservation des Espèces et des Populations Animales (CEPA) as full members and the Zoological Society of London (ZSL) and Gemeinschaft Deutscher Zooförderer (GDZ) as partners. The main goal of this conservation project is to preserve primates in the Upper-Guinean rainforest, but it also has a number of specific interim targets to achieving this long-term goal: the establishment of a "Centre for Endangered Primates" in Ghana at the partner zoo in Accra, the provision of environmental education to local peoples, the provision of training to wildlife guards, the protection of habitat through providing adequate resources for local wildlife officials, and the improvement of captive breeding through improved knowledge.

Three main species are used as focal animals: the Roloway monkey (*Cercopithecus diana roloway*), the white-naped mangabey (*Cercocebus atys lunulatus*), and Miss Waldron's colobus (*Procolobus badius waldroni*). While the Roloway monkey and the mangabey are maintained in European zoos and utilized by the WAPCA partner zoos as flagships for the project, Miss Waldron's colobus has special status as being the first monkey declared extinct in the new millennium (Oates *et al.* 2000). Although reports still suggest that small isolated groups may exist, none have been reported from surveys since the late 1970s. The use of Miss Waldron's colobus in this context promotes a powerful message, to both potential donors in the developed world and local peoples directly impacting on species in the wild, namely that extinction is ongoing and will continue unless action is taken.

The establishment of the center at Accra Zoo will fulfil Article 9 of the Convention on Biodiversity, in that Article 9 gives clear mandate for the use of *ex situ* conservation action through maintenance and breeding of captive populations of threatened taxa, but it also proposes that this is best carried out (where possible) in the country of origin. While the center will initially fulfil a rescue role it can then be used in pursuit of the educational goals of the project, providing further information on the maintenance of these species in captivity. The building of the center is financed through the WAPCA zoo alliance, which also pays for a staff member to be based at Accra Zoo to help manage the project. The project is still at an early stage but has the benefit of a number of base founders, which should ensure continuity and long-term financial support, vital if conservation efforts such as these are to prosper.

WHY HAVE THESE COALITIONS SUCCEEDED?

Several factors have made these coalitions successful to date, and these factors can be identified, modified, and incorporated to suit most coalition structures. Enlisting and gaining support, clear coalition objectives, ownership, focused champions to drive progress, continuity, conflict resolution, communication, and targeted conservation spend are all essential components. In addition to the comments below a useful investigation of conservation alliances through the Biodiversity Support Program (WWF, The Nature Conservancy and World Resources Institute funded by USAID) has been produced by Margoulis *et al.* (2000).

Enlisting and gaining support

Attempting to persuade a bird park to raise money for primates would seem problematic. Yet this was done very successfully during the EAZA lion tamarins campaign. To do so, it was important to widen the inclusivity of the campaign objectives. This was done by targeting the campaign at the Atlantic rainforest in totality, ensuring that species from different taxa were represented. Special focus was given to any species that were part of European captive breeding programs, while lion tamarins were utilized as flagship species. Although it is fair to say that perhaps some of the target species were "more equal than others" enlarging the focus of the campaign brought in more coalition members and aided in instilling commitment to the objectives.

The Bushmeat Campaign utilized powerful images, invoking empathy within both the zoo community and the public, who contributed monies and/or their support to the petition. Such powerful, compelling stories are an area where zoos can excel, using an effective method to garner support, the "front of house" providing a stage to promote *in situ* conservation action. The Bushmeat Campaign also provided an example of where it is prudent for coalitions to gain political support to "sponsor" activities, as in the case of the Irish MEP Proinsias de Rossa and the Bushmeat Campaign.

Coalition objectives

Clear objectives are vital in both gaining support and general effectiveness. This is particularly important for zoos so that they can focus their fundraising and also give tangible explanations to their visitors as to how money raised is spent. It allows clearer interpretation opportunities for zoos, particularly important if education of the public is to be regarded as one of the most important aspects of the function of the modern zoo.

Equal partners

Zoos participating in coalitions should be able to contribute to a focused campaign, regardless of their level of contribution. This allows smaller zoos with lower budgets to fulfil their conservations aims, and financially more able zoos to spread their funds across a number of projects. It is the overall result rather than the individual contributions that ultimately matters, the "whole being more than the sum of the parts." In addition, larger, more experienced partners in coalitions must ensure that smaller partners feel that they are actually *partners*, not simply convenient donors.

Champions

All coalitions benefit enormously from having a number of "champions" and the influence of people such as Jeremy Mallinson (DWCT) and Devra Kleiman (SNZP) in furthering the conservation cause of lion tamarins has been hugely significant. Bryan Carroll of Bristol Zoo Gardens has provided a vital ongoing focus for the Bushmeat Campaign in Europe. Conservation campaigns need their champions, whether this is a single zoo initiative or part of a wider coalition. Projects need a driving force and these individuals may be crucial to finance provision. This also requires zoos to allow these champions to devote time to what may be construed as individual passions, within the context of institutional needs, and allow ideas to develop. However, these champions at some point will want to hand over the reins and without the benefit of multiple partners it is possible that there may be nobody to succeed them.

Continuity

A project reliant on a single or small number of zoos is highly dependent on the fortunes of that zoo (and in some cases of one individual within it). Multiple partners provide an obvious advantage in this area, providing a higher level of insurance to any individual project. Given that the time scales of conservation projects may be entirely unpredictable (though an initial project planning process should put in place clear indications of when a foreign conservation NGO plans to hand over the project to in-country colleagues), staying power is an important facet of conservation. Restoration of habitats may take years, decades in some cases. Behavioral change in local populations that impact positively on conservation outcomes is also an inherently slow process. Multiple partners cater for the changes in fortune of individual partners. Zoos must learn to make long-term commitments, not change projects without good reason.

Conflict

One potential disadvantage of coalitions is the conflict that could arise in terms of the conservation action targeted and actions taken to reach that target. How and what initiatives are supported can easily become the source of disagreement. For the Rainforest Campaign the potential of this problem was mitigated by using an established route for the funding: the LTBF and an established and respected decision-making process, the ICCM. The strong zoo representation on the committee and the strength of the ICCM (which was promoted in the Campaign literature) helped to add value to the campaign or at least assisted in enlisting participation. The potential for conflict can be mitigated from the beginning by clearly establishing goals before partners enter into coalitions.

Again the issue of equal partners must be raised as institutional rivalries can blight future conservation progress. One problem that has been somewhat apparent in the EAZA campaigns has been the view from some zoos that, since they are already carrying out individual projects or are in separate partnerships with other conservation NGOs, fundraising for other projects (projects identified for funding by the campaign organizing group in any one year) is in competition with their institutional aims. However, this attitude is at odds with the entire purpose of a common, coalition-based approach to conservation problems. If individual zoos are serious about their engagement in EAZA-type cross-cutting campaigns, individual institutional "showboating" should be set aside when greater conservation impacts and public education (a core mission of zoos) can be achieved.

Communication

Good communication is essential to maintaining support amongst partners, however loose the coalition, and relates very strongly to the possibility of conflict affecting projects. Communication can take many forms, and initiatives such as e-groups and newsletters aid communication. Despite there being clear goals, to which all partners agree, projects may ultimately fail if coherent communication is lacking. This is particularly important in relation to less experienced partners, who may be providing financial support but not personnel and thereby rely on feedback from "larger" partners. If the less experienced partner does not receive full and clear messages, and in turn feels unable to state their own concerns, relationship difficulties can fester and escalate. It should not be underestimated that such communication is a skill that requires attention and time.

Conservation spend

The Atlantic Rainforest Campaign provided an avenue for zoos, regardless of their size, financial status or previous conservation experience, to commit funds and feel equal as partners. Whilst it is true that the total sums of money collected may be dwarfed by the monies spent on zoo exhibitory, and could potentially have been provided by a single donor, in reality this is unlikely to happen and may not benefit zoos and long-term conservation aims in as powerful a way. From an individual zoo perspective, being seen by their visiting public to be supporting X, Y, and Z conservation projects rather than just project X will not only benefit the zoo, but educate the public, who may only make one zoo visit in a year, about the diversity of conservation activities carried out by zoos. Multi-zoo projects therefore allow zoos to diversify their conservation spend without diluting the impact on a single project.

IS THERE MORE ZOOS CAN DO?

Given the holistic nature of zoos and the fact that they have an identifiable, fixed, and accessible physical base, zoos in the developing world may be able to bring a new type of visible partnership that may impact upon conservation needs. If such zoos can partner with local and international conservation NGOs, information concerning conservation activities could be communicated to a far larger audience (N. Lindsay, pers. comm.) – the zoo visitor.

Zoos should also look more widely at what kind of partnerships can be established that have potential benefit for conservation. Despite a keen focus on habitat restoration in recent years, zoos have formed relatively few partnerships with botanical gardens, many of which have extensive field programs. Many botanical gardens would welcome closer links to zoos in relation to conservation output (S. Sharrock, pers. comm.) and this form of non-traditional synergy could bring new avenues for action. Working with the commercial sector for conservation is also a relatively neglected area.

Education is a core mission area for zoos: the majority of the 1200 or so WAZA zoos have at least some education staff on site. Incorporating education partnerships at field sites, involving local or international NGOs or other institutions can bring added value to projects, bringing a training aspect, for example, to a project whose main focus may be habitat protection or ecological monitoring (Brewer 2002).

CONCLUSIONS – COALITIONS VERSUS SOLO ZOO

In answer to the question as to whether it is better for zoos to pool their resources or go it alone, it would seem that both have a role in different circumstances. Solo zoo initiatives can not only provide finance but are well suited to a technical support role. However, they must ensure that they can provide for the long term. Coalitions can provide both finance and a technical role and should ensure greater continuity, allowing conservation activities, which may take many years to come to fruition, to prosper, and gain local support. Whilst it may, in some circumstances, be easier to get projects off the ground more quickly by going it alone, no project is truly individual. Relationships have to be made with in-country authorities and local peoples if field conservation is to take place. With additional emphasis, turning these relationships into real partnerships may ultimately be of greater benefit to all. Zoos have an enviable track record of working together in *ex situ* breeding programs. By applying the same principles of openness, trust, and community to field conservation, zoos can have even greater impact. In an ever more crowded world, working with others will be inevitable and perhaps the most effective action if species and habitats are to be maintained, for their intrinsic value and their benefit to us all.

References

Balmford, A., Moore, J. L., Brooks, T., Burgess, N., Hansen, L. A., Williams, P., & Rahbek, C. (2001). Conservation conflicts across Africa. *Science*, **291**, 2616–2619.

Berkes, F. (2004). Rethinking community-based conservation. *Conservation Biology*, **18**(3), 621–630.

Brewer, C. (2002). Outreach and partnership programs for conservation education where endangered species conservation and research occur. *Conservation Biology*, **16**(1), 4–6.

Byers, O. & Seal, U. S. (2003). The Conservation Breeding Specialist Group (CBSG): activities, core competencies and vision for the future. *International Zoo Yearbook*, **38**, 45–53.

Chapin, M. (2004). A challenge to conservationists. *World Watch Magazine*, Nov/Dec, 17–31.

Clark, T. W. & Brunner, R. D. (1996). Making partnerships work in endangered species conservation. *Endangered Species Update*, **13**(9), 1–5.

Clark, T. W., Mattson, D. J., Reading, R. R., & Miller, B. J. (2001). Interdisciplinary problem solving in carnivore conservation: an introduction. In *Carnivore Conservation*, eds. J. L. Gittleman, S. M. Funk, D. W. Macdonald, & R. K. Wayne. Cambridge: Cambridge University Press.

Cowlishaw, G., Mendelson, S., & Rowcliffe, J. M. (2005). Structure and operation of a Bushmeat Commodity Chain in southwestern Ghana. *Conservation Biology*, **19**(1), 139–149.

Davies, G. (2002). Bushmeat and international development. *Conservation Biology*, **16**(3), 587–589.

Fa, J. E., Currie, D., & Meeuwig, J. (2003). Bushmeat and food security in the Congo Basin: linkages between wildlife and people's future. *Environmental Conservation*, **30**(1), 71–78.

Hart, L. A. & O'Connell, C. E. (2000). Human conflict with African and Asian elephants and associated conservation dilemmas. Unpublished Report. Centre for Animals in Society, School of Veterinary Medicine and Ecology, University of California at Davis.

Jackson, R. & Wangchuk, R. (2000). People-wildlife conflicts in the trans-Himalaya. Management Planning Workshop for the Trans-Himalayan Protected Areas, Symposium sponsored by the Wildlife Institute of India, US Fish and Wildlife Service Himalayan Biodiversity Project and the International Snow Leopard Trust. Leh, Ladakh, 25–29 August 2000.

Kleiman, D. G. & Mallinson, J. J. C. (1998). Recovery and management committees for lion tamarins: partnerships in conservation planning and implementation. *Conservation Biology*, **12**(1), 27–38.

Lewis, M. (2005). Indian science for Indian tigers: conservation biology and the question of cultural values. *Journal of the History of Biology*, **38**(2), 185–207.

Mallinson, J. J. C. (1991). Partnerships for conservation between zoos, local government and non-governmental organisations. In *Beyond Captive Breeding: Re-Introducing Endangered Mammals to the Wild*, ed. J. H. W. Gipps. Symposium. London: Zoological Society of London, pp. 57–74.

Margoulis, R., Margoulis, C., Brandon, K., & Salafsky, N. (2000). *In Good Company: Effective Alliances for Conservation.* Washington, D.C.: Biodiversity Support Program.

Metrick, A. & Weitzman, M. L. (1998). Conflicts and choices in biodiversity preservation. *Journal of Economic Perspectives*, **12**(3), 21–34.

Nadkarni, N. M. (2004). Not preaching to the choir: communicating the importance of forest conservation to nontraditional audiences. *Conservation Biology*, **18**(3), 602–606.

Oates, J. F., Abedi-Lartey, M., McGrew, W. S., Struhsaker, T. T., & Whitesides, G. H. (2000). Extinction of a West African red colobus monkey. *Conservation Biology*, **14**(5), 1526–1531.

Refisch, J. & Koné, I. (2005). Impact of commercial hunting on monkey populations in the Ta region, Côte d'Ivoire. *Biotropica*, **37**(1), 136–144.

Romero, C. & Andrade, G. I. (2004). International conservation organizations and the fate of local tropical forest conservation initiatives. *Conservation Biology*, **18**(2), 578–580.

Rose, A. L. (2001). Social change and social values in mitigating bushmeat commerce. In *Hunting and Bushmeat Utilization in the African Rain Forest*, eds. M. Bakarr, G. Fonesca, R. A. Mittermeier, A. B. Rylands, & S. Walker. Washington, D.C.: Conservation International.

Schwartzman, S. & Zimmerman, B. (2005). Conservation alliances with indigenous peoples of the Amazon. *Conservation Biology*, **19**(3), 721–727.

Sekerak, C. M., Eason, T. H., & Small, C. R. (2002). Using partnerships to address human black bear conflicts in central Florida. *Proceedings of the Annual Conference of the Southeastern Association of Fish and Wildlife Agencies*, **56**, 136–147.

Steinmetz, R., Chutipong, W., & Seuaturien, N. (2006). Collaborating to conserve large mammals in Southeast Asia. *Conservation Biology*, **20**(5), 1391–1401.

Thibault, M. & Blaney, S. (2003). The oil industry as an underlying factor in the bushmeat crisis in Central Africa. *Conservation Biology*, **17**(6), 1807–1813.

Thirgood, S., Woodroffe, R., & Rabinowitz, A. (2005). The impact of human-induced conflict on human lives and livelihoods. In *People and Wildlife: Conflict or Coexistence*, eds. R. Woodroffe, S. Thirgood, & A. Rabinowitz. Cambridge: Cambridge University Press, pp. 13–26.

Walker, S., Jordan, M., & Molur, S. (2005). *Conservation of small mammals in south Asia – a case study in international zoo partnerships.* In *Proceedings of the 58th Annual Meeting, The World Zoo and Aquarium Conservation Strategy – the Key to a Sustainable Future*, eds. P. Dollinger. San Jose, Costa Rica, 16–20 November 2004, pp. 152–154.

WAZA (2005). *Proceedings of the 58th Annual Meeting, The World Zoo and Aquarium Conservation Strategy – the Key to a Sustainable Future*, ed. P. Dollinger. San Jose, Costa Rica, 16–20 November 2004.

20

The conservation mission in the wild: zoos as conservation NGOs?

ALEXANDRA ZIMMERMANN AND ROGER WILKINSON

THE CONSERVATION ROLES OF ZOOS

The role of zoos and aquariums in conservation is complex, not only because it is multi-faceted, but also because it varies from institution to institution and changes over time in response to the values of the public, its scientific peers, and its critics. Traditionally, the contributions of zoos and aquariums (hereafter referred to collectively as "zoos") to conservation have involved their expertise in breeding threatened species, their responsibility to educate and influence large numbers of adults and children, and the opportunity to conduct a wide range of scientific studies, including the development of veterinary medicine, on a vast diversity of otherwise inaccessible animals (Hutchins and Conway 1995).

These traditional roles are played out *ex situ*, away from the original habitats of the animal species in question. No other conservation sector has a comparable capacity and infrastructure to pursue these functions to the same extent. But although they are fundamentally and uniquely the domain of zoos, zoos are criticized for not focusing their efforts sufficiently on working with habitats, species, and people *in situ*. The recent emphasis by zoos on supporting conservation projects in the wild is in part a response to critiques about the relevance and/or efficacy of their *ex situ* activities. As a result, the hallmark of a "good" zoo is generally proclaimed by lay-people and conservation scientists alike to be the participation in or support of, conservation work in the field. Having a portfolio of *in situ* activities has become a common and esteemed activity for many zoos around the world and seems to be a prerequisite for a zoo to call itself a conservation organization, or a conservation-missioned zoo (Miller *et al.* 2004).

With such a wide range of both opportunity and peer expectation, what exactly is the conservation niche of a zoo? Zoos endeavor to make significant contributions to education, from school children to university students, but are not acclaimed as institutions of learning (see Chapters 4, 5 and 9). Zoos carry out a wide range of fundamental and original research, but are not regarded as establishments of training and science. Zoos are uniquely skilled in the husbandry of endangered species, at coordinating complex meta-populations across the globe, and have on several occasions saved critically reduced populations from certain extinction (see Chapters 10 and 11) yet they are criticized for not focusing on the roots of the problems and working *in situ* as "proper" conservation organizations should (Balmford *et al.* 1995, Chapter 16). Zoos by their very nature maintain animals in confined spaces and as a consequence face a continuous onslaught from animal welfarists despite investing enormous effort, research, and funds into behavioral enrichment and driving forward the frontiers of intellectual ethical debates (see Chapter 7). Zoos also play very important roles in society and culture, which we rarely discuss in our papers and conferences. And zoos, as they acknowledge that the survival of biodiversity requires conservation *in situ*, are trying at the same time to define the role that they can play in this additional area of work – yet so far without being fully accepted into the global community of conservation organizations.

In fact, a few zoos have become so good at conservation work in the field that one is no longer sure whether to think of them as zoos or as conservation non-Governmental organizations (NGOs). At the moment, this is the exception rather than the rule, but some believe that this change in emphasis is an evolution in progress that is gathering great momentum.

Essential policies and structures are in place, and the new, revised *World Zoo and Aquarium Conservation Strategy* (WZACS) ". . . calls on all zoos and aquariums to increase their work in support of conservation in the wild" (WAZA 2005). In Europe, the EU Council Directive 1999/22/EC relating to the keeping of wild animals in zoos includes a requirement to contribute to conservation (EC 1999), although it does not explicitly state that it requires direct *in situ* conservation by zoos (Rees 2005, Wehnelt and Wilkinson 2005). In the UK, however, a serious commitment to *in situ* conservation work is required of large zoos and a proportionate contribution of smaller zoos (DEFRA 2003), and examples are provided as benchmarks and guidance (DETR 2000, DEFRA 2006).

UNDERSTANDING THE NICHE

How much and how widely do zoos and aquariums actually contribute to *in situ* conservation? How do they become involved in *in situ* conservation projects and what drives them to do so? What is the extent of their work, in what types of projects do zoos engage, and what is the nature and level of their stake in these projects? Do most zoos support *in situ* conservation projects financially or are they able to offer specialized skills to the field on which other organizations depend?

To understand better the niche of zoos in the global conservation community and to make recommendations for the future, a clearer picture of what zoos currently do is required. A study was carried out that examined the true extent to which zoos and aquariums participate in conservation in the wild. A questionnaire sent to zoos and aquariums around the world was used to gather both quantitative and qualitative baseline information about their *in situ* conservation activities. This study differed from the databases and surveys carried out by various zoo associations such as WAZA, EAZA, and BIAZA (Stevenson 2001, WAZA 2004b, EAZA 2005), in that it did not collect information about individual projects, but rather about the overall activities, preferences, and patterns of the conservation missions of zoos around the world.

The objective of this study was to measure the *activities* of zoos, not their ultimate conservation impact. Measuring conservation activities is the necessary precursor to the much more challenging task of measuring conservation impact, i.e., the real contribution that an activity is making toward the conservation of a given species or habitat. Evaluating the *impact* of zoos is a challenge still ahead that has been explored by few studies (e.g., see Chapter 21). Whether measuring activities or impact, it is important for zoos and zoo associations to evaluate their contribution to conservation, in order to be able to respond to critics who argue that zoos' conservation activities are public relations stunts and that their conservation efforts are superficial and ineffective (Bartos and Kelly 1998, Hewitt 2000, Scott 2001).

A GLOBAL SURVEY OF ZOOS AND CONSERVATION

A web-based questionnaire survey was designed in which respondents were asked questions about financial input, scope and type of conservation work, decision making, prioritizations, opinions, and perceived trends. In order to maximize chances of an acceptable sample size and increase the willingness to respond, questions were short and closed, response options were pre-grouped into bands, Likert scales were used for opinion questions, and a reminder was sent two days before expiry of the reply deadline. The survey was also anonymous – none of the questions asked could reveal the identity of the respondent organization.

The questionnaire was sent by email to the chief executives or conservation departments of zoos and aquariums worldwide. A list of institutions and contact email addresses was compiled from the websites of the associations of zoos and aquariums and zoo directors' associations, as well as independent web-based zoo databases, namely: World Association of Zoos and Aquaria (WAZA 2004a), Australasian Regional Association of Zoological Parks and Aquariums (ARAZPA 2004), European Association of Zoos and Aquaria (EAZA 2004), Association of Zoos and Aquariums (AZA 2004), African Association of Zoos and Aquaria (PAAZAB 2004), South East Asian Zoos Association (SEAZA 2004), Conservation Breeding Specialist Group (CBSG 2002), Diretório de Zoológicos Brasileiros (BDT 2004), Asociación Ibérica de Zoos y Acuarios (AIZA 2004), Verband Deutscher Zoodirektoren (VDZ 2004), Danish Association of Zoological Gardens (DAZA 2004), the Federation of Zoological Gardens for Great Britain and Ireland, now BIAZA (FZG 2004), Canadian Association of Zoos and Aquariums (CAZA 2004), and Zoos Worldwide (Anon 2004).

Our sampling frame therefore was that of all zoos and aquariums in the world, whether charitable, for-profit or governmental, regardless of size and of whether the institution itself professes to engage in conservation. Theoretically, each representative had an equal chance of being included in the sample and therefore the sampling would be random. Nevertheless, there are some biases implicit in this survey methodology, linked mainly to the web-based method: only zoos with access to email had the chance to participate (although it was possible to source the email addresses of nearly 800 zoos), and only zoos with internet access would have the chance to reply. Web-based surveys do not necessarily generate higher response rates than paper-based or telephone surveys (Thorpe 2002), but the international scope of this survey made this the method of choice. Furthermore, due to financial constraints, the survey was sent out in English only; and, finally, although the survey instructions stated that it should be completed by the chief executive or the conservation department (if applicable), it was not possible to control for which individual in the organization filled out the survey, so some subjectivity was inevitable. However, the questionnaire was pilot-tested prior to release in order to check for individual subjectivity and a

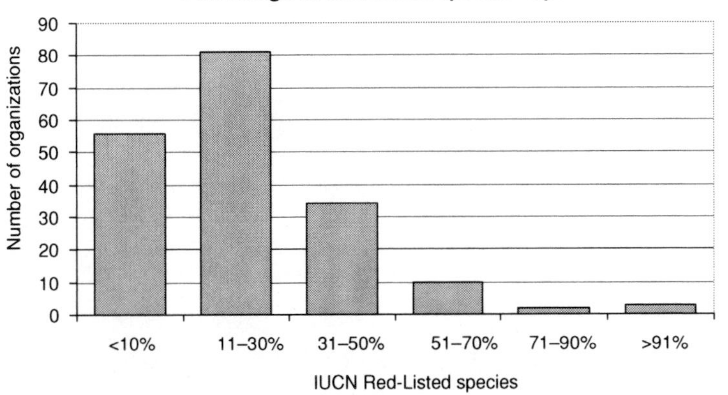

Figure 20.1 Percentages of threatened species kept by the responding organizations

good consistency of responses across several people in a single organization was found.

INSTITUTIONAL PROFILES

The survey was sent to 725 zoos and aquariums in 68 countries, of which 190 institutions in 40 countries responded. This constitutes a response rate of 27%, or approximately 16% of the WAZA network of the world's zoos. Of the respondent organizations, 81% ($n = 153$) were zoos, the remainder were aquariums ($n = 16$), safari or theme parks ($n = 8$), and specialized organizations or "other" ($n = 13$). Similar proportions of the respondent organizations were privately owned (28%), trusts or registered charities (25%), publicly owned (19%), and governmentally owned (21%). The remaining 7% defined themselves as "other."

For the majority (72%) of institutions, fewer than 30% of their species were listed in an IUCN category of threat (Figure 20.1) (IUCN 2004). This proportion of threatened species held by zoos is highly relevant as it echoes the argument that the *ex situ* conservation value of zoos, in terms of keeping threatened species whether for possible reintroduction or as "insurance populations," is not reflected by the species they keep in reality. Although Leader-Williams *et al.* (Chapter 16) have found a significant increase in the number of threatened taxa held in zoos since 1993, for most zoos, conservation relevance (defined by either reintroduceability or degree of

Table 20.1
Matrix showing relationship between income (in bands) and expenditure on conservation (in bands): number of zoos which fall into each respective band ($\chi^2 = 33.88$, $df = 12$, $P < 0.001$)

		Conservation expenditure				
	EUR/USD	<1k	1–10k	10–100k	100k–1M	>1M
Income	10–100k	–	8	1	–	–
	100k–1M	7	22	5	–	–
	1–10M	6	21	28	11	2
	>10M	–	4	14	11	4

endangerment) is still not the top criterion for most zoos in their stock selection, or "collection planning," processes (Balmford *et al.* 1996).

CONSERVATION MISSION AND EXPENDITURE

The majority (89%) of the respondent organizations had mission statements, and of these 77% specifically included biodiversity conservation in their missions. Of the respondents, 82% ($n = 156$) said they participated in conservation activities *in situ*. The majority (62%) of zoos received fewer than 500 000 visitors per year. Each organization expressed its income and its conservation expenditure in categorical scales (or "bands") (see Table 20.1), ranging from €10 000 to greater than €10 million for income, and €1000 to > €1 million for conservation expenditure. An analysis of the relationship between these bands showed that income was significantly related to their conservation expenditure ($\chi^2 = 33.88$, df $= 12$, $P < 0.001$). The total conservation expenditure by the 156 organizations which had field activities [calculated as $\Sigma\ (n \times M)$ where $n =$ the number of zoos and $M =$ the median of each scale] was 49 million US dollars or Euros per year.

Interestingly, over half (55%) of zoos generated the majority (>75% or more) of their conservation expenditure from their own income (i.e., gates and earnings) rather than from externally sourced funds (i.e., grants) (Figure 20.2). Indeed almost a third (31%) of zoos relied *entirely* on their own incomes to fund field projects and had no additional funding sources. This result could be interpreted as being indicative either of how important many zoos regard their conservation mission, or of the fact that many zoos do not have either the capacity or priority to raise funds for conservation from external sources. Encouragingly, the results also showed that conservation

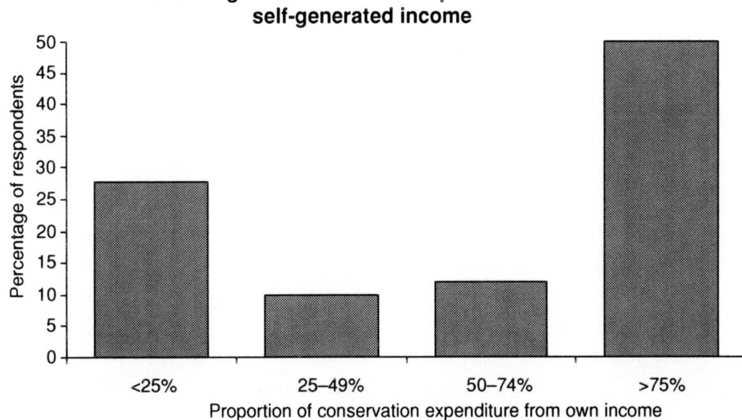

Figure 20.2 The percentage of zoo's conservation expenditure which originates from self-generated income (i.e., tickets and sales in the zoo)

is not only the realm of zoos in the comparatively richer countries in Europe, North America, and Australia. Conservation expenditure did not vary proportionally between zoos in high-GDP versus low-GDP countries (Figure 20.3). This appears to be consistent with other studies that have quantified the *in situ* conservation work of low-GDP regions. For example, one review of Latin American zoos showed that 16 zoos were participants in 56 field projects in 11 countries (Matamoros Hidalgo 2002). Similarly, in South Asia, the Indian-based Zoo Outreach Organization works to foster links and collaboration between a large number of zoos, rescue centers, and field-based NGOs (ZOO 2005).

Financing and fundraising for conservation is the simplest way for zoos to contribute to *in situ* conservation. The sums donated are, according to both this study and the literature, not insignificant. For example, a review of British and Irish zoos for the years 1997–2000 showed that a total of over £5 million (around €9.5 million) was spent by zoos on field conservation (WAZA 2005). Collaborative fundraising can generate even more impressive results. Through the annual EAZA Conservation Campaigns, in which its members cooperate to raise funds and increase conservation awareness, zoos are able to raise significant sums each year through their themed campaigns, (e.g., Atlantic Rainforest Campaign: €280 000, Tiger campaign: €750 000 over 2 years, Shellshock campaign: over €370 000; Anon 2006, EAZA 2006). Nevertheless, while a few zoos spend large amounts of money on conservation, other studies have shown that, proportionate to their

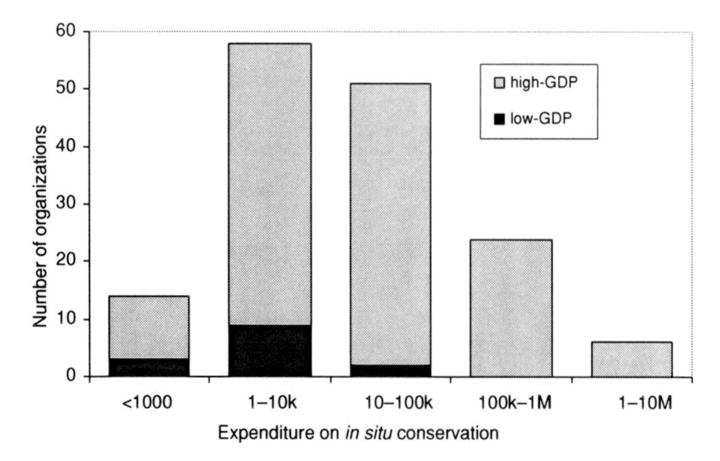

Figure 20.3 Expenditure on *in situ* conservation by zoos, showing proportion of zoos located in high-GDP and low-GDP countries, where low-GDP is defined as <$10 000 per capita. The frequency distribution of expenditure is consistent regardless of GDP category ($\chi^2 = 10.07$, df = 4, $P < 0.05$)

overall budgets, zoos spent surprisingly little: a study of American zoos in 1999 by Bettinger and Quinn (2000) revealed an average expenditure of only 0.1% of operating budgets, even when zoo-based research and staff time were included in the definition of "conservation expenditure."

INITIATIVE AND ALLIANCES

However, financing and fundraising alone do not necessarily confer conservation credibility upon a zoo. Zoos also need to have an active involvement in the conservation projects they support, otherwise they differ little from, for instance, a "green" corporation that regularly donates funds to conservation and environmental programs. This raises the question of how zoos become involved in conservation in the field in the first place. The respondents of this study replied as follows: 40% receive requests for help from other organizations, 35% approach existing projects and offer assistance, and 25% initiate, set up, and run their own projects. The levels of initiation and ownership of conservation projects therefore vary a great deal, and are related in part to the levels of in-house conservation expertise that zoos have.

Regardless of whether they prefer to fund or initiate conservation projects, zoos appear to be good collaborators, often participating in coalition strategies (see Chapter 19) to become involved with conservation and

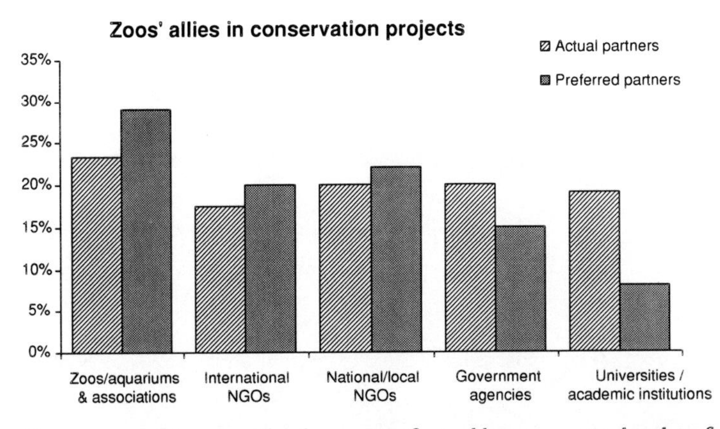

Figure 20.4 Alliances with other sectors formed by zoos – actual and preferred

fulfil their conservation missions. When asked with which types of organizations they collaborate in their conservation projects, no significant preference for one group over another was revealed: 23% of zoos work with other zoos/aquariums and associations, 20% work with national or local NGOs, 20% work with government agencies, 19% with academic institutions, and 18% with international NGOs. When asked about their preferences for collaboration (which *one* of these groups they preferred as their ally in collaborative conservation projects), they ranked zoos/aquariums and zoo associations as their highest preference (29%), followed by national or local NGOs (22%) and international NGOs (20%), government agencies (15%) and, perhaps surprisingly, in last place, academic institutions (8%) (Figure 20.4). This result might be explained as follows: organizations like to work with peer organizations, so zoo–zoo collaborations make sense. National or regional NGOs can be more appreciative of a zoo's contributions, and the zoo may have a larger stake in a collaboration with a small NGO, than in collaborations with large international NGOs or government agencies where the zoo's role may be minor and recognition inadequate from the zoo's point of view. The most interesting result here though is that academic institutions are the least preferred partner for zoos. A qualitative follow-up study would be useful to understand the reason for this perception. We can hypothesize that it relates to the fact that many zoos offer their facilities to universities for research projects but the return is often minimal in terms of benefits for the zoos. However, those zoos wishing to professionalize their research and conservation capacities in-house, whether for academic research or for applied conservation in the field, could

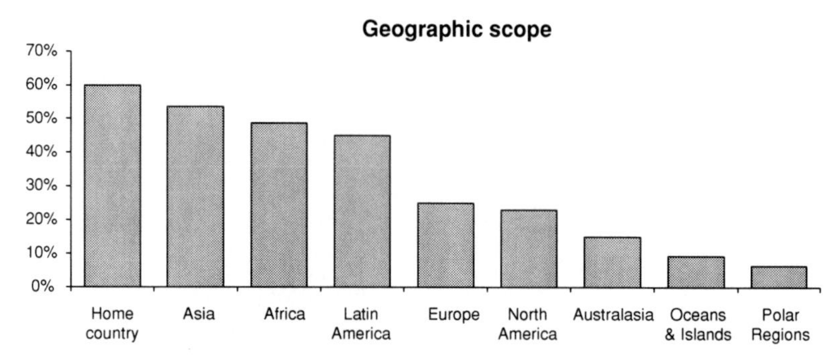

Figure 20.5 Geographic scope of the conservation work carried out by zoos

do so by fostering collaborations with universities and attracting their top graduates.

IN SITU CONSERVATION ACTIVITIES AND SPECIALIZATIONS

Having identified to what extent zoos and aquariums initiate or get involved in conservation projects, the survey asked the respondents about the type of participation they favor. A financial role prevailed slightly (40%) over an advisory role (33%) and over a technical role (27%). In contradiction to this, however, when asked in what level of participation they engage, purely financial support was only ticked by 0.6% ($n = 2$) organizations. All others described their participation as including roles of investigator (12%), participant (33%), advisor (31%), and leader (23%). This indicates that while zoos most commonly assist conservation projects through donation of funds, they almost exclusively do not perceive this to be their only role in these projects.

When it comes to the scope and orientation of their work, the respondents in this survey operate their *in situ* conservation programs foremost near home. Of the organizations, 60% had projects in their own countries, closely followed by projects in the high-biodiversity/high-threat regions of Asia, Africa, and Latin America; 50% of organizations had projects on three or more continents (Figure 20.5). Their thematic orientations favor biological/ecological projects and those that link directly into their *ex situ* work such as assistance with breeding in-range, re-stocking, and reintroduction. They also support capacity-building and training, protected area management, and conservation medicine. The social side of conservation (e.g., community-based conservation) appears to be gaining popularity among

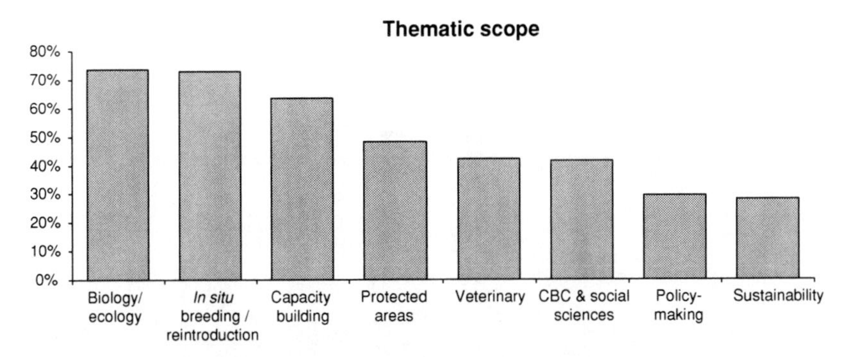

Figure 20.6 Thematic scope of the conservation work carried out by zoos

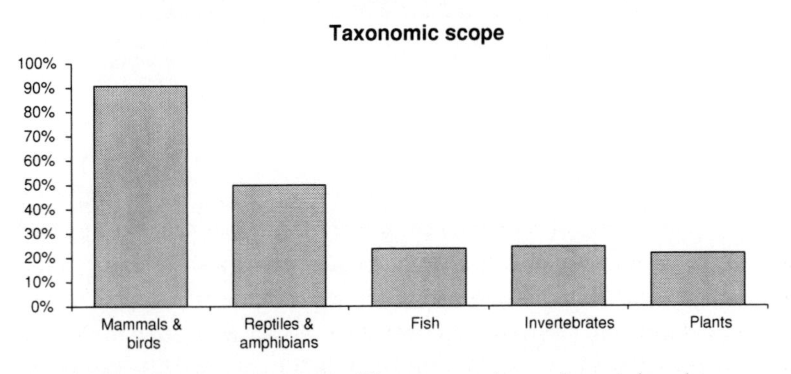

Figure 20.7 Taxonomic scope of the conservation work carried out by zoos

zoos but wider issues such as policy-making and sustainable use are considered the realm of other organizations (Figure 20.6). In terms of species focus, zoos work largely with mammals and birds *in situ*, which directly mirrors their preferences for *ex situ* species holdings (see Chapters 9 and 16) (Figure 20.7). This wide spread of particularly geographic and thematic participation in conservation projects is remarkable and leads on to the question of how these compare with the activities of non-zoo-based conservation NGOs.

A COMPARISON OF ZOOS WITH NON-ZOO-BASED CONSERVATION NGOS

To compare conservation activities between zoos and NGOs, a subset of those zoos in this sample which had the highest annual conservation expenditure was examined more closely. Six zoos had annual expenditures of

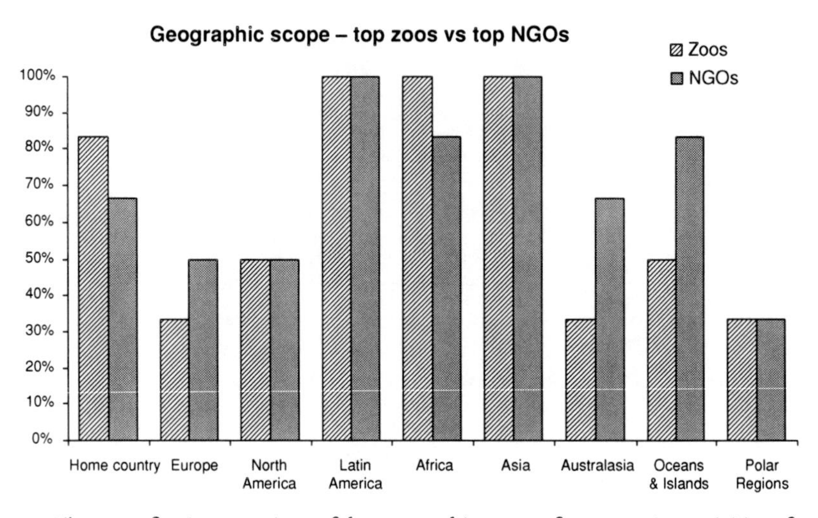

Figure 20.8 A comparison of the geographic scope of conservation activities of zoos compared with NGOs

over €1 million, so for this study these can be considered to rank very high in their league of conservation activity. They were then compared with an equal number of conservation NGOs which also rank very high within their sector in terms of conservation expenditure and scope: Conservation International, The Nature Conservancy, World Wide Fund for Nature (WWF-International), World Wildlife Fund (WWF-US), Wildlife Conservation Society (International Programs), Flora and Fauna International, and the World Conservation Union (IUCN). The most recent annual reports of these NGOs served as the source data.

A few patterns emerged when geographic scope, thematic scope and taxonomic scope were compared. The zoos carried out more work in their home countries than did the NGOs, but both focused largely on high-biodiversity regions. NGOs were also more active in Australasia, oceans and small islands ($\chi^2 = 1.45$, df $= 8$, NS) (Figure 20.8). In terms of thematic scope there were no meaningful differences between the sectors, except that zoos carried out much more veterinary-related conservation work than did NGOs ($\chi^2 = 0.93$, df $= 7$, NS) (Figure 20.9). In terms of species orientation in conservation projects of zoos and NGOs, zoos concentrated more on higher vertebrates than on lower vertebrates, invertebrates, and plants, while NGOs had a much more even coverage of species-focus within their activities ($\chi^2 = 1.36$, df $= 4$, NS) (Figure 20.10). The latter, again, reflects the species holdings preferences of zoos described by Balmford *et al.* (1996) and Leader-Williams *et al.* (Chapter 16).

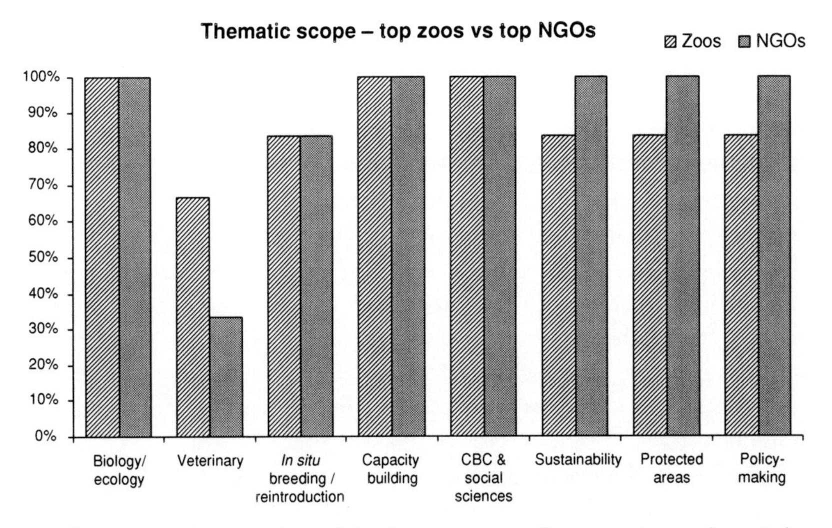

Figure 20.9 A comparison of the thematic scope of conservation work carried out by zoos and NGOs

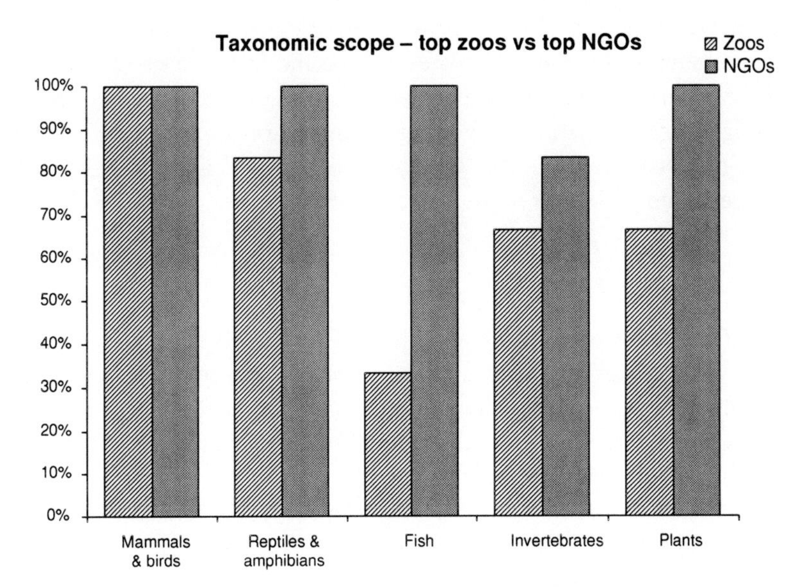

Figure 20.10 A comparison of the taxonomic scope of conservation work by zoos and NGOs

These comparisons and calculations were made to gain an approximate idea of relative niches, yet it would be interesting re-investigate these questions with more precision by comparing zoos with international NGOs of equal conservation expenditure (rather than by sector rank). While some zoo professionals may argue that zoos *are* essentially conservation NGOs, there remain marked differences that separate them into very distinct subsectors. One is that, unlike most NGOs, many zoos are not registered charities, so they fall into a different category legally. The other is that zoos draw the public to their physical base (and build conservation, education, and all their other activities around that foundation), while NGOs, lacking such physical presence, have to create a virtual identity with which to attract their membership and supporters and to build upon the idealism and values they promote.

The third difference is a question of professional capacity. Zoos pursue a very multi-faceted role and the portfolio of staff expertise is entirely different from that of NGOs. This gives zoos a competitive handicap when it comes to sheer numbers of professional conservation staff. Conservation activities are often carried out by curators or other staff who often lack the time, terms of reference, and advanced training of conservation science to pursue conservation projects. In this survey it emerged that only a third (34%) of all zoos interviewed had *any* staff dedicated specifically to conservation, and, of these, most (86%) had fewer than five FTEs (full-time equivalents). Encouragingly, however, 23% of all zoos/aquariums had some *in situ* conservation staff (i.e., staff hired and paid wholly or partly to work at the location of the conservation project for more than 40% of their time).

CAPACITY BUILDING AND CONSERVATION EXPERTISE

The facts about conservation staffing are particularly interesting because they allude to an important strategic direction that zoos need to explore if they wish to become leaders in field conservation. Zoos need to attract more professional conservation scientists into their staff teams. Young professionals and graduate students seeking a career in conservation should begin to see zoos as attractive employers for a career in conservation. While more and more zoos now have conservation scientists on staff, the shift from donating to other people's projects to being in charge of their own initiatives is still a challenge for many zoos.

However, given the variety of expertise that a zoo requires in order to fulfil the expectations of its supporters and the demands of its ambitious mission, as well as to earn the respect of its conservation peers, are zoos spreading themselves too thin? Is it realistic to expect zoos to excel at education,

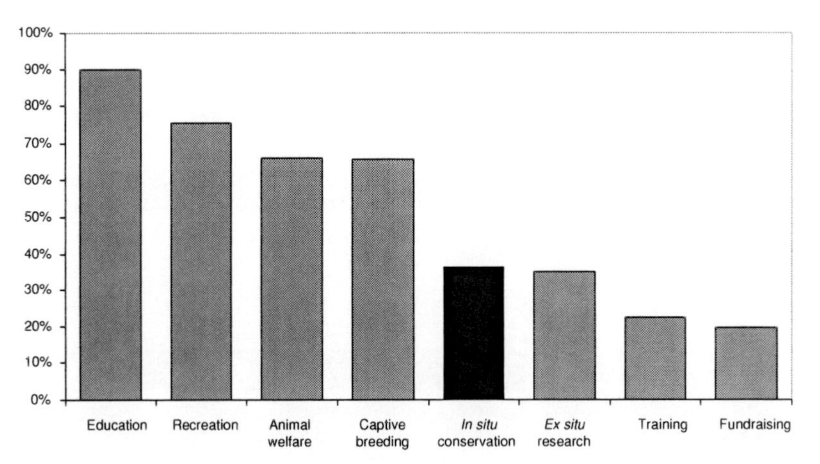

Figure 20.11 What types of conservation-related work do zoos, in their own opinion, do best?

awareness-raising, research, captive breeding, reintroduction, welfare, and *in situ* conservation? Meeting such multi-disciplinary terms of reference, while reflective of the complex characteristics of the conservation field, is perhaps not necessary. For many zoos, a niche defined by a sub-set of these disciplines in which it is particularly adept may be a better choice than trying to become "full-service" NGOs. Most zoos, however, are still in the process of identifying their niche within the wider conservation NGO community. Unlike conservation NGOs, which are established and planned entirely around their conservation missions, zoos have evolved into their new *in situ* roles through a combination of societal value changes and external pressure. While the critiques and evaluations are necessary to drive this evolution further, many zoos are trying hard to meet the expectations of the scientific and public community and already do more than is generally accredited to them.

Zoos know their current limitations. When they were asked what they considered the strongest skills in their own organizations, *in situ* conservation did not feature in the top ranks (Figure 20.11). However, by far their main motivation for being active in *in situ* conservation was their mission (74%) as opposed to other reasons, such as expectations of the public, peer pressure from other zoos, fund-raising incentives, attracting more visitors, or legal requirements.

Nearly half (46%) of zoos considered conservation to be *very important* in the philosophy of their organizations, while 21% rated it even more strongly,

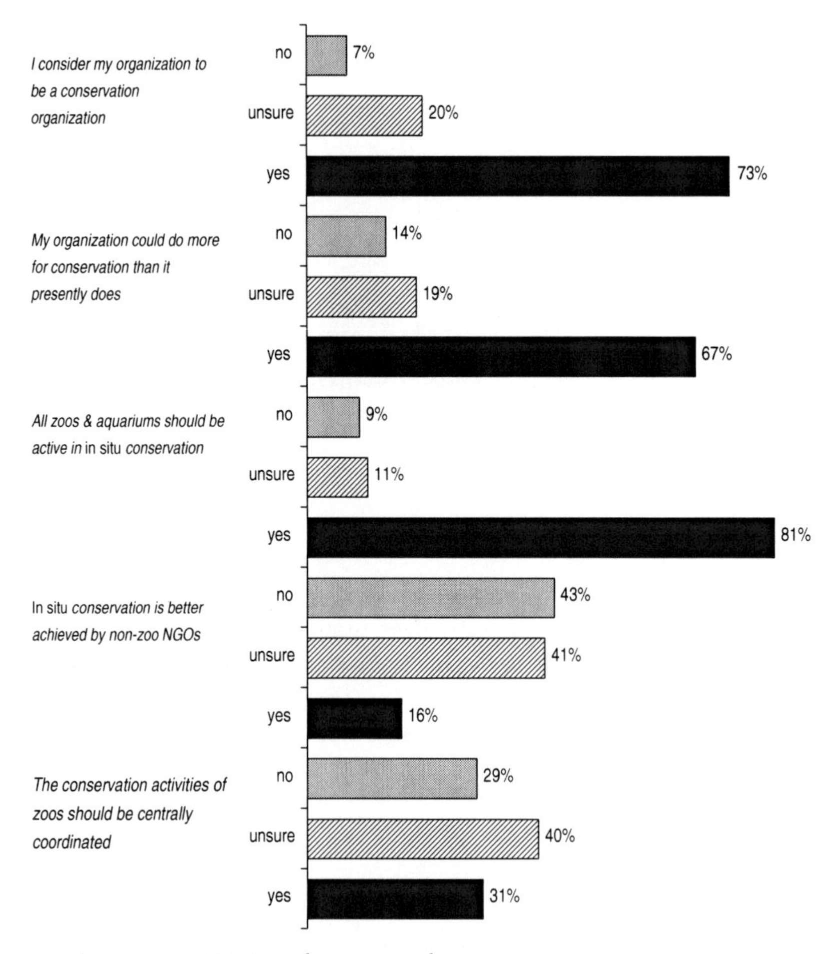

Figure 20.12 Opinions about zoos and conservation

as *central*. Respondents were also asked their opinion about five statements regarding their conservation work. These opinions, shown in Figure 20.12, show an overall strong trend of conservation awareness and valuation, but also an interesting perception of independence. The majority of respondents considered their zoo to be a conservation NGO (73%), but that they had not reached their potential for conservation input (67%) and conservation work was something in which all zoos should be active (81%). The opinions about whether zoos or non-zoo-based NGOs were better at conservation were more mixed (43% no, 41% unsure), and the response regarding whether zoo conservation should be coordinated by central organizations (e.g., associations) was also divided (40% unsure).

Whatever their priorities and mission strategies, the important bottom line is that, if a zoo identifies its main mission as being *in situ* wildlife conservation, it must be accountable to that endeavor (Miller *et al.* 2004). Important, too, is the appropriateness of the type of contribution a zoo makes. For some, a lack of in-house conservation scientists may mean that financing the projects of others is the most responsible form of participation. For others, a lack of funds may mean that sharing skills or labor with other partners may be the right thing to do.

CONCLUSIONS

The results of this study provide a global overview not only of the scope and type of conservation projects in which zoos and aquariums engage, but also of their strengths and potential for the future. The conviction that zoos of the 21st century must embrace direct *in situ* conservation as part of their *raison d'être* has a wide following.

A handful of zoos have already made the leap from being zoos that do conservation to becoming conservation organizations that run zoos. The rest still have some way to go. While a number of species would no longer exist today if it were not for the efforts of zoos (e.g., California condor, Mauritius kestrel, black-footed ferret, Guam rail) (Snyder *et al.* 1996), if in the future zoos want to share the responsibility for saving humpback whales, controlling wildlife trade, studying emerging diseases or mitigating human–wildlife conflicts, they need to shift their focus even further to the wild, and develop more targeted individual and collective strategies that will transform them into ever more effective *Catalysts for Conservation.*

ACKNOWLEDGEMENTS

This study relied on the generous help and support of several people, in particular: Martin King, Scott Wilson, Mike Jordan, Olivia Walter and Christian Rohlff. Grateful thanks also to all of the survey respondents for their time and participation.

References

AIZA (2004). *Quiénes Somos?* Asociación Ibérica de Zoos y Acuarios. www.aiza.org.es.

Anon (2004). Zoos Worldwide – zoos, aquariums, animal sanctuaries and wildlife parks worldwide. www.zoos-worldwide.com.de.

Anon (2006). 21st Century Tiger. http://www.21stcenturytiger.org/ [accessed 01/08/2006].

ARAZPA (2004). Membership directory. Australasian Regional Association of Zoological Parks and Aquaria. www.arazpa.org.au/Membership_MemberList.htm.

AZA (2004). AZA zoo and aquarium directory. American Zoo and Aquarium Association. www.aza.org.

Balmford, A., Leader-Williams, N., & Green, M. J. B. (1995). Parks or arks: where to conserve threatened mammals? *Biodiversity and Conservation*, **4**, 9–607.

Balmford, A., Mace, G. M., & Leader-Williams, N. (1996). Designing the ark: setting priorities for captive breeding. *Conservation Biology*, **10**(3), 719–727.

Bartos, J. M. & Kelly, J. D. (1998). Rules towards best practice in the zoo industry: developing key performance indicators as benchmarks for progress. *International Zoo Yearbook*, **36**, 143–157.

BDT (2004). *Diretório de Zoológicos Brasileiros*. Base de Dados Tropical. Campinas SP Brazil. www.bdt.org.br.

Bettinger, T. & Quinn, H. (2000). Conservation funds: how do zoos and aquariums decide which project to fund? In *Proceedings of the AZA Annual Conference*, St. Louis, Missouri, pp. 88–90.

CAZA (2004). Members' directory. Canadian Association of Zoos and Aquariums. www.caza.ca.

CBSG (2002). *Global Zoo Directory*. Apple Valley, MN: Conservation Breeding Specialist Group, Species Survival Commission, IUCN. www.cbsg.org.

DAZA (2004). *Medlemmer*. Danish Association of Zoological Gardens. http://www.dazaportal.dk/medlemmer.htm.

DEFRA (2003). Zoo Licensing Act 1981. Department for Environment, Food and Rural Affairs (UK). http://www.defra.gov.uk/wildlife-countryside/gwd/zoo.htm.

DEFRA (2006). Zoos Forum guidance on the Zoo Licensing Act's requirement for zoos to participate in conservation measures. Annex to Chapter 2. In *The Zoos Forum Handbook*. London: Department for Environment, Food and Rural Affairs (UK). http://www.defra.gov.uk/wildlife-countryside/gwd/zoosforum/handbook/index.htm.

DETR (2000). *Secretary of State's Standards on Modern Zoo Practice*. London: Department of the Environment, Transport and the Regions, H.M. Government.

EAZA (2004). *Membership*. European Association of Zoos and Aquaria. www.eaza.net.

EAZA (2005). EAZA *in situ* conservation database. Data collection CDROM, Second Edition. European Association of Zoos and Aquaria. http://www.eaza.net.

EAZA (2006). *EAZA Annual Conservation Campaigns*. http://www.eaza.net/news/frameset_1news.html?page=1news. [Accessed 01/08/2006.]

EC (1999). Council Directive 1999/22/EC of March 29, 1999 relating to the keeping of wild animals in zoos. Official Journal of the European Communities. 09/04/1999.

FZG (2004). *Federation of Zoos Contact Report*. London: The Federation of Zoological Gardens for Great Britain and Ireland.

Hewitt, N. (2000). Action stations – zoo check is go. *Wildlife Times*, 17.
Hutchins, M. & Conway, W. (1995). Beyond Noah's ark: the evolving role of modern zoological parks and aquariums in field conservation. *International Zoo Yearbook*, **34**, 117–130.
IUCN (2004). *The IUCN Red List of Threatened Species*. Gland: World Conservation Union.
Matamoros Hidalgo, Y. (2002). *In situ* conservation programmes of Latin American and Caribbean zoos. *WAZA Magazine*, **4**, 8–11.
Miller, B., Conway, W., Reading, P., Wemmer, C., Wildt, D., Kleiman, D., Monfort, S., Rabinowitz, A., Armstrong B., & Hutchins, M. (2004). Evaluating the conservation mission of zoos, aquariums, botanical gardens and natural history museums. *Conservation Biology*, **18**(1), 86–93.
PAAZAB (2004). African Association of Zoological Gardens and Aquaria. www.paazab.org.
Rees, P. A. (2005). Will the EC Zoos Directive increase the conservation value of zoo research? *Oryx*, **39**, 128–131.
Scott, S. (2001). Captive breeding. In *Who Cares for Planet Earth? The Con in Conservation*, ed. B. Jordan. Brighton: Alpha Press, p. 72.
SEAZA (2004). *Membership List*. South East Asian Zoo Association. www.seaza.org.
Snyder, N. F. R., Derrickson, S. R., Beissinger, S. R., Wiley, J. W., Smith, T. B., Toone, W. D., & Miller, B. (1996). Limitations of captive breeding in endangered species recovery. *Conservation Biology*, **10**(2), 338–348.
Stevenson, M. (2001). Report on review of *in situ* projects supported by members of the Federation of Zoos. Unpublished report of the Federation of Zoological Gardens for Great Britain and Ireland, London.
Thorpe, S. W. (2002). Online student evaluation of instruction: an investigation of non-response bias. Conference presentation at the 42nd Annual Forum of the Association for Institutional Research, Toronto, Canada. www.airweb.org/forum02/550.pdf [accessed 25/07/2006].
VDZ (2004). *Zoos im VDZ*. Verband Deutscher Zoodirektoren e.V. www.zoodirektoren.de.
WAZA (2004a). Membership. Liebefeld-Bern: WAZA. www.waza.org.
WAZA (2004b). The WAZA Network links *ex situ* breeding with *in situ* conservation. Liebefeld-Bern: WAZA. http://www.waza.org/conservation/projects.
WAZA (2005). *The World Zoo and Aquarium Conservation Strategy: Building a Future for Wildlife*. Liebefeld-Bern: WAZA.
Wehnelt, S. & Wilkinson, R. (2005). Research, conservation and zoos: the EC Zoos Directive – a response to Rees. *Oryx*, **39**, 132–133.
ZOO (2005). Zoo Outreach Organisation: about us. http://www.zooreach.org/aboutzoo.htm [accessed 26/11/05].

21

Measuring conservation success: assessing zoos' contribution

GEORGINA M. MACE, ANDREW BALMFORD, NIGEL LEADER-WILLIAMS, ANDREA MANICA, OLIVIA WALTER, CHRIS WEST, AND ALEXANDRA ZIMMERMANN

INTRODUCTION

Interest in auditing conservation projects, and therefore in ways of measuring conservation success, has been growing. This has partly been driven by increased pressure from conservation funders to be shown tangible progress as a result of their investments (Christensen 2003), but has also developed as a consequence of conservation organizations' own recognition that with the limited resources available for conservation, they need to evaluate successful from unsuccessful approaches, and even be able to assess their relative own cost-effectiveness (Possingham 2001, Salafsky *et al.* 2002, Pressey *et al.* 2003, CMP 2004).

The project that we report on here was a collaborative effort to identify methods to score the conservation value of projects that zoos support. Our interest in this work stemmed from two processes. First, the new *World Zoo and Aquarium Conservation Strategy* (WAZA 2005) states that the major goal of zoos and aquariums will be to integrate all aspects of their work with conservation activities. In consequence, zoo governing bodies and the national and regional zoo associations are all interested in gathering information about the contributions that zoos make to conservation. Second, as zoos take their conservation responsibilities seriously and begin to invest substantially in their own and other organization's projects, there is increasing interest in ensuring that they are doing the best they can with limited resources. But on what basis can a zoo manager decide where best to place rather limited conservation spend? How can these internal and external assessments weigh the relative benefits of investing in different kinds of conservation? For example, what is the difference in conservation impact that could result from allocating similar resources to a zoo-based education program focusing on species conservation, a small local field-based recovery project, or a contribution to some major, multi-organization project in a globally threatened hotspot? We wanted to find a way to make a more robust assessment of the value of different projects, both for more useful reporting by zoos, and to guide them in making sensible choices of projects to support.

Conservation project databases held by the zoo associations tend simply to gather information on the number and/or costs of conservation projects. But these metrics are derived unsystematically, and are often far removed from measures of conservation success. For example, while one zoo director might choose to include expenditure on animals in the collection that are the focus of conservation breeding programs as conservation spend, another might only report on unrestricted resources spent on field conservation. Miller *et al.* (2004) suggest eight questions that collection-based institutions (including zoos) might use to assess their own contributions toward a conservation mission. These include financial and organizational commitments of various kinds, but all the measures are of in-house processes, rather than assessments of the consequences of conservation spend for wild species and habitats. We wanted to test their presumption that evaluating outcomes will take too long, and that therefore simply assessing whether or not an organization actually lives by its mission statement is the best that can be done in the short term.

The issue of conservation project assessment is of course considered more widely by other conservation organizations, and some common systems that will allow different organizations to share best practice are now in place (CMP 2004). In general, classification of the stages in project development, and the identification and naming of different approaches to project implementation have all proved easier to accomplish than has the assessment of outcomes, especially once the evaluation seeks to be broad-based. While there are several examples of species- and project-specific evaluations (e.g., Clark *et al.* 1994, Jepson 2004), we could find very few examples of project-scoring systems that have got to the stage of producing results (but see examples in Belokurov *et al.* 2003, Salafsky and Margoulis 1999). For this work, by focusing only on projects run by zoos, we hoped to be able to make faster progress in developing and implementing an explicit and repeatable scoring system.

As well as developing a system, we undertook a pilot study to assess the practicality and the reliability of our proposed method. In particular we used a pilot study to see how straightforward it was to assess a real set of zoo-based conservation projects, to examine the consistency of scores from different assessors, and to see whether project leaders consistently scored projects differently from independent assessors. Finally, we were interested to see the congruence between how the system scored projects compared to subjective impressions about what were more or less successful conservation projects.

METHODS

Framework for assessment

The first requirement for evaluation is to agree on activities that can qualify as conservation projects, and this requires an explicit definition of what conservation is. For our purposes here we define conservation as the persistence of wild habitats and wild species. Conservation projects therefore include actions that directly enhance the persistence of wild habitats and species. This definition is quite challenging since measuring success at this level requires an assessment of the impact of a project, not just whether the work was funded, whether the planned activities were undertaken, or indeed whether any particular outputs were successfully delivered. Instead we are proposing to assess the consequences of the project – and examine what difference it made to the conservation status of a target species or habitat.

Project evaluation can take place at many points along the project delivery cycle, from inputs (money, equipment or expert time), to activities (workshops, training courses, survey), to outputs (reports written, people trained, fences built, etc.), and finally to impacts (a measurable improvement in the conservation status of the species or habitat). Much current project monitoring is focused on inputs and outputs, and there are relatively few examples of attempts to measure the longer-term impact of a project. There are good reasons for this. Outputs are tangible products that will often have had a deadline for production; impacts can be far harder to assess objectively, and it may take time before the final consequences can be fairly assessed. Nevertheless, because most conservation projects of zoos can be explicitly linked to the conservation status of wild species or their habitats, we felt that this was the right way to evaluate their success.

The diversity of zoo projects also adds to the complexity of any system for assessing effectiveness. The range in the scope and scale of zoo conservation projects is vast, just as are the size, style, and mission of different zoo organizations. The assessment system needs to deal with very local projects of modest scale, as well as major international efforts to save species, habitats, and ecosystems. Additionally, whereas some organizations choose to design and implement their own projects, others might simply donate resources to projects run by others. Finally, the means by which conservation is done can be highly variable. Many different kinds of activities contribute to conservation on the ground; ranging, for example, from education and training programs to habitat restoration, and from ecological surveys of target species to land purchase. How can we fairly contrast projects that consist of such different activities?

One approach developed by Salafsky and Margoluis (Margoluis and Salafsky 1998, Salafsky and Margoluis 1999) is based around assessing the importance of different threats affecting a system, and then measuring the impact of different interventions on reducing those threats. The threat-reduction approach allows many different activities (e.g., direct protection, management and restoration, policy and advocacy, etc.) to be measured against a common standard (threat reduction). Along rather similar lines, Jepson (2004) describes the application of a common structure for scoring diverse projects for elephant conservation derived from a conceptual framework of the components considered necessary in order for conservation to be successful. They then evaluated projects according to the extent to which they supported these components.

We adopt a comparable approach here, but seek to calibrate all projects against a common standard of how influential the project was (regardless of what kind of activity it involved) for conservation. We ask how much the project improved the status of a target species or habitat. To do this across many different types of project we divide our assessment of impact into three sub-components which can vary according to the type of activity that the project involved. Thus, the overall impact score for a project is a function of the importance, the volume, and the effect of a project.

Overall impact $= f$(project importance, project volume, project effect).

Put simply, importance, volume, and effect are measures of the conservation significance of the target, the scale of the intervention, and its outcome, respectively.

The **importance** measure reflects how influential or significant the target of the project was for conservation. This measure seeks to distinguish projects that are strongly focused upon one place, species or activity, from those that might have a more general or dispersed focus, and projects that are focused on targets of wide significance compared to those of more restricted relevance.

The **volume** measure reflects the scale of the project. The way in which volume is best measured depends upon the nature of the project – for a habitat re-creation project it might be the area to be restored, whereas the volume of a training program might be the number of people successfully trained.

The **effect** measure assesses how far the project was successful in terms of the conservation-related outcomes on which it was focused. Clearly, this measure will depend to a large degree on the time elapsed. Given the complexities of project implementation as well as the response times characteristic of biological systems, it is unlikely that scores on this measure will be high in the short term. Nevertheless this is a measure of high relevance for the overall impact score. Continued investment in projects over the long term, at least for the purposes of monitoring project outcomes, will therefore be a necessity for projects to score highly on this measure.

Project types

The assessment of importance, volume, and effect need clear guidelines to encourage consistent assessment. This becomes a much simpler task when the projects are all of a similar nature. For example, it is relatively straightforward to compare the volume of two projects directed at habitat restoration, but almost impossible to come up with measures of volume that could be applied to both a habitat restoration project and a training program.

An initial survey of projects in the British and Irish Association of Zoos and Aquariums (BIAZA, formerly Zoo Federation) conservation projects database suggested that the following list would include most kinds of projects that zoos support (Table 21.1):

- Enhancing public education and awareness: changing attitudes, informing and inspiring (**Education**).
- Enhancing the capacity of people who are in a position to manage species or habitats, change legislation or change behavior (**Training**).
- Undertaking research and monitoring related to species and habitats (**Research**).
- Undertaking direct action to enhance species viability and persistence (**Species**).
- Undertaking direct action to enhance habitat quality, viability, and persistence (**Habitat**).

Table 21.1
The five project types used in the project scoring scheme, and examples of each type in zoos and in field-based projects

	Examples in zoo	Examples *in situ*
Education	Zoo education – talks and interpretation	Sign posts on the edge of reserves
	Coloring-in books for school children	Coloring-in books for school children
	Informative poster and leaflets	Radio shows with environmental messages
Training	Training for *in situ* managers and rangers	Workshops/conferences for scientists and policy-makers: e.g. ranger training
	Training in husbandry, survey or other techniques relevant to wild population	Production of species identifying aids for CITES officials
Research	Investigations of husbandry, veterinary care, reproductive biology	Investigation into ecology and behavior, social attitudes, threat and status surveys, reproductive biology
Species	Managed breeding programs, EEPs, SSPs, JMSPs, etc.	Stock from managed breeding programs released into native habitat
		Translocation of animals
Habitat	n/a	Supporting the costs of a protected area
		Habitat restoration/management, creating fire breaks
		Compensation schemes for domestics taken by wild animals

EEPs, European Endangered Species Programs; JMSPs, Joint Management of Species Program; SSPs, Species Survival Plans.

Table 21.2
The criteria, scores and explanations for scoring projects according to project category. The final impact score for a project will be a function of Importance, Volume, and Effect. The scores for each measure are given as A–E or A–D to emphasize that these need not be simply ranked ordinally. Some measures should possibly scale in increasing steps or even exponentially

	Project assessment measures		
General question to be addressed	Importance	Volume	Effect
	How influential/significant was/is the target (people, species, habitat, policy) of the project for conservation?	How many/much of the target (people, species, habitat, policy) were/was addressed by the project?	How did/does this project affect relevant conservation outcomes? (Do not take importance or volume into account)
EDUCATION (Zoo)	Influence of the target group: A: low (untargeted) B: moderate (children) C: high (school teacher, media people) D: very high (PM, community leader, large land owners, business leader, religious leader) **Target**: people (zoo visitors) **Influence**: extent to which they can influence relevant policy or practice, now or in the future	Number of people who received the education: A: 0–10 B: 11–100 C: 101–1000 D: 1001 +	Effect of the project: A: no discernible effect B: marginal improvements C: improvement D: substantial improvements N: in its first phase **Effect**: a measured change in awareness or behavior that is likely to have beneficial outcomes for conservation compared to no project

EDUCATION (wild)	Influence of the target group: A: low (untargeted) B: moderate (children) C: high (school teacher, media people) D: very high (PM, community leader, large land owners, business leader, religious leader) **Target**: people (local community, tourists, visitors) **Influence**: extent to which these people influence relevant policy or practice, now or in the future	Number of people who received the education: A: 0–10 B: 11–100 C: 101–1000 D: 1001 +	Effects of the project: A: no discernible effect B: marginal improvements C: improvement D: substantial improvements N: in its first phase **Effect**: a measured change in awareness or behavior that is likely to have beneficial outcomes for conservation compared to no project
TRAINING (captive)	Influence of people trained: A: low (front line: zoo keeper) B: moderate (Manager: supervisor) C: high (Curator: Director) D: very high (Senior management: decision-maker, senior civil servants) **Target**: people (involved directly or indirectly in conservation outcomes) **Influence**: the extent to which these people can influence relevant policy or practice, now or in the future	Number of people who received the training program: A: 0–10 B: 11–100 C: 101–1000 D: 1001 +	Effect of the project: A: no discernible effect B: marginal improvements C: improvement D: substantial improvements N: in its first phase **Effect**: a documented change in attitude or behavior affecting relevant conservation policy or practice

(*cont.*)

Table 21.2
(*Cont.*)

General question to be addressed	Project assessment measures		
	Importance	Volume	Effect
TRAINING (wild)	Influence of the person trained: A: low (front line: goat herder) B: moderate (Supervisor: game scouts, rangers) C: high (middle-ranked conservation personnel) D: very high (decision-makers: large land owners, park wardens, senior civil servants) **Target**: people (involved directly or indirectly in conservation outcomes) **Project**: directed education (in wild) **Influence**: the extent to which these people influence relevant policy or practice, now or in the future	Number of people who received the training program: A: 0–10 B: 11–100 C: 101–1000 D: 1001 +	Effect of the project: A: no discernible effect B: marginal improvements C: improvement D: substantial improvements N: in its first phase **Effect**: a documented change in attitude or behavior affecting relevant conservation policy or practice
RESEARCH	Significance of the research target: A: low/ negligible B: moderation/ local C: high/ national or regional D: very high/ international **Target**: the subject of the investigation: species, policy, habitat **Significance**: the relative importance of this target in relation to global priorities	The cost of the project (£): A: 1–100 B: 101–1000 C: 1001–10 000 D: 10 001–100 000 E: 100 000 +	Effect of the project: A: no relevance B: marginal relevance C: considerable relevance D: clear relevance **Effect**: the potential relevance of the research project for conservation outcomes, compared to no project

SPECIES	Significance of the target: A: not threatened B: Nationally/regionally*** threatened (UK or EU) only C: Globally lower risk (NT, DD) D: Globally threatened (EW, CR, EN or VU) **Target**: species in zoo or wild **Significance**: how endangered is this species globally, regionally, nationally, etc.	The proportion of the global population targeted by the project: A: 0–1% B: 2–10% C: 11– 50% D: 50% +	Effect of the project: A: no discernible effect B: marginal improvements C: improvement D: substantial improvements N: in its first phase **Effect**: a documented change in the conservation status of the species within the focus of the project, compared to no project
HABITAT	Significance of the target area: A: negligible/ no biodiversity/ no global loss B: some evidence of decline/local importance (e.g., local nature reserves) C: many areas are in decline/ national or regional importance (e.g., SSSI) D: under global threat/international importance (e.g., CI hotspot, endemic bird areas) **Target**: an area **Significance**: the importance of the target habitat globally, regionally, nationally, etc.	The area targeted by the project (hectares): A: 1–1000 B: 1001–10 000 C: 10 001–100 000 D: 100 000–1 000 000 E: 1 000 000 +	Effect of the project: A: no discernible effect B: marginal improvements C: improvement D: substantial improvements N: in its first phase **Effect**: a documented change in the overall conservation status of the habitat, compared to no project, within the area of the project

Although at first sight it might seem important to distinguish whether these five project types are undertaken in the zoos (*ex situ*) or in the field (*in situ*), in fact we do not think this will be significant for the outcome of the scoring process. Table 21.1 provides a fuller account of the kinds of projects that fall into each of the project types by giving examples of each, both in zoo and *in situ*. The distinction between Training and Education appears subtle but in practice the two are differentiated by whether or not the subject is receiving formal input relevant to their professional activities (Training) or whether the input is more general (Education).

Project scoring methods

For assessing a project it is necessary to define the *target* for the work. For Education and Training projects the target is people (at whom the education or training is directed). In the case of Species projects and Habitat projects the target is self evident: either a species (or set of species) or a place. In the case of Research projects the target might be any one of a range of things that the research is specifically investigating: a species, a habitat or place, or a particular policy.

Measures of importance for the scoring system can be derived from the relative influence of the target on the conservation issue. Thus, educating a high-level decision-maker and changing his/her attitudes to some conservation issue has the potential to be far more influential on a *per capita* basis, and therefore more important for conservation than, for example, educating primary school children. For Habitat and Species projects, the measure of importance is related to the significance of that area or species for conservation. Thus, habitats and species that are globally recognized for their level of threat, extent of decline or rarity will score more highly for Importance than habitats or species that are of local significance only. In the Research projects category, measuring importance is more difficult, and we suggest that it is simply scored according to the relative importance of this target in relation to global priorities. This will be context specific: research on a policy of international significance (for example, the impact of the illegal wild bird trade) would score more highly than research on illegally held wild birds in Cambridgeshire. Equivalently, research on the habitat requirements of some critically endangered species will score above research on a local population of a widespread species. Measures of importance for each project type therefore depend on the target and the relative influence of that target (see Table 21.2 for details).

Once the target is defined, measures of volume can be quite easily derived. Thus the volume of Education and Training projects is a simple multiple of the number of people, for Habitat projects it is an assessment of the area, and for Species projects it is the proportion of the global population of that species targeted by the project. As with assessment of Importance, the measure of volume for Research projects is harder to define. In general the volume of a Research project will vary according to the effort expended, the thoroughness and inclusiveness of the study methods, and the breadth of the enquiry. In general we recommend the cost of the project as a surrogate for all this, but we recognize the potential for serious distortions caused by technological and operational costs affecting some kinds of research. In applying the scoring table, assessors need to be aware of these potential biases, and take them into account where possible. For example, the same kind and quality of research into the status of a species off the coast of the UK and in Antarctica should have the same volume score. But in reality, the costs in Antarctica are likely to be orders of magnitude higher, purely for operational reasons.

Finally, the score for the Effect of a project also depends upon the project type. For Education projects it is assessed as a measured change in the awareness or behavior of the target people. For Training programs, the effect measure is similar, but here the effect is determined by changes in the trainees' attitude, skills or behavior. As a consequence of the training a successful project will have equipped people with new skills that allow them to address the conservation issues within their sphere of expertise more successfully. Note that both these measures will ideally be based on some pre- versus post-project assessment of the quality and/or extent of the change in awareness, attitude or behavior (see Chapter 9). For Species and Habitat projects the measure of effect is the change in the conservation status of the target population or area. Again, ideally some standard measures can be used to compare the conservation status before and after the project (e.g., population size, population growth rate, IUCN threat category for species; habitat quality, integrity or extent for habitats). For Research projects our measure of effect is the relevance of the information obtained for conservation. The reason that the effect of Research projects is scored so differently is due to the nature of research. By itself, research does not contribute to conservation outcomes; by definition research aims to produce new understanding which (in the case of conservation research) might then lead to improved management and conservation. The success of a research project could perhaps be measured in terms of what it set out to achieve. By this yardstick, well-executed projects of low conservation relevance could be rated very highly, even though they have a negligible impact on conservation, unless the findings are of high conservation relevance. Instead, we assume (perhaps unrealistically) that all research outcomes eventually feed into conservation practice, and on this basis the best measure of effect is the relevance of the findings for conservation. To take an example, a successful conservation research project on the taxonomic status of a species of African monkey could achieve all it set out to do, and a new nomenclature and quantitative measures of its relationship to other species published and disseminated. But this research may have little impact on the conservation of that species compared to say a study of illegal hunting of the species that identified the source and motivations of the hunters.

Although the five project types can be found both in zoo and *in situ*, and the basic elements of the work are common to both, as we move to the specifics of the scoring system it becomes clear that it is helpful to distinguish the two in the case of Training and Education projects, as the targets of the project are rather different in the two situations.

A PILOT STUDY

In order to test the system, we designed a small pilot study based around a set of real conservation projects. Altogether we scored 27 projects contributed by five zoo-based organizations (North of England Zoological Society, Wildlife Conservation Society, Zoological Society of London, Durrell Wildlife Conservation Trust, and Marwell Preservation Trust). The projects were chosen by these organizations from their own conservation project portfolios. At each organization we asked an expert on the project from among their staff to first complete a short questionnaire about the project. This contained basic information, such as the aims, location, cost, and timescale, but also had some questions that indirectly addressed the variables for scoring in Table 21.2. Importantly, we did not reveal to these experts what the scoring matrix consisted of, nor indeed any details beyond the general information that we were testing a method for assessing conservation projects.

Once the form had been completed by the expert, four of us collectively acted as assessors, and interviewed the expert about the project. Based on the material in the written questionnaire, we discussed details, again without revealing what it was that we were scoring, until all four assessors had independently completed the score sheet for the project. In practice, there was little disagreement about what the project type was, but many projects had elements of more than one type. Hence at the start of each project discussion the assessors agreed among themselves about what the project

Table 21.3
The 41 projects in the pilot study, classified according to project type. There were 27 different projects but 7 of these included more than one type of activity

	Frequency	%
Education (Zoo)	–	–
Education (Wild)	2	4.9
Training (Zoo)	1	2.4
Training (Wild)	5	12.2
Research	12	29.3
Species	11	26.8
Habitat	10	24.4

types were, so that scoring could proceed for each one in turn. Once we had completed our scoring, we revealed the scoring system to the expert, who was then invited to assess their project(s) using the same system.

The way in which scores for different measures should be scaled against the A to E categories, and how they should be used in combination to provide an overall project score are complex questions that require more work. For the pilot study we chose simply to adopt the simplest approach and scored A = 1; B = 2; C = 3; D = 4; and E = 5 for all measures, and the total project score as Importance × Volume × Effect.

Of the 27 projects we assessed, 7 projects contributed to more than one project type, so that overall we obtained 41 project scores.

RESULTS

Project profile

Most of the projects we assessed were Research, Species, and Habitat projects (Table 21.3). There were fewer in Education and Training, and only one Training project, and none in the Education category, was zoo-based (the latter is the focus of a separate paper – Chapter 9). Our projects sample included local- to regional-scale projects; the cost of projects ranged from a few hundred to several million dollars; and the length of the zoo's support from over 15 years, to just a few months (Table 21.4). Of the projects we examined, most were supported by several institutions, with a median contribution from the focal zoo of 35% of total cost; only 11 of the 41 projects were 100% supported by a single zoo.

Table 21.4

Profile of projects assessed. The table shows the minimum, median, and maximum for the length of time the zoo had been involved in the project, the percentage of the total project costs supported by the zoo to date, and the mean annual cost to the zoo to date

	Minimum	Median	Maximum
Duration of zoo's involvement (years)	0.17	6.00	16.00
% contribution from zoo	5	35	100
Annual cost to zoo ($/year)	754	43 077	12 300 000

Table 21.5

Mean scores for each of four independent assessors for each measure, and the scores for project experts across 25 projects

	Importance	Volume	Effect
Independent 1	3.48	2.88	3.00
Independent 2	2.68	2.84	2.80
Independent 3	3.40	2.88	2.68
Independent 4	3.48	3.00	3.20
Project expert	3.20	2.76	3.04

Consistency of scoring

Of the 41 projects, only 25 were scored by all 4 independent assessors and the project expert (assessor 2 did not score 16 of the projects); the analysis here is restricted to that subset. In order to analyze the score, the categories A–E (Table 21.2) were converted to numerical values of 1–5. Thus higher scores are better across the measures.

First we examined the overall consistency of scoring among assessors. The mean scores for each assessor are given in Table 21.5. There was no significant difference among the assessors in their mean scores for Volume (Friedman rank sum test, $\chi^2_4 = 1.65$, $P = 0.8$) or Effect (Friedman $\chi^2_4 = 9.29$, $P = 0.054$), although the scores in the latter case only just miss statistical significance. There was however a clear significant difference among assessors in the Importance scores (Friedman $\chi^2_4 = 20.84$, df $= 4$; $P < 0.001$). Among the measures, Importance is often clear-cut, so this was surprising but this result seems to be largely due to assessor 2 sometimes inadvertently scoring Importance the wrong way around. These inconsistent Importance scores led to significant variation among assessors (Friedman $\chi^2_4 = 11.99$,

Table 21.6
Correlations among assessors for each measure. The statistic given is Rho from Spearman rank correlations with the significance level given by the number of asterisks

	Independent 2	Independent 3	Independent 4	Project Expert
Importance				
Independent 1	0.44*	0.89***	0.80***	0.60**
Independent 2		0.38	0.27	0.48*
Independent 3			0.83***	0.52**
Independent 4				0.57**
Volume				
Independent 1	0.97***	0.97***	0.89***	0.90***
Independent 2		0.95***	0.86***	0.86***
Independent 3			0.87***	0.87***
Independent 4				0.77***
Effect				
Independent 1	0.72***	0.79***	0.54**	0.31
Independent 2		0.74***	0.29	0.23
Independent 3			0.50*	0.43*
Independent 4				0.01
Impact (I×V×E)				
Independent 1	0.82***	0.90***	0.82***	0.82***
Independent 2		0.85***	0.61**	0.70***
Independent 3			0.78***	0.83***
Independent 4				0.68***

* $P < 0.05$.
** $P < 0.01$.
*** $P < 0.001$.

$P = 0.017$) in the overall impact score (Importance × Volume × Effect), but this is again due to assessor 2, rather than any systematic differences among other assessors or the project expert.

Correlations of individual project scores across the different assessors and the expert are generally high, especially for the Volume score, and for the Importance score (except for independent assessor 2). Not surprisingly the correlations are weaker, though still always positive and often significant for the Effect score (Table 21.6). If the overall impact of the project is calculated simply as Importance × Volume × Effect, then the correlations are all highly significant (Table 21.6). There is no indication here that the project expert scored any of the variables differently to the independent assessors.

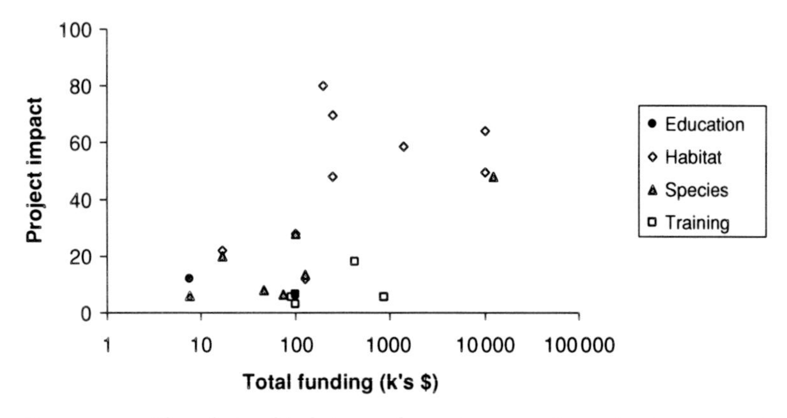

Figure 21.1a The relationship between the project score (Impact = Importance × Volume × Effect) and total project cost to date

Characteristics of successful projects

Since assessor 2 scored differently from the other three assessors, we built an overall impact project score for each project based on the average of the marks from the three consistent assessors. This provided us with a full set of 41 projects to investigate the scoring system. The scoring system used here gave significantly different overall impact scores for the different types of projects (Kruskal–Wallis $\chi^2_4 = 20.3$, $P<0.01$). Habitat projects ranked highest followed by Research projects, then Species projects, with Education and Training projects ranked lowest. However, it is important to note that the scoring system here is arbitrary and not calibrated for different activities. At this stage the impact scores can be usefully compared within types of project but probably not between them.

These scores were then used to examine the relationship between conservation impact, and duration and funding. Information on funding was lacking for some projects, thus reducing the pool of available projects to 33. Furthermore, since the amount of funding received by a research project was used as its measure of volume, Research projects were excluded from this analysis to avoid circularity, leaving us with 23 projects. Whereas project duration (Spearman's $\rho = -0.065$; $P = 0.485$) was not correlated with project impact, total funding (Spearman's $\rho = 0.519$; $P = 0.011$) as well as annual funding contributions (Spearman's $\rho = 0.542$; $P = 0.008$) showed significant positive relationships. (Figure 21.1a, b). This figure also illustrates the different impact scores characteristic of different project types (see above); more significant than the difference between project types are the differences within project type resulting from total financial investment.

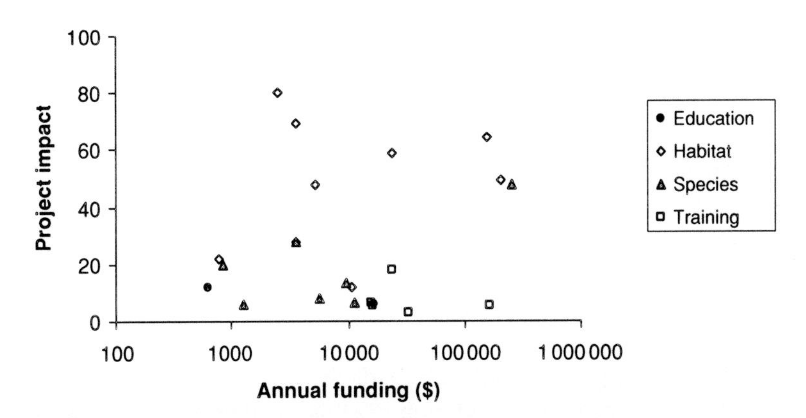

Figure 21.1b The relationship between the project score (Impact = Importance × Volume × Effect) and annual funding for all projects scored by three consistent independent assessors. Different project types are distinguished by symbols as shown in the key

DISCUSSION

Our method for scoring project impact turned out to be reasonably easy to use in practice. All of the assessors, including the project experts who only had a few minutes to pick up the system, grasped the concept quite quickly and were able to apply it across a range of very different projects. Allocating scores to each of the measures of Importance, Volume, and Effect was often very straightforward, although at times it required more difficult judgements. In general, once the project type was agreed, the scores for Importance and for Volume were straightforward to determine – these scores are largely based on factual information. It is the assessment of Effect that is the most difficult as well as the most subjective. Given this, it is reassuring to note that our system showed relatively consistent scores for Effect across assessors, as well as between the expert and independent assessors.

While it seems that the basic method and structure for our scoring system is sound, significant work remains to be done on the details of the scoring system. We labeled the scores A–E to indicate that the numerical scores and their scaling are as yet undetermined. For our numerical tests we converted these to simple ordinal values, but this is almost certainly inappropriate, and may alone be responsible for the consistently different impacts found among project types (Figure 21.1a). What alternative weights could be used? Some scores reflect quantifiable aspects of the project (e.g., the number of people trained, the area on which a habitat project is focused, etc.). However, these quantities may not always be appropriate as weights for scoring. In other cases where the attributes are qualitative, simple surveys or experiments could be used to determine appropriate scaling. For example, it would be possible to find out how much people value the conservation of a globally endangered species above that of a locally common form (see chapter 9); and it would be possible to weight the political influence of alternative targets of Education and Training programs by the conservation budgets over which they have influence.

Further work would certainly be needed if it were felt to be relevant or necessary to compare scores across different project types. This is a more difficult exercise than calibrating scores within project types, because it involves making a judgement about the relative importance of different activities. One approach to determining this could be through an analysis of a large number of successful conservation projects, looking at the contribution of different kinds of activities. The problem with this approach is that it may not yet be feasible, and it may prove to be difficult to generalize between projects that involve very different circumstances of time, place, economic and political context, biological and environmental conditions, etc. An alternative would be to simply sidestep the issue, and instead of attempting to score different project types on a single measure (effectively trying to value apples against oranges), to accept that there are five different scales (i.e., value apples and oranges in their own right). In this case, the scoring system could only be used to compare projects within project types, never between them.

Finally, the Research project category poses some particular problems in assessment and in relation to the other project types. The impacts of research may be removed in time, and hard to trace, so the assessment of Effect is always difficult. As discussed above, the choice of a Volume measure proved difficult, and our solution to use the costs of the project may prove to be a distortion of the scheme as a whole. Research can form a part of all the other projects, and is often a necessary precursor to conservation project design. It seems to us that there are two possible solutions: to live with this difficulty but be particularly careful not to compare scores between Research and other project types, or to exclude altogether projects that only include research activities. We also found few Education projects to assess, but noted that the scoring system for these was quite difficult. In this case, more formal procedures for evaluating effect will certainly be needed (see Chapter 9).

Although the projects we studied were quite varied, it is certain that there were some important gaps. We probably saw fewer unsuccessful projects than might be expected by chance. Unsuccessful projects probably fall into two categories: those that immediately fail or fade after a very short period, and those that are persisted with over many years, despite there being no evidence of any progress. Organizations may be less willing to air such projects publicly, and indeed the information on them may not be so easy to obtain. For a true calibration of a scoring system, however, it is important to include projects that fail or that take a very long time to succeed.

Our pilot study also suggests some factors that need to be considered in implementing any such scheme. First, although there was no systematic bias in the scores of the project experts, there were often differences. In any system such as this, we recommend that at the very least there is an independent audit of project scoring, if not independent assessors. Leaving all scoring up to the project leaders would introduce an unacceptably arbitrary element to the scores. Second, the process of being able to "interview" the project leaders proved especially useful as the necessary details were often not apparent in answers to questionnaires. We suggest that this may not always be necessary, but might be useful for resolving difficult cases, as well as for monitoring and maintaining the operation of the scheme overall.

Based on our pilot study, we recommend proceeding with the development of a system such as this. Reassuringly, it seems that an assessment of impacts – surely the most relevant measure of the success of a conservation project – is possible.

ACKNOWLEDGEMENTS

We are especially grateful to the various people who helped us by providing project information and acting as project experts in the pilot study – in particular Andrea Fidgett, Tim Woodfine, Matthew Hatchwell, Sarah Christie, Glyn Davies, Heather Hall, Nick Lindsay, Paul Pearce-Kelly, and Tim Wacher. We are also grateful to the North of England Zoological Society, the Wildife Conservation Society, the Zoological Society of London, the Durrell Wildlife Conservation Trust, Marwell Preservation Trust and BIAZA for staff support, and for giving us access to their projects.

References

Belokurov, A., Dudley, N., Higgins-Zogib, L., Hockings, M., Lacerda, L., Stolton, S., & management (2003). *Tracking Effectiveness in Forest Protected Area Management*. Gland: WWF.

Christensen, J. (2003). Auditing conservation in an age of accountability. *Conservation in Practice*, 4, 12–19.

Clark, T. W., Reading, R. P., & Clarke, A. L. (1994). *Endangered Species Recovery*. Washington, D.C.: Island Press.

CMP (2004). Open standards for the practice of conservation. Conservation Measures Partnership. www.conservationmeasures.org.
Jepson, P. (2004). Auditing conservation: reflections and findings from an independent audit of Asian elephant conservation. *Loris*, **23**, 29–34.
Margoulis, R. A., & Salafsky, N. (1998). *Measures of Success: Designing, Managing and Monitoring Conservation Projects*. Washington, D.C.: Island Press.
Miller, B., Conway, W., Reading, R. P. *et al.* (2004). Evaluating the conservation mission of zoos, aquariums, botanical gardens, and natural history museums. *Conservation Biology*, **18**, 86–93.
Possingham, H. P. (2001). *The Business of Biodiversity: Applying Decision Theory Principles to Nature Conservation*. Fitzroy: Australian Conservation Foundation.
Pressey, R. L., Cowling, R. M., & Rouget, M. (2003). Formulating conservation targets for biodiversity pattern and process in the Cape Floristic Region, South Africa. *Biological Conservation*, **112**, 99–127.
Salafsky, N. & Margoluis, R. A. (1999). Threat reduction assessment: a practical and cost effective approach to evaluating conservation and development projects. *Conservation Biology*, **13**, 830–841.
Salafsky, N., Margoluis, R., Redford, K. H., & Robinson, J. G. (2002). Improving the practice of conservation: a conceptual framework and research agenda for conservation science. *Conservation Biology*, **16**, 1469–1479.
WAZA (2005). *The World Zoo and Aquarium Conservation Strategy: Building a Future for Wildlife*. Liebefeld-Bern: WAZA.

22

Conclusion: the future of zoos

MATTHEW HATCHWELL, ALEX RÜBEL, LESLEY A. DICKIE, CHRIS WEST, AND ALEXANDRA ZIMMERMANN

The *World Zoo Conservation Strategy* (IUDZG/CBSG 1993) was the product of an evolutionary process going back 200 years (Wemmer 1995, Kisling 2001, Conway 2003, Knowles 2003), but at the same time a revolutionary document that challenged the international zoo community to re-examine the thinking underlying the very existence of its member institutions. The strategy spoke of the "responsibilities" of the world's zoos and aquariums "towards the conservation of the variety of global wildlife," and called on the zoo community to focus its "conservation potential . . . primarily at supporting the conservation of natural habitats and ecosystems."

The first chapter of WZCS traced the evolution of zoos from menageries through zoological parks to conservation or environmental resource centers. The diagram reproduced in Figure 22.1 was accompanied by the following text, which is worth citing in its entirety:

> Zoos are rapidly evolving to serve in multiple ways as conservation centres. The horizontal arrows indicate that professional capacities of concern and subjects communicated to the public in earlier phases of zoo development are now vital services to conservation. As conservation centres, zoos must additionally address sustainable relationships of humankind and nature, explain the values of ecosystems and the necessity of conserving biological diversity, practise the conservation ethic throughout zoo operations, and cooperate within the world zoo network and with other conservation organizations. Immersion exhibits involve zoo visitors in the environmental circumstances of the animals, and such experiences are conducive to favourable reception by visitors of strong conservation messages.

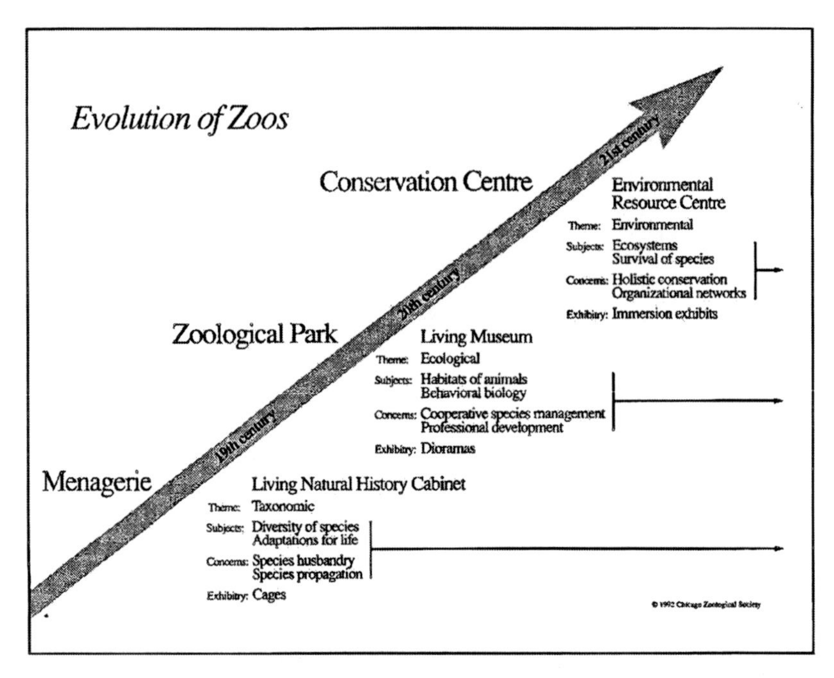

Figure 22.1 The evolution of zoos (from IUDZG/CBSG with permission)

What is most remarkable about this passage, besides its prescience, is the tone of ambition and aspiration in the language. In this summary chapter, we set out to update the discussion, which has been continued in the meantime by other authors (e.g., Hutchins and Smith 2003), about exactly what the attributes of a "conservation centre" might be. Earlier chapters of this book have identified a number of characteristics which can be grouped under three broad headings:

- Effecting behavior change among zoo visitors and decision-makers in ways that may contribute to *ex situ* or *in situ* conservation
- Establishing linkages between *ex situ* zoo-based operations and *in situ* conservation actions, and
- Contributing directly to *in situ* wildlife conservation.

While the traditional function of zoos, inherited from menageries, as leisure destinations where the public can experience animals "in the flesh," must surely remain as a *sine qua non* of the New Zoo paradigm as well, these three functions can be seen as the defining traits of the conservation centers which the authors of WZCS in 1993 saw as the next stage in the evolution of zoos.

EFFECTING BEHAVIOR CHANGE AMONG ZOO VISITORS AND DECISION-MAKERS

Despite adopting the education of the public as a key part of their role over a century ago (Chapter 4), it is clear now that zoos and aquariums have not been as rigorous as they might have been in demonstrating that visitors undergo any significant change in awareness or understanding of wildlife conservation issues as a result of their visits. In future, zoos must (1) learn how to detect such changes if they occur, (2) use that knowledge to design new exhibits or adapt old ones and develop associated interpretation materials and tools so that they are more effective in getting a conservation message across to the public, and (3) assess whether the increased awareness and/or understanding results in any change of behavior by zoo visitors. Broadly speaking, museums have made more progress than zoos in tackling the challenges outlined above, although with some notable exceptions such as Brookfield Zoo, Illinois (Saunders and Myers 2001, Saunders 2003). In addition, evaluation of the efficacy of exhibits is now becoming more commonplace (Hayward and Rothenberg 2004).

In order to detect any changes in knowledge or attitudes about wildlife, zoos must acquire a deeper understanding of visitor psychology (Pekarik 2004, Bickert and Meier 2005). Rather than trying to measure changes in awareness among adults resulting from a single zoo visit (Dierking *et al.* 2004), for example, they may have to change their focus to detecting and measuring the longer-term impacts on children of multiple family visits to the zoo. Positively changed attitudes have been measured in college-level students after taking conservation electives (Caro *et al.* 2003), but can the same response be elicited by a single visit to a zoo? What are the behavioral changes that zoos wish to bring about on the part of their visitors in order to increase their conservation impact? What is the overall effect of a zoo visit in terms of bringing about those changes? Are some exhibits more effective than others in changing visitor behavior? What cultural norms (Schultz and Zelezny 2003) does the public bring with them to a zoo and what does that mean when it comes to getting messages across in an effective way (Bitgood *et al.* 1993)?

Clearly, the answers to these questions are likely to have profound implications for the design of zoo and aquarium exhibits (Gwynne 2000). Certainly a good exhibit must inspire, which means that it should appeal to visitors on both intellectual and emotional levels (Vining 2003). But zoos in future must be more deliberate about *how* they inspire their visitors, distinguishing for example between the ways in which children and adults learn.

Do children learn primarily from their own experience or do they depend on the adults they are with to interpret it for them and to fill in key details? In addition, the dynamic of moving from the emotional connection that may be generated with an individual animal to a generalized behavioral change is equivocal. Can zoo visitors, once a caring response has been evoked for an animal, link that to caring for whole ecosystems (Vernon 2003)? These are important questions to which we have only partial answers, at best.

In many cases, zoos will find it easiest to influence the behavior of the general public and of decision-makers by addressing wildlife conservation needs in their own backyards. Zoo employees may apply their professional expertise to conducting research on locally threatened or endangered species or ecosystems, to raising awareness and mobilizing public support of such issues, or to adopting particular sites or ecosystems as foci for conservation actions. Zoos are also well placed to set themselves up as models when it comes to "green" practices: using natural filtration systems to treat polluted water, recycling waste materials, reducing or offsetting carbon emissions, encouraging staff and visitors to use public transport and leave their cars at home, reducing water and electricity consumption, and so on. We know now that human-induced climate change is the greatest threat that exists to wildlife on a global scale (Thuiller *et al.* 2006).[1] Zoos are well placed to get that message across and provide information on what the public can do to tackle the problem.

Just as zoos can be models in terms of reducing their environmental footprints, so too must they be models when it comes to applying the highest possible standards of animal care and welfare (Anon 2005a, Cuaron 2005). We undermine our own conservation objectives if we are perceived to treat the animals in our care with anything less than the consideration and respect that our graphics demand of our visitors. Failure to do so can attract the unwelcome attention of animal rights activists (Jamieson 1995, Regan 1995), further reducing our ability to communicate our conservation message effectively to the general public. Most fundamentally, many of us believe that we owe it to other animal species, as fellow travelers on Planet Earth, to treat them humanely. That includes catering for their psychological well-being as well as for their physical requirements. Among the latter, obvious provisions include suitable accommodation, diets and medical care, all adapted to the ecology of the species in question. Psychological requirements include addressing the social needs of animals by enabling

[1] Although Lewis (2006) cautions conservationists to not ignore the immediate threats of ongoing habitat destruction.

them, where appropriate, to form the types of social groups in which they would live in the wild, and designing behavioral enrichment programs that encourage animals to practise skills that they would normally use in the wild and forestall the repetitive behaviors frequently developed by animals living in unsuitable enclosures (Shepherdson *et al.* 1998, Swaisgood and Shepherdson 2005). In cases where captive animals do show aberrant behaviors, zoos should explain, articulately and honestly, what is happening. To quote Juniper (1998), "If an animal is showing stereotypic or abnormal behaviour, then don't the public warrant at least an explanation why the animal is doing that?"

Finally, zoos in future must be more alert to the broader conservation context in which they exist and the role they can play in influencing the debate. We know now, for example, that the best zoo exhibits can highlight the interdependence of humans and animals, demonstrating that biodiversity conservation and human poverty alleviation are two sides of the same coin and not polar opposites as they are frequently portrayed (Chapter 14). In the context of that global policy debate, it is not constructive for zoos (or any other conservation organizations) to present human activities solely in terms of the threats they pose to wildlife. By presenting complex issues in terms that are accessible to the general public, as advocates for conservation, zoos may be able to influence the actions of entire communities. The Masoala exhibit in Zürich has alerted the whole of Switzerland to the importance of conservation in Madagascar (Bauert *et al.* 2006). Future zoo exhibits could prompt local decision-makers to take action on other issues of global import such as climate change. In all cases, the role of zoos can be summarized by the mantra attributed to René Dubois: "Act locally, think globally."

In summary, under the rubric of effecting behavior change among zoo visitors and decision-makers, the following lessons emerge for the zoos of the future:

1. Learn to demonstrate and measure their impacts on zoo visitors and any behavior changes on the part of those visitors that may occur as a result.
2. Develop a more rigorous approach to inspiring zoo visitors to care about wildlife conservation.
3. Become models and catalysts for conservation change in their own neighborhood.
4. Define and adopt the highest possible standards of animal care and welfare.
5. Emphasize the links between humans and biodiversity at all levels.

ESTABLISHING LINKAGES BETWEEN *EX SITU* ZOO-BASED OPERATIONS AND *IN SITU* CONSERVATION ACTIONS

The second function that emerged from the Catalysts symposium in 2004 upon which this volume is based, as a focus for zoos of the future, is the systematic development of links between *ex situ* zoo-based operations and *in situ* conservation actions.

As a first step, regardless of their rhetoric in the past, zoos must be realistic about the true potential for the reintroduction of captive-bred animals by zoos to contribute to *in situ* conservation, and correspondingly honest about the purposes and utility of captive breeding as a conservation tool. While breeding endangered species in captivity for eventual reintroduction to the wild may be an appropriate strategy in some cases, it is likely to be among the more expensive options available to conservation managers (Lindsay *et al.* 2005).

Although there have been high-profile successes (including, for example, the Golden Lion Tamarin), the focus of this "breed and release" ethos on larger, predominantly mammalian, species may be misplaced in all but the most dire circumstances. Conversely, a partial solution to the extinction crisis threatening more than 30% of the Amphibia (Stuart *et al.* 2004) may lie in captive breeding. Once their ecology has been studied and understood, the majority of threatened amphibian species can be kept in captivity relatively inexpensively. Captive-bred animals display high fecundity rates and few, if any, behavioral problems that would make return to the wild difficult (Bloxam and Tonge 1995). It is abundantly clear that without such possibilities hundreds of species of amphibian are vulnerable to extinction within a few decades. A similar case can be made for invertebrates (Pearce-Kelly 1994, Ferguson and Pearce-Kelly 2005) and for some fish species (Andrews and Kaufman 1994) which are amongst the most cost-effective species for breed and release programs (Pearce-Kelly 1994).

While it is important to bear such examples in mind, when it comes to the charismatic mammals that have been the bread-and-butter of zoos up to now, preventing the extinction of wild populations in the first place is likely to be a cheaper and more effective option than reintroduction. The costs associated with setting up and running reintroduction projects in developing countries are on a par with those of entire protected areas, which protect many more animals as well as their habitats. It must be recognized that reintroduction projects such as that of black-and-white ruffed lemurs to Madagascar's Betampona reserve (Britt *et al.* 2002, 2003, 2004, Britt and Iambana 2003) or of western lowland gorillas to the Bateke Plateaux

National Park in Gabon have brought benefits for protected area management which increase dramatically the overall conservation impacts of the reintroduction itself (Beck *et al.* 1994). Unless such linkages are put in place, however, where mammals are concerned, zoo-based expertise and experience in animal handling are likely to be more useful to conservation in the context of translocations of animals from one site to another (e.g., to re-populate habitats where the species has been extirpated for one reason or another) than through reintroductions. Similarly, zoo-based skills in the genetic management of small populations of captive animals (Anon 2005b) may be more useful to conservation when applied to the management of small populations in the wild than to captive breeding and reintroduction programs, although the proper genetic and demographic management of captive populations remains an essential task for zoos (Wiese *et al.* 1994).

The key requirements for zoos wishing to increase their involvement in such field-based activities are, first, for them to broaden their thinking about the potential applications of their scientific expertise to *in situ* species conservation, and, second, to develop the links into *in situ* conservation efforts that would make their skills available to conservation managers on the ground. At one extreme, those linkages may be so close and so numerous that an exhibit such as the Masoala exhibit in Zürich is perceived, at least by the zoo-going public, as an outpost of the site on which it is based. That sort of twinning relationship between exhibits and protected areas may not be within the reach of all zoos, but there are still individual links that can be established. Examples cited elsewhere in this volume include:

- *In situ* conservation projects initiated by zoos in parallel with the development of new zoo exhibits.
- Membership of associations of like-minded zoos that have agreed to develop cooperative programs in particular countries or targeted at particular species.
- Participation in annual fundraising campaigns organized by regional associations of zoos.
- Conceptual links between zoo exhibits and protected areas explored through graphics and interpretive materials.
- Multi-year, zoo-based fundraising for conservation of specific sites or species.
- Visits to *in situ* conservation projects by zoo donors, members and staff.
- Scientific expertise on small population management provided by zoo staff to *in situ* conservation.

- Support to protected areas in development of visitor centers and interpretive graphics.
- Support to developing countries on issues of wildlife health and epidemiology.

There are certainly other types of linkage with *in situ* wildlife conservation programs that imaginative zoos around the world will develop in coming years. Whatever their nature, it is vital that the relationship between *in situ* and *ex situ* brings benefits to both parties. Otherwise it will fail in the long run. In return for funding from zoos, for example, field sites must agree to provide regular and high-quality information and visual materials which the zoos can use to maintain the interest of zoo-goers and donors. Without it, their support will falter and ultimately die away. Examples include the monthly field reports sent to zoos that support the Lion Tamarins of Brazil Fund and the Madagascar Fauna Group. Like human relationships, the ones between zoos and conservation programs on the ground must be a two-way street.

In the same way that two partners each have to work out their respective roles in a relationship, zoos and their collaborators in the field also need to work out their respective strengths. In that respect, there is little doubt that wildlife conservation organizations have failed so far to realize, let alone capitalize upon, the capacity of zoos to reach vast numbers of people and therefore to communicate the conservation message to a public upon whom, at present, they are having little impact. This is particularly true, for example, when compared to the success of animal welfare organizations in the United Kingdom, whose ability to galvanize an emotional response from the public makes them more successful in fundraising terms than conservation organizations that rely almost exclusively on rational, intellectual arguments to attract support.

Another reason why zoos must insist on receiving regular reports from field programs with which they are involved is that they have to get better at demonstrating and quantifying the impacts of their activities, and therefore of these linkages, on *in situ* conservation. Modern zoos, like most other organizations, are expected to be able to demonstrate that they are achieving their institutional goals. As those goals focus more and more on conservation in the wild, new tools must be developed to monitor the effectiveness both of zoo exhibits as vehicles for conservation messages and of zoo-sponsored field programs as direct conservation interventions. The science of evaluating the conservation impact of zoos is still in its infancy. That will have to change if – as articulated in the 2005 *World Zoo and*

Aquarium Conservation Strategy – zoos wish to be accepted as *bona fide* members of the worldwide conservation community. It must be said, however, that zoos are not alone among conservation organizations in finding it difficult to measure their impacts (Jacobson 1991, Kleiman *et al.* 2000, Saterson *et al.* 2004, Ferraro and Pattanayak 2006). The fact that demonstrating the benefits of conservation education in rural communities is especially challenging (Struhsaker *et al.* 2005) echoes the problems faced in zoo education and interpretation.

CONTRIBUTING DIRECTLY TO *IN SITU* WILDLIFE CONSERVATION

There is a clear overlap between *ex situ* – *in situ* linkages and the third set of activities that emerge from this volume as a focus for zoos of the future: direct contributions by zoos to *in situ* wildlife conservation.

Some of those opportunities for direct involvement by zoos in conservation have already been mentioned. Zoos are especially well placed to raise money from the millions of visitors who pass through their gates every year and probably there is more that most of them can do in future to maximize the funds they can raise for *in situ* conservation programs, especially if linkages are in place as discussed in the previous section. Some have gone a step further by resolving to allocate a minimum of 10% of their annual budget to *in situ* conservation activities.

Another course of action pursued by some zoos is to develop their own field programs as an alternative to funding projects set up and managed by others (Chapter 20). For some organizations, it may be important to be able to lay claim to the exclusive credit for the achievements of a particular project, and to reap the associated benefits in terms of advertising, public relations and fund raising. Others have decided that establishing and running field conservation programs lies beyond their core competency as zoos and that they are better off working in partnership with local conservation NGOs or the large international conservation organizations. Certainly there are management expenses associated with setting up and running conservation projects which are reduced when zoos are prepared to support *in situ* projects managed by other agencies as an alternative to establishing their own programs.

In cases where zoos decide to fund *in situ* programs operated by other organizations, the broader the range of linkages they can develop the more likely it is that they will be perceived and treated as conservation partners rather than just donors. Funding is obviously a vital part of the equation, but it is just one among many potential linking mechanisms between zoos and field programs.

An innovative proposal for direct *in situ* interventions by zoos, which was made over 15 years ago but has not so far been tested in practice, is the creation of extractive reserves. This controversial idea was proposed in part as a solution to the growing problem of finding reliable legal sources of animals for zoo collections in cases where captive-bred animals are not available (Conway 1998, Redford 1998). The main principle behind extractive reserves is that local communities own and control the harvesting of natural resources – in this case wild animals for export to zoos. It is a particularly attractive form of land use in that, in theory at least, the natural habitat remains intact and indeed contributes to improving the livelihoods of local people. Rather than keeping people out of a given area, extractive reserves are supposed to permit people to manage and benefit from it without destroying it.

Again in theory, a well-managed extractive reserve could provide a legal and sustainable source of animals for zoo collections as well as for commercial exports. In practice, their success would require broad consensus on the part of the community involved, as well as rigorous law enforcement and technical supervision, neither of which is easy to guarantee in remote, developing country contexts. It is a contentious proposal in that there is growing consensus in the international zoo community that zoo animals should be sourced to the greatest possible extent from captive populations (WAZA 2005), and from ones that are palatable to the public, but that is not possible in all cases, and so many people believe that it is a proposal worth testing under carefully controlled conditions (Redford 1998). If it is successful, it would be another way in which zoos could contribute to *in situ* wildlife conservation.

To summarize, this volume proposes a number of ways in which zoos can engage directly in *in situ* conservation. These approaches may be combined very effectively with the development of the zoo-based linkages mentioned in the previous section.

- Raise funds for field conservation programs operated by partner organizations.
- Develop their own "proprietary" field programs by increasing in-house conservation capacity.
- Explore the use of extractive reserves to source animals not available from captive-bred sources.

As in other areas, at the same time as thinking about how to increase their impacts on *in situ* wildlife conservation, zoos must also design and implement ways to evaluate those impacts. As discussed elsewhere, it is a problem that zoos have in common with other conservation organizations, but it is especially important for zoos if they want to demonstrate their own value as relative newcomers to the conservation community.

A framework for change

No matter how zoos and aquariums evolve from their current state, WAZA, in particular through its *World Zoo and Aquarium Conservation Strategy* (WZACS), is likely to constitute the global institutional framework for change. Rather than defending the *status quo*, WAZA must make sure that WZACS does prove to be the "blueprint for urgent local and collective action by zoos and aquariums worldwide" that it aspires to be. Where WZCS, despite its aspirational tone, stopped short of defining *in situ* wildlife conservation as the *raison d'être* of zoos, WZACS goes one very significant step further and defines it as the "overarching mission" of WAZA members. Some zoos are indeed heading rapidly toward that goal. On a global scale, the vision has yet to be expressed in the missions of zoos as a whole, let alone translated into concrete actions on a worldwide scale or into changed public perceptions, but it already represents a significant step forward as an expression of collective intent.

This is the context too in which the *Catalysts for Conservation* symposium, from which this volume emerged, was held. The results of the Catalysts meeting played an important role in defining the agendas for the 2004 and 2005 WAZA annual meetings and have spawned over time an open-ended forum – complementary to and coordinated with the processes of WAZA and CBSG (the Conservation Breeding Specialist Group of IUCN) – for promoting the WZACS vision. WAZA as a professional association has limited operational capacity except through the actions of its members. The ongoing Catalysts Initiative, as the open-ended forum has become known, represents an effort by some of those members to harmonize their actions in order to move the WZACS agenda forward.

One international zoo community, one agenda

A criticism that has been leveled at WAZA, WZACS, and the Catalysts Initiative alike is that they are elitist undertakings whose ambitions are beyond all but the wealthiest institutions in the most developed Western countries. It is certainly true that there are relatively few zoos that can afford to build state-of-the-art, climate-controlled exhibits faithfully reproducing the habitats and fauna of exotic far-off lands – or even the personnel and travel budgets that it takes to participate in the international meetings where collaborative projects are frequently conceived. Nonetheless, neither the measures proposed by WZACS nor those discussed in this volume should be considered the domain of just a privileged few. The zoo community must act together if it genuinely wishes the public to view its individual institutions – however rich or poor – as vehicles for conservation. The step-wise approach to zoo accreditation discussed below may be the way ahead for some less wealthy institutions. For others, guidelines and support mechanisms must be put in place to help needy zoos in both developed and developing countries implement the WZACS provisions. We must do this as a community because the poor image of the "bad" zoos tarnishes that of our entire industry (Maple 1995).

Zoos possess a range of attributes that position them to play a unique role in effecting *in situ* wildlife conservation. That role is recognized in the language of the Convention on Biological Diversity signed in Rio in 1992, which is the highest legal instrument governing biodiversity conservation on the global scale. Those unique characteristics of zoos are evoked and explored individually elsewhere in this volume. The most obvious of them is that zoos as visitor attractions in both developing and developed countries have the potential not only to educate the general public about wildlife conservation issues but also to bring about behavior change on a massive scale. No other conservation organizations, as a group, have the same ability to get their message across to some 600 million people a year, or to evoke the emotions that direct encounters with animals can inspire and channel them towards conservation action. As we have seen in other chapters, much remains to be done in coming years to develop the role of zoos as agents for behavior change and to demonstrate and quantify the impacts they can have, but there is growing consensus now that this is the direction in which the zoo community should move.

A first step that any zoo can take is to bring about and demonstrate behavior change in its own operations and in the actions of its own employees. As we have seen elsewhere in this volume, zoos can set themselves up as models to be emulated in one respect or another, and incentives can be provided for zoo staff to get involved in conservation issues "in their own backyard." Such actions are within the reach of all zoos.

An important incentive to encourage all zoos to move in these directions could be provided by the introduction of "stepwise" accreditation schemes which supplement, or even replace, the current all-or-nothing approach.

Credit could thus be given to zoos from the moment that they decide to step onto the first rung of the accreditation ladder. Each subsequent step is clearly defined and leads toward a final rung of "full accreditation" which can be redefined and supplemented over time as standards and public expectations rise. In this way, even the much-derided "roadside zoos" could potentially be brought into accreditation schemes in a way that is unthinkable under the current, all-or-nothing system in use by most zoo associations. To ensure steady progress by zoos within the scheme, timetables and fee structures can be set to motivate them to move on to the next step within an agreed timeframe. According to the number of milestones passed, zoos could be awarded star ratings as the basis for publicizing their achievements to the public.

The European Association of Zoos and Aquariums (EAZA) already uses a simple form of the stepwise accreditation system to accommodate candidate institutions that do not meet the standards required for full or associate membership but are expected to do so with 1 or 2 years. Similarly, zoos that are under construction or working toward compliance with full EAZA accreditation criteria can be awarded Candidate for Membership status. Institutions in this category have access to technical support provided by EAZA members to help them toward full accreditation (www.eaza.org).

Probably the main reasons why institutions in any field of endeavor pursue accreditation are to distinguish themselves among their peers and competitors as applying "best practice," to comply with regulations established for the industry in question, and to enhance their credibility with the public at large. With their audience of 600 million or more visitors a year, zoos and aquariums worldwide have great potential as public advocates for conservation. Anything that increases an institution's credibility with the public is especially important when it comes to influencing decision-makers, whether at local, national or international level. As long as their campaigns are compatible with their charitable status, when that applies, individual zoos as potential leaders of public opinion on local wildlife conservation issues have considerable power to influence planning decisions that may affect locally threatened species. By acting collectively, they can influence legislation and decision-making at national and even at international levels. Advocacy, however, requires zoos to act outside their traditional comfort zone by acquiring and applying political skills in addition to the technical ones which, as scientific institutions, they have worked so hard to develop. In combination with the emotional responses that good zoo and aquarium exhibits can evoke from their visitors, a well-planned advocacy campaign conducted by a scientifically credible and politically astute zoo or zoo association is hard for politicians and other decision-makers at any level to ignore.

It will be instructive, in this context, to monitor the effects of the new Manifesto for Zoos (Regan 2005) that has been developed by a consortium of nine British zoos in partnership with the British and Irish Association of Zoos and Aquariums (BIAZA). The Manifesto "constitutes the first attempt ever to establish the overall value and true 'public good', actual and potential, available to British society through the progressive UK zoos." It concludes in particular that, "with the right encouragement and discussion, Zoos can partner more effectively and more consistently with Government to deliver even greater social and environmental benefit." Several Australian zoos have already joined forces to develop a similar document. Such initiatives, developed as part of systematic advocacy campaigns, could raise considerably the profile and effectiveness of zoos in the conservation policy arena.

Changing public perceptions

Two very substantial challenges underlie the specific issues raised by this volume. The first is for individual institutions to translate the objectives of the *World Zoo and Aquarium Conservation Strategy* of 2005 into concrete actions. It is a massive task that will demand great creativity and leadership on the part of the zoo directors and executive staff all over the world. Once that first challenge has been confronted and it is clear that a true transformation is under way, the second one – to change public perceptions of what modern zoos are about – will have to be confronted. It may prove to be a problem that takes care of itself as the conservation mission of zoos becomes more overt. Most likely, it will have to be helped along by proactive communications campaigns. One of the strongest messages to emerge from this volume, however, is that – even as public perceptions change – it is vital that zoos learn to demonstrate and measure their own effectiveness as conservation institutions in ways that they can market to the zoo-going public, their donors, the decision-makers, and even to the broader community of conservation organizations. Until that happens, their potential as catalysts for *in situ* conservation is likely to remain under-estimated, under-funded, and under-developed.

If, on the other hand, these challenges can be overcome, we believe that zoos can develop a new global identity that fills a vacant niche in the broader community of conservation organizations. Decisions made by the current generation of humans on Earth are likely to influence the future of our planet, and of all the organisms on it, to a greater degree than those made by any previous one (Saunders *et al.* 2006). Zoos as advocates for conservation have the unique ability to inspire hundreds of millions of people every year to make the right decisions with regard to their own place, role, and responsibilities in this unique and glorious world. If there was ever a time for zoos to nail their colors to the mast as passionate and effective advocates of wildlife conservation and ecologically sustainable economic development, this is it.

References

Andrews, C. & Kaufman, L. (1994). Captive breeding programmes and their role in fish conservation. In *Creative Conservation: Interactive Management of Wild and Captive Animals*, eds. P. J. S Olney, G. M. Mace, & A. T. C. Feistner. London: Chapman and Hall, pp. 338–351.

Anon. (2005a). Ethics and animal welfare. In *Building a Future for Wildlife: The World Zoo and Aquarium Conservation Strategy*. Bern: WAZA, Chapter 9.

Anon. (2005b). Population management. In *Building a Future for Wildlife: The World Zoo and Aquarium Conservation Strategy*. Bern: WAZA, Chapter 4.

Bauert, M. R., Furrer, S. C., Zingg, R., & Steinmetz, H. W. (2006). Two years of experiences running the Masoala Rainforest ecosystem at Zurich Zoo, Switzerland. *International Zoo Yearbook*.

Beck, B. B., Rapaport, L. G., Stanley Price, M. R., & Wilson, A. C. (1994). Reintroduction of captive-born animals. In *Creative Conservation: Interactive Management of Wild and Captive Animals*, eds. P. J. S. Olney, G. M. Mace, & A. T. C. Feistner. London: Chapman and Hall, pp. 265–286.

Bickert, I. & Meier, J. (2005). Customer expectation and self-conception of zoological gardens – results of a visitor survey. *Zoologische Garten*, **75**(3), 202–208.

Bitgood, S., Formwalt, D., Patterson, D., & Zimmerman, C. (1993). The Noah's Ark dilemma: zoo visitor's ratings of how much animals are worth saving. *Journal of the International Association of Zoo Educators*, **27**, 41–43.

Bloxam, Q. M. C. & Tonge, S. J. (1995). Amphibians: suitable candidates for breeding-release programmes. *Biodiversity and Conservation*, **4**, 636–644.

Britt, A. & Iambana, B. R. (2003). Can captive-bred *Varecia variegata variegata* adapt to a natural diet on release to the wild? *International Journal of Primatology*, **24**(5), 987–1005.

Britt, A., Welch, C., & Katz, A. (2002). The release of captive-bred black and white ruffed lemurs into the Betampona Reserve, Eastern Madagascar. *Re-introduction News*, **21**, 18–20.

Britt, A., Welch, C., & Katz, A. (2003). Can small, isolated primate populations be effectively reinforced through the release of individuals from a captive population? *Biological Conservation*, **115**(2), 319–327.

Britt, A., Welch, C., Katz, A., Iambana, B., Porton, I., Junge, R., Crawford, G., Williams, C., & Haring, D. (2004). The re-stocking of captive-bred ruffed lemurs (*Varecia variegata variegata*) into the Betampona Reserve, Madagascar: methodology and recommendations. *Biodiversity and Conservation*, **13**(3), 635–657.

Caro, T., Borgerhoff Mulder, M., & Moore, M. (2003). Effects of conservation education on reasons to conserve biological diversity. *Biological Conservation*, **114**, 143–152.

Conway, W. (1998). Zoo reserves: a proposal. In *Proceedings of the AZA Annual Conference*, Tulsa Zoo & Living Museum, Tulsa, Oklahoma. Chicago, IL: AZA, pp. 54–58.

Conway, W. (2003). The role of zoos in the 21st century. *International Zoo Yearbook*, **38**, 7–13.

Cuaron, A. D. (2005). Further role of zoos in conservation: monitoring wildlife use and dilemma of receiving donated and confiscated animals. *Zoo Biology*, **24**(2), 115–124.

Dierking, L. D., Adelman, L. M., Ogden, J., Lehnhardt, K., Miller, L., & Mellen, J. D. (2004). Using a behaviour change model to document the impact of visits to Disney's Animal Kingdom: a study investigating intended conservation action. *Curator*, **47**(3), 322–343.

Ferguson, A. & Pearce-Kelly, P. (2005). Captive management of the Fregate Island giant tenebrionid beetle, *Polposipus herculeanus*. *Phesuma*, **13**, 25–42.

Ferraro, P. J. & Pattanayak, S. K. (2006). Money for nothing? A call for empirical evaluation of biodiversity conservation investments. *PLoS Biology*, **4**(4), 482–488.

Gwynne, J. A. (2000). Conservation by design. In *AZA Annual Conference Proceedings*, Disney's Wild Animal Kingdom, pp. 277–282.

Hayward, J. & Rothenberg, M. (2004). Measuring success in the "Congo Gorilla Forest" Conservation Exhibition. *Curator*, **47/3**, 261–282.

Hutchins, M. & Smith, B. (2003). Characteristics of a world-class zoo or aquarium in the 21st century. *International Zoo Yearbook*, **38**, 130–141.

IUDZG/CBSG (IUCN/SSC) (1993). *The World Zoo Conservation Strategy: The Role of the Zoos and Aquaria of the World in Global Conservation*. Brookfield, IL: Chicago Zoological Society.

Jacobson, S. K. (1991). Evaluation model for developing, implementing and assessing conservation education programs: examples from Belize and Costa Rica. *Environmental Management*, **15**(2), 143–150.

Jamieson, D. (1995). Wildlife conservation and individual animal welfare. In *Ethics on the Ark: Zoos, Animal Welfare and Wildlife Conservation*, eds. B. G. Norton, M. Hutchins, E. F. Stevens, & T. L. Maple. New York: Smithsonian Institution Press, pp. 69–73.

Juniper, P. (1998). Soapbox – keepers versus the animal rights. How can we get across the right message to ensure that the public perceive us well? *Ratel*, **25**(6), 224–226.

Kisling, V. N. (2001). *Zoo and Aquarium History: Ancient Animal Collections to Zoological Gardens*. Boca Raton: CRC Press.

Kleiman, D. G., Reading, R. R., Miller, B. J., Clark, T. W., Scott, J. M., Robinson, J., Wallace, R. L., Cabin, R. J., & Felleman, F. (2000). Improving the evaluation of conservation programs. *Conservation Biology*, **14**(2), 356–365.

Knowles, J. M. (2003). Zoos and a century of change. *International Zoo Yearbook*, **38**, 28–33.

Lewis, O. T. (2006). Climate change, species-area curves and the extinction crisis. *Royal Society Philosophical Transactions Biological Sciences*, **36**(1465), 163–171.

Lindsay, P. A., Alexander, R., Du Toit, J. T., & Mills, M. G. L. (2005). The cost efficiency of wild dog conservation in South Africa. *Conservation Biology*, **19**(4), 1205–1214.
Maple, T. (1995). Towards a responsible zoo agenda. In *Ethics on the Ark: Zoos, Animal Welfare and Wildlife Conservation*, eds. B. G. Norton, M. Hutchins, E. F. Stevens, & T. L. Maple. New York: Smithsonian Institution Press, pp. 20–30.
Pearce-Kelly, P. (1994). Invertebrate propagation and re-establishment programmes: the conservation and education potential for zoos and related institutions. In *Creative Conservation: Interactive Management of Wild and Captive Animals*, eds. P. J. S. Olney, G. M. Mace, & A. T. C. Feistner. London: Chapman and Hall, pp. 329–337.
Pekarik, A. J. (2004). Eye-to-eye with animals and ourselves. *Curator*, **47/3**, 257–260.
Redford, K. (1998). The role of extractive reserves in wildlife conservation. In *Proceedings of the AZA Annual Conference*, Tulsa Zoo & Living Museum, Tulsa, Oklahoma. Chicago, IL: AZA, pp. 103–105.
Regan, J. (2005). The manifesto for zoos. Unpublished report. John Regan Associates, Ltd.
Regan, T. (1995). Are zoos morally defensible? In *Ethics on the Ark: Zoos, Animal Welfare and Wildlife Conservation*, eds. B. G. Norton, M. Hutchins, E. F. Stevens, & T. L. Maple. New York: Smithsonian Institution Press, pp. 38–51.
Saterson, K. A., Christensen, N. L., Jackson, R. B., Kramer, R. A., Pimm, S. L., Smith, M. D., & Wiener, J. B. (2004). Disconnects in evaluating the relative effectiveness of conservation strategies. *Conservation Biology*, **18**(3), 597–599.
Saunders, C. D. (2003). The emerging field of conservation psychology. *Human Ecology Review*, **10**(2), 137–149.
Saunders, C. D. & Myers, G. (2001). Using conservation biology as a model for thinking about conservation psychology. *Population and Environmental Psychology Bulletin*, **27**(2), 7–8.
Saunders, C. D., Brook, A. T., & Myers, O. E. (2006). Using psychology to save biodiversity and human well-being. *Conservation Biology*, **20**(3), 702–705.
Schultz, P. W. & Zelezny, L. (2003). Reframing environmental messages to be congruent with American values. *Human Ecology Review*, **10**(2), 126–136.
Shepherdson, D. J., Mellen, J. D., & Hutchins, M. (1998). *Second Nature: Environmental Enrichment for Captive Animals*. New York: Smithsonian Institution Press.
Struhsaker, T. T., Struhsaker, P. J., & Siex, K. S. (2005). Conserving Africa's rain forests: problems in protected areas and possible solutions. *Biological Diversity*, **123**, 45–54.
Stuart, S. N., Chanson, J. S., Cox, N. A., Young, B. E., Rodrigues, A. S. L., Fischman, D. L., & Waller, R. W. (2004). Status and trends of amphibian declines and extinctions worldwide. *Science*, **306**, 1783–1786.
Swaisgood, R. R. & Shepherdson, D. J. (2005). Scientific approaches to enrichment and stereotypies in zoos animals: what's been done and where should we go next? *Zoo Biology*, **24**(6), 499–518.
Thuiller, W., Broennimann, O., Hughes, G., Alkemade, J. R. M., Midgley, G. F., & Corsi, F. (2006). Vulnerability of African mammals to anthropogenic climate

change under conservation land transformation assumptions. *Global Change Biology*, **12**(3), 424–440.
Vernon, C. (2003). Moving guests from caring *about* the oceans to caring *for* the oceans. Unpublished discussion paper, Strategic planning workshop Monterey Bay Aquarium.
Vining, J. (2003). The connection to other animals and caring for nature. *Human Ecology Review*, **10**(2), 87–99.
WAZA (2005). *The World Zoo and Aquarium Conservation Strategy: Building a Future for Wildlife*. Liebefeld-Bern: WAZA.
Wemmer, C. (1995). *The Ark Evolving: Zoos and Aquariums in Transition*. New York: Smithsonian Institution.
Wiese, R. J., Willis, K., & Hutchins, M. (1994). Is genetic and demographic management conservation? *Zoo Biology*, **13**, 297–299.

Bibliographic resource: zoos and conservation

SCOTT WILSON AND ALEXANDRA ZIMMERMANN

The body of literature about the conservation role of zoos is extensive and growing as the debates and developments within this topic continue to evolve. We have compiled a bibliography of this literature which may be useful to professionals, managers, researchers and students, whether based in a zoological organization or interested in the work of zoos and aquariums.

The bibliography is a comprehensive listing of all published and selected unpublished papers and reports since 1960 containing the keywords *zoo* or *aquarium* in connection with keywords such as *conservation, biodiversity,* ex situ, in situ, *reintroduction, field, breeding, research, education, learning, attitudes, visitors, outreach, role, mission,* and others.

Rather than include this bibliography in print in this volume, we have created an online version, which is updated annually. The compilers welcome any new contributions (or correction of any errors it may contain). To contribute to this project please contact: conservation@chesterzoo.org.

The *Zoos & Conservation Bibliography* can be found at: **www.chesterzoo.org**.

Index

Lightning Source UK Ltd.
Milton Keynes UK
UKOW04f2114020316

269463UK00001B/120/P